Environment and Man

W · W · NORTON & COMPANY · INC · *NEW YORK*

Richard H. Wagner

ASSISTANT PROFESSOR OF BOTANY
THE PENNSYLVANIA STATE UNIVERSITY
AND CONSULTING ECOLOGIST

Environment and Man

SECOND EDITION

The photograph on the title page is by Jim Page.
All uncredited photographs in the book are by the author.

THIS BOOK IS PRINTED ON RECYCLED PAPER

ISBN 0 393 09317 4

1 2 3 4 5 6 7 8 9 0

". . . To the wide world and all her fading sweets . . ."

SHAKESPEARE, *Sonnet 19*

CONTENTS

2 Natural Traumas Become Pollutants

3 Man Makes New Traumas

4 The Biotic World and Man

PREFACE

TRADITIONALLY, many colleges and universities have advised their students to take at least one course in general biology, on the premise that their graduates ought to have some idea of their fundamental structure and of the physiological processes that govern their lives. Such courses on biological form and function can be lively, vital, and relevant if the instructor relates his material to the needs and interests of his students. All too frequently, however, they give short shrift to, or beg altogether, the critical question of the interaction of man with the lives and activities of other organisms and with the environment itself.

Because of our sudden awareness of the magnitude of man-environment problems, efforts have been made on many campuses to fill this curricular gap with courses that recognize man's tremendous impact on the natural world and analyze it from various points of view. Although there were scores of books covering the many individual aspects of the man-environment field, as well as symposia, anthologies, and exposés, there were no balanced texts that pulled all the threads together. The first edition of this book was written to fill that need.

Since the first edition of *Environment and Man* went to press, there have been enormous changes both in the scope of environmental problems and in our reactions to them. Although the great furor of public concern has subsided as anticipated, governmental agencies— federal, state, and local—have finally been stirred into action. As a result, many problems that seemed intractable or unavoidable in 1970 are being dealt with and reduced to manageable proportions. To keep abreast of these sometimes dramatic changes, it became necessary to revise *Environment and Man* somewhat sooner than had been anticipated. There are three new chapters, the sequencing of some other chapters has been altered, and many parts of chapters have been extensively reworked to incorporate new information. Further, reading lists have been updated where appropriate, and the appendices retained to provide a guide to the reader who wishes to take an active role in what is increasingly being recognized as a continuing and long-term struggle of heroic proportions.

As with the first edition, it was not my intention to write another basic ecology text, for many fine ones already exist, but to provide an introduction to man-environment problems that presupposes no background in the sciences, one that any college student or other interested reader could read with understanding and profit.

Consequently, although basic ecologic principles have been mentioned wherever necessary, there has been no attempt to underline them or develop them systematically. To do so in this book would double its length and significantly change its focus. Instead, I have begun preparation of a companion book which will examine the application of ecological theory to practical problems. In the interim, for those who wish a fuller treatment of ecological principles than has been provided by *Environment and Man,* I suggest using as a supplement one of the several fine paperbacks currently available: Storer's *Web of Life* (New American Library, 1968), Melillo's *Ecology Primer* (Pendulum Press, 1972), Boughey's *Fundamental Ecology* (Intext, 1971), or Emmel's *Ecology and Population Biology* (W. W. Norton, 1973).

The central theme of this book is that man's relationship to his environment has passed through several phases. In the beginning, man was shaped by an environment which acted as a selecting agent and controlled the evolution of his present features. Throughout this period man, like the other animals, remained in equilibrium with his environment. But then something happened that changed the face of the world, completely and irreversibly: man developed culture and thereby shattered this equilibrium. From that time on man has exerted an ever-increasing influence upon his environment.

Initially, he simply increased the amplitude of already existing environmental traumas. Smoke from natural fires, or polluted water from animal carcasses or mineral seeps, for example, was magnified in proportion to man's numbers and cultural diversity. My discussion of these natural traumas has been arranged in a sequence reflecting man's increasing technological involvement and seriousness of effect.

More recently, new traumas have been introduced by technology, which has synthesized compounds totally new to the environment, such as DDT, and has produced a great many new forms of known substances like beryllium and asbestos.

Man has also profoundly affected the organisms that share the earth with him, both by introducing exotic species in new locations and by driving other species to extinction. His biotic effects have not, of course, been limited to the lower species, for environmental warfare has long been applied by man to man.

Finally, man's rampant technology has placed him, once again, under the direct selective influence of an environment of his own making: the city.

The book concludes with a discussion of man's ultimate problem—finding the balance of resources and population that will enable all of earth's organisms to coexist in an environment prejudicial to none.

While the order of the chapters reflects my own approach to man and environment, worked out over several years' experience teaching a course in this area, others may prefer a different sequence. The comprehensive scope of the book should allow the sections or even individual chapters to be read and used in any order. To provide this flexibility, a full index is included. As with the first edition, this book is printed on recycled paper.

In preparing the revision of this book I had the gracious assistance of Drs. William Niering and Peter Stuttard, who reviewed Chapter 16; Dr. Gerald Garvey, who read Chapter 8; and Dr. Arthur Westing, who examined Chapter 17.

RICHARD H. WAGNER

Easton, Maryland
August, 1973

1

Man in the Landscape

Forest fires can be enormously destructive when fire is ignored as a natural part of many forest environments. (*Wide World Photos*)

1

ENVIRONMENT AND THE EVOLUTION OF MAN

LIQUIDS HAVE NO shape of their own; they assume the configuration of their container. Although the time scale is quite different, the form of a plant or animal is also shaped by its environment, the "container" in which it exists. This fitting into an environment is called evolution. When a species conforms to its environment perfectly, evolution is complete. Completed evolution is rare in nature, however.

Man also evolved to conform with his environment, and for the greater part of his existence on earth has constantly been molded by that environment. While man has existed on earth for only a few million years, at most, the thread of his evolution goes far back in geological time.

The earth's crust has undergone extreme changes during its 4.5-billion-year-long history. Surface temperatures have varied from that of molten rock to a hundred or more degrees below zero; seas have formed, receded, and reformed over what is now dry land; mountains have arisen from ancient sea beds, have eroded away, and risen again. In spite of these extreme changes, complex and delicately balanced forms of organized living matter have developed. These organisms are so fragile that they may be quickly disorganized into nonliving matter by relatively slight internal changes in chemical composition, temperature, or pressure. How, then, do organisms survive the widely varying environmental conditions on earth?

The key to survival is not resistance to change, but meeting change with change. Even though a single species must inevitably be modified or become extinct, continued life is assured through evolution.

If you were to view the evolution of life on earth as a thirty-minute film, you would see wave after wave of new species evolving, filling the environment with a diversity of life forms, and then receding—sometimes totally, but occasionally leaving a few of the best-adapted species behind. It is humbling to note that man's existence on earth would flash by in the last 3.5 seconds of the film!

THE NICHE CONCEPT

Every animal except man has a well-defined habitat, a place where it is generally found and to which it has become adapted: trout "like" cold-water streams; a wood thrush, heavy woods; a woodchuck, weedy pastures. Within this habitat each species has a unique life style not shared by any other species. This complex of activity within a given habitat is called a niche. A species in a niche undergoes adaptation that is specialized; no two species occupy the same niche. Under experimental conditions, the best-adapted of two organisms occupying a niche outcompetes the other and causes its disappearance. As the complexity of the habitat increases, the number of niches increases, and with this, the number of species filling those niches. In an undisturbed habitat in a reasonable state of equilibrium, most, if not all, niches are occupied. As habitats change, the niches also change. If an animal cannot adapt to the changes and so evolve along with its niche, it becomes extinct; its place is usurped by a species which has evolved to fit that particular niche.

By the beginning of the Tertiary period, 65 million years ago (Figure 1.1), conifers had been replaced by broad-leaved trees, in just this way. A rich forest covered large areas that today are too rough, too dry, or too cold to support forests. This early Tertiary forest, by giving ample opportunities for new forms of species to evolve and survive, was responsible to a great extent for the rapid evolution of the most recent wave of animals—the mammals.

THE RISE OF THE PRIMATES

One group of mammals, the primates (forerunners of man), was quick to exploit the new environment of broad-leaved trees. Primates (Figure 1.2) are difficult to characterize because they do not share obvious primary characteristics, such as the hooves of the grazers or the specialized teeth of the carnivores (meat-eaters). Perhaps the simplest way to describe primates is to say that compared with other animal groups, they are unspecialized.

Once an animal becomes specialized, its specialization either leads to success, if the environment is stable, or to extinction, if the environment is rapidly changing, for evolution tends to be irreversible. The early primates were able to adapt to a changing environment because they retained their generalized features. If the primates had lost toes, as did the horse, they could not have regained them for use in tree climbing. This lack of specialization allows man, the most ad-

PERIODS AND EPOCHS	DURA-TION	TIME PRIOR TO PRESENT DAY	DISTINCTIVE LIFE
	(MILLIONS OF YEARS)		

PHANEROZOIC
CENOZOIC ERA

Quaternary	1 or 2	1 or 2	Modern man
Recent			
Pleistocene	1 or 2	1 or 2	Stone Age man
Tertiary			
Pliocene	9 or 10	12	Great variety of mammals
			Elephants widespread
Miccene	13	25	Flowering plants in full development
			Ancestral dogs and bears
Oligocene	15	40	Ancestral pigs and apes
Eocene	20	60⎫	Ancestral horses, cattle, elephants, and
Paleocene	10	70⎭	primates

MESOZOIC ERA

Cretaceous	65	135	Extinction of dinosaurs. Mammals and
			flowering plants slowly appear
Jurassic	45	180	Dinosaurs abundant
			Birds and mammals appear
Triassic	45	225	Flying reptiles and dinosaurs appear
			First corals of modern types

PALEOZOIC ERA

Permian	45	270	Rise of reptiles and amphibians
			Conifers and beetles appear
Carboniferous			
Pennsylvanian	35	305⎫	Coal forests
Mississippian	45	350⎭	First reptiles and winged insects
Devonian	50	400	First amphibians
			Earliest trees and spiders
			Rise of fishes
Silurian	40	440	First spore-bearing land plants
			Earliest known coral reefs
Ordovician	60	500	First fish-like vertebrates
Cambrian	100	600	Abundant fossils first appear

PRECAMBRIAN

Late Precambrian	1000	2000	Scanty remains of primitive inverte-brates: sponges, worms, algae, bacteria
Earlier Precambrian	1500	3500	Rare algae and bacteria back to at least 3000 million years for oldest known traces of life

Fig. 1.1 • Geologic time is divided into eras and periods of varying duration and remoteness from the present. The Tertiary period is particularly important to our study of man-environment interactions because of the development at that time of the broad-leaved forests. (*After Holmes*)

Fig. 1.2 • (a) This baboon shows clearly some of the primate characteristics that allow binocular vision and the opposable thumbs and big toe that permit grasping. (*Courtesy of the American Museum of Natural History*) (b) The brooding quality of this orangutan is not a true measure of either the species or the primate group: the eyes, hands, and feet, however, show its affinities with the primates. (*Freelance Photography Guild*)

vanced primate, to occupy a tremendous range of ecological niches.

The earliest primate species for which we have evidence is a squirrel-like animal who lived on the ground. A combination of competition and predation on the ground, together with increasing evolutionary opportunity in the expanding Tertiary forest, probably led such primates to shift their habitat from the ground into the trees.

Living in a tree, as any tree climber knows, requires two basic abilities: an ability to grasp branches and an ability to perceive three dimensions, so that movement can be rapid and sure. The danger of falling is a constant hazard for tree dwellers (squirrels *do* fall out of trees occasionally). It is not surprising then, considering the special qualities of an arboreal environment, that animals invading this new environment began to adapt to it.

The Opposable Thumb Leads to New Occupations

Primates characteristically have five digits terminating the fore- and hindlimbs. These digits are the prime contact the animal has with its environment; consequently, they are usually the first part of the animal to undergo adaptation in response to changes in its environment. When a five-toed animal grasps an object, the digits tend to separate into two groups. If the big toe or thumb is able to move so that it opposes the other digits, the hand or foot can lock firmly onto the object being grasped, making movement from branch to branch easy (Figure 1.3).

With a set of four appendages that were able to grasp effectively, the position of primates in the treetops was secure—literally and figuratively. A walking animal like the horse needs all four limbs for support; therefore, it has little opportunity for the evolution of other, independent functions for the limbs. The opposable toe made it possible for a primate to hold on to a branch with its hindlimbs while its forelimbs were freed for other activities. This freedom of one set of appendages to explore, feel, and touch opened a new avenue of evolutionary possibilities.

Snout Length as a Function of Survival

Yet the tactile sense in itself was incomplete without a sensor to evaluate the impressions collected. Here the second major aspect of the arboreal environment—its three-dimensional quality—placed great emphasis on a visual system that could perceive depth. The ground-dwelling primate had a long snout with an eye positioned on each side, precluding stereoscopic vision. After several million years of tree life a reduction in snout size permitted stereoscopic vision, giving an accurate representation of the three-dimensional environment to which these primates were becoming adapted.

A ground-dwelling mammal's snout serves two important functions: the long, powerful jaws secure and deliver food to the mouth, and the nose serves as an important environmental sensor. Because four legs are needed and adapted for support and locomotion, objects of interest must be picked up with the jaws and are rarely manipulated with the paws. The first two or three feet of air above the ground are usually rather moist, and carry scents and odors; the closer to the ground the animal, the more important is his nose as a sense organ.

A primate abandoning the ground for the trees would certainly have found the dry upper air much freer of odors and his nose correspondingly less useful. Procurement of food by the snout also became less important as the forelimbs no longer performed a strictly supportive function, but began to supplement and finally to supplant the jaws as a means of selecting food and placing it in the mouth.

These two factors allowed the snout to diminish in size and become less important in function, without adversely affecting survival.

NATURAL SELECTION

Adaptive change has, so far, been rather casually considered—forelimbs grasped, vision became stereoscopic, snouts shortened. How did these changes occur? Certainly animals weren't able to make these

Fig. 1.3 • By curling its fingers around an object and applying opposing pressure with its thumb, a primate can grasp firmly and securely. (*Freelance Photography Guild*)

changes consciously; even man can manipulate the environment to his ends only at times, and cannot—at least now—make intentional structural improvements in his physical adaptation to the environment.

Mutations

Every living cell of an organism contains information in its nucleus that governs the activities of that cell, either singly or in concert with other cells. But the reproductive cells are especially important because they produce the egg and sperm which, in higher organisms, are the only link between generations. The potential for change or evolution is greatly enhanced when the information-containing material (DNA) of the egg or sperm is altered. Such alterations of DNA are called mutations and the environmental factors which cause them, mutagens. Some of the more important mutagens are radiation (see Chapter 9), chemicals (see Chapter 12), temperature extremes, and ultraviolet light.

About 20 percent of the individuals in any generation carry recessive or concealed mutations. This genetic load appears to be increasing through medical advances that have "cured" more hereditary diseases and hence enabled the carriers to breed more frequently, and through technological change, which exerts new selectional pressures. George III of England, for example, suffered most of his life from porphyria, a mutation in which porphyrin, a breakdown product of hemoglobin, is excreted, coloring the urine wine red. For some reason, porphyria sufferers are extremely sensitive to barbiturates. Had barbiturates been available in George III's time he might well have died at a rather early age, radically altering the course of American history.

In man, a mutation causing the malfunction of one lung or kidney is not necessarily disastrous, since most humans are born with a pair of each of these organs. However, were the mutation to involve a change in the heart, or perhaps a block in an important biochemical pathway, the effect of the mutation would be more critical (Figure 1.4).

Fig. 1.4 • Spinus bifida is a mutation that randomly occurs in man from time to time. The spine divides near the base into two branches, seen here in an infant as a bulge at the base of the spine (in profile to the left, face-on on the right.) (*R. B. Salter, The Textbook of Disorders and Injuries of the Musculo-Skeletal System,* © *1970 The Williams & Wilkins Co., Baltimore*)

Occasionally a mutation permits an organism to interact with its environment in a more efficient manner; hence, the mutant organism proliferates. Far more often, however, mutations result in an organism less able to cope with its environment.

Suppose you had a color television, but felt it could stand some improvement. One approach would be to expose the complicated circuitry and splash solder around at random. There is a remote possibility that you could somehow improve the quality of the picture, but the chances are overwhelmingly greater that you would make such a mess of things that the set would not work at all. If you had an infinite number of television sets and millions of years to experiment, sooner or later you would probably make an improvement. Such are the odds against a successful mutation.

Selection Enhances Favorable Mutations

Mutations move from an individual to future generations only by way of the egg and the sperm. If a mutation impairs the offspring's ability to cope with its environment, the animal may fail to mate or even die before it can reproduce. As a result, unfavorable mutations are lost. Changes that are neither favorable nor unfavorable can be carried indefinitely, simply because there is no selection pressure to get rid of them. Finally, any mutation that even slightly increases the ability of an organism to survive or to produce more offspring will be transmitted throughout the species population.

To see how mutation is related to evolution, consider snout length gain. Snout length is determined by a complex interaction of genetic and environmental controls. Because of slight variations in the environment during development or perhaps minor mutations, no two individuals of a species are born with exactly the same snout length.

Since having a long snout means less effective stereoscopic vision, a primate with a long snout would be at a greater disadvantage in his arboreal habitat than a primate with a short snout and less likely to survive long enough to reproduce; hence, short-snouted primates would tend to replace longer-snouted primates.

Over a period of evolutionary time all organisms are subject to random mutational change. While the environment does the selecting, it selects differently for each species; a mutation that is useful to one animal could be a disaster for another. A spider monkey without a tail, for example, would be as inconvenienced as a man with one. While mutations producing monkeys without tails rarely occur, when they do, the aberrant individual does not proliferate; hence, monkeys have tails and men do not.

THE EMERGENCE OF MAN

Had the Tertiary period and its forest continued up to the present, man might never have evolved. Instead, there probably would have been a normal diversity of primates, well-adjusted to their arboreal world. This, however, was not to be. The luxuriant early

Tertiary forest that extended across the continents from coast to coast resulted from the free movement of moist air from the oceans over the interior of the continents, providing enough rain to support the forest. About the middle of the Tertiary period, old mountain chains were rejuvenated, and new mountains like the Sierra-Nevada–Cascade complex, the Andes, and the Himalayas were uplifted. The rising mountains deflected the moisture-laden air and the interior of the continents became drier.

As the humid ocean air was forced up and over the mountains, it was cooled, losing much of its water on the slopes facing the ocean. The air that flowed down the inland slopes became hot and dry. The higher the mountains, the drier the land in the lee of the mountains, until that land could no longer sustain the Tertiary forest. This effect, known as rain shadow (Figure 1.5) resulted in the desert-like conditions in the Great Basin of the United States, the Gobi desert north of the Himalayas, and the dry plains east of the Andes in Argentina. Climates were changing, and the plants and animals dependent on certain climatic conditions were obliged to change also.

Fig. 1.5 • As moisture-laden systems move in off the Pacific, they are forced to rise first over the coast ranges in California, and then over the Sierras. To the lea of these mountain ranges the precipitation is markedly less. This phenomenon is the rain shadow effect. (*From Daubenmire, R., 1959.* Plants and environment, *2nd ed. John Wiley, New York, p. 91*)

The rich and diverse forest that had always required a generous supply of moisture could not adapt, and as the forest died out, a savannah—a mixture of grasses and small trees—gradually took its place. Today the remnants of the Tertiary forest (Figure 1.6) are to be found in scattered patches throughout the tropical and temperate zones of the earth; in between are deserts, grasslands, and savannahs (Figure 1.7).

Descent from the Trees as a Selection Pressure

What happened to the tree-adapted primates as their habitat changed? Some managed to stay with the retreating forest and are found today in its remnants—the orangutan in southeast Asia, the chimpanzee and gorilla in central Africa, lemurs in Madagascar, and various monkeys in the tropics of Central and South America. But some were able to survive the loss of trees and return to the ground their ancestors had abandoned millions of years before. It is within this group of primates that we again pick up the thread leading to man.

The descent from an arboreal to a grassland-savannah niche was not sudden: no primate could have adapted to an abrupt disappearance of the sheltering forest. But as the distance between trees increased, more and more time, of necessity, was spent on the ground.

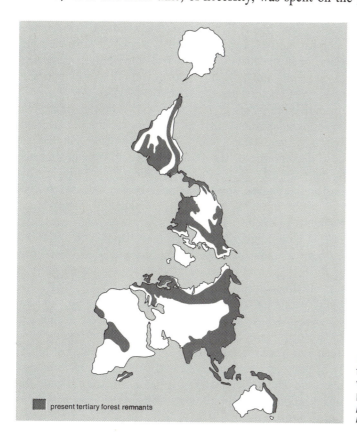

present tertiary forest remnants

Fig. 1.6 • The remnants of the Tertiary forest are now scattered around the world. This forest once covered a large part of the earth's surface. (*Dymaxion projection copyright © 1954 by R. Buckminster Fuller*)

Fig. 1.7 • A savannah is a grassland with scattered trees. These are acacia trees in east Africa. (*Photo by Alice Mairs*)

This opened a new area of selective pressure, which had an effect on appearance and activities of certain primates. For example, once on the ground it became important for primates to be able to see over the grass to avoid predators and to communicate with other primates. This encouraged standing upright and further reinforced the functional separation of forelimbs and hindlimbs. Whatever tail was present in the early primates became less important on the ground, and hence could be lost without adversely affecting survival.

Toward the end of the Tertiary period, when the environment began to stabilize again, there were several stocks of advanced primates—tailless or short-tailed, bipedal, possessing stereoscopic vision, with shortened snouts and smaller, less powerful jaws, and opposable thumbs giving great manipulative power to the hands.

The primitive brain was rapidly changing at the same time that the more obvious adaptations, first to an arboreal, then to a grassland environment, were taking place. As the eyes were emphasized and the nose deemphasized, the brain responded with greater development in the portion associated with seeing. The interplay between hands and eyes was of paramount importance in the further development of the brain. Species with manipulatable hands found new environments to explore, new food to eat, while their eyes, seeing the world in three dimensions, guided their hands with a deftness impossible in other animals. The brain responded to these new stimuli by further growth in size and complexity, which in turn stimulated manual dexterity.

Emergence of man

In qualitative terms, brain development, stimulated by hand-eye interaction, seems to have moved through three stages: seeing and doing, remembering what was seen and done, and imagining what might be seen and done. Seeing and doing is common to most higher animals; remembering what was seen and done is characteristic of many species and indicates an ability to learn and be trained. But imagining what might be seen and done is limited to man and a few species of apes. Once this final stage was reached, self-sustaining intellectual evolution became possible. After an unusually short evolutionary period, man emerged. But why did only man evolve to this degree, leaving his fellow primates so far behind?

The answer lies in the comparative specialization of the higher primates. Many highly specialized species have survived through long periods of time, but they have not usually evolved into a variety of new species. Generalized species have the greatest evolutionary potential because they can respond to the greatest number of environmental options. As specialization removes these options, a species inevitably approaches an evolutionary dead end, becoming extinct in periods of environmental change. Because they are relatively unspecialized, sparrows and rodents are rapidly evolving, whereas ostriches and elephants, which are quite specialized, have reached this evolutionary dead end.

In the late Tertiary period there were three major primate stocks. One of these stocks reverted to the functional quadripedalism seen today in the baboon (Figure 1.8). Another specialized in a different direction evolving the flattened thorax, more erect posture, and modified arms that enabled movement from tree to tree by swinging (brachiation) typified by the modern gibbon (Figure 1.9). A third stock developed a terrestrial bipedalism involving certain spinal modifications which allowed weights to be lifted vertically, a flattened chest making possible a shift in the body's center of gravity so that it lay over the pelvis, a tail modified to form the floor of the pelvis, and a head balanced on the neck.

 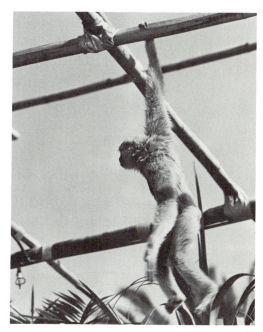

Fig. 1.8 • (*left*) Although baboons live on the ground, they are not true bipeds like man because the knuckles of their forelimbs are usually in contact with the soil. (*San Diego Zoo Photo*)

Fig. 1.9 • (*right*) With its long spidery arms the gibbon brachiates or swings easily through its environment—in this instance, the roof of its cage in a zoo. (*San Diego Zoo Photo by Ron Garrison*)

This last stock early gave rise to chimpanzees and gorillas that never became fully independent of trees nor achieved a true bipedal stance; they walk knuckles-down on the ground during the day and return to the trees at night for shelter. But one group of this third primate stock developed the hand-eye-brain correlation before excessive specialization limited future evolutionary potential—from that stock, man emerged. Evidence that several forms of that stock existed at the same time has recently been uncovered by R. E. Leakey. Apparently 2.5 million years ago a small, rather graceful biped, *Australopithecus africanus* (Figure 1.10) lived in east Africa along with a more advanced species, *Homo erectus*. At some later date *Australopithecus* became extinct and was replaced by *Homo erectus*, who was replaced in turn by *Homo sapiens*. Whether these extinctions and replacements were an early manifestation of man's inhumanity to man or rather a peaceful miscegenation, the fossil record, unfortunately, does not reveal. It would be intriguing to know.

Fig. 1.10 • (a) The late Dr. L. S. B. Leakey, noted British archeologist, discovered many australopithecine remains in Tanzania, east Africa. The chimpanzee skull on the left gives an idea of the relative size of the skull.

(b) This artist's conception prepared under Dr. Leakey's supervision gives a glimmer of how *Australopithecus* might have looked. (*Wide World Photos*)

CULTURATION: A TRANSITION

Up to the appearance of *Australopithecus*, evolutionary changes came rather quickly and markedly affected the features of the protomen of the day. After *Australopithecus*, outward changes were more subtle. Neanderthal man—or even the more primitive *Homo erectus*—might pass unnoticed on a city street today if dressed and shaven. Why the apparent turning point in man's evolutionary change? The answer, in a word, is culture.

Development of the Use of Tools

There are suggestions that *Australopithecus* had tools—crude pebbles, perhaps, but purposefully used. Many animals use tools in one way or another, but only men, and in rare instances monkeys and apes, use them to defend themselves against predators or to kill their prey.

The consistent use of tools is of great significance in the effect of environment on man's evolution. For the first time an animal could use its brain to meet the challenges of its environment, instead of having to make gross structural adaptations. That is why further changes in the outward appearance of man have been small. When man first started living in the Arctic, he did not have to grow fur, he wore clothes; when he first flew, he did not have to develop wings by mutation, he used a tool. With culturation, that is, the achievement of culture, man was freed from most of the direct selective effects of his environment.

Culturation, however, had its own effects which further refined and modified the crude australopithecine stock. Monkeys and apes have rather impressive canine teeth (Figure 1.11). Neither early nor contemporary man has prominent canines. Why? Monkeys and apes use their canines for defense, exposing them in threatening grimaces, biting or slashing, if necessary. This action, however, requires a dangerous proximity to be effective. By using tools, sticks, or stones, early man could fend off attackers without having to risk injury to himself in close combat. The large canine teeth became expendable, and if by random mutation they were reduced in size, survival was not threatened.

As the defense function of teeth and a powerful jaw was superseded by a brain directing hands that used tools, the massive musculature supporting the jaw was no longer necessary for survival. The ridges to which these muscles were attached, and which gave the skulls of early men such a formidable appearance, also became unnecessary. Mutations producing more delicate skulls, no longer selected against, spread throughout the population.

Later, man learned to control fire and use it to his advantage, just as he learned to use natural objects as tools. Cooked food is softer and more easily chewed than raw food, again lessening the importance of large, powerful teeth and jaws. In addition, new varieties of food became available through the use of tools and fire, broadening still further the tendency of early man to dietary generalization. The development of tools and the control of fire did more to make man truly human than any other cultural or biological advance.

ENVIRONMENT AND THE EVOLUTION OF MAN

monkey

man

Fig. 1.11 • As man evolved physically and culturally, he no longer used his teeth as a weapon, and hence the jaw receded and the size of the canines was drastically reduced.

HUMAN ADAPTATIONS

Although culturation seemed to remove man from some specific selective forces, environmental selection continued. If man had remained in the place where he evolved, the task of relating adaptation to environment would be relatively simple. But man has been on the move since the early days of his evolution; a primitive hunting and gathering culture must cover enormous areas of territory to survive. Climatic and cultural changes also stimulated people to migrate far from the site of their evolution and the environment responsible for their particular adaptations.

Despite the difficulty of unraveling the connection between environment and adaptation, some explanations can be pieced together from circumstantial evidence. Two adaptations are discussed in the next section: one—pigmentation—is very obvious; the other— sickle cell anemia—became apparent only through recent scientific study.

Pigmentation and Vitamin D Synthesis

Human skin, as well as the skin of most vertebrates and many invertebrates, contains a pigment known as melanin, which is formed from the amino acid tyrosin. Skin color is determined primarily by the amount of melanin it contains; albinos, for example, have no melanin, and their skin looks pink because of blood-filled capillaries near the skin's surface. Among the human population, skin tones range from pink to very dark brown. If we look at the distribution of pigment types we see that the darkest pigmentation is in the tropics, generally, and the lightest skin is found in the high latitude temperate areas. The two extremes of the gradient might be represented by a Nigerian on the one end and a Laplander on the other, with a North African representing an intermediate degree of pigmentation.

How can we explain the lighter pigmentation of people living nearer the poles? One hypothesis involves pigmentation and vitamin D synthesis. The skin contains a precursor substance called 7-dehydrocholesterol, which under the influence of ultraviolet radiation from sunlight is converted into vitamin D. Since egg yolks and fish liver oil are the only common foods with much vitamin D, the sunlight-skin reaction, until quite recently, was the most important source of this important vitamin. Vitamin D controls the absorption of calcium from the intestine and the deposition of this mineral in the bones. If there is less than 0.01 milligram per day of vitamin D available from both sunlight and the diet, not enough calcium is absorbed, and the result is brittle and deformed bones. In children this condition is known as rickets (Figure 1.12); in women it may result in pelvic deformities leading to difficulties in childbirth. Too much vitamin D also causes problems: greater than 2.5 milligrams a day may lead to kidney stones, calcification in the joints, and deposits in the aorta and other large blood vessels. The body cannot regulate the amount of vitamin D absorbed from food, or eliminate a possible toxic dose. However, control of vitamin D from sunlight is achieved by the rate of photochemical synthesis in the skin. The

Fig. 1.12 • Today rickets has become quite rare in the developed countries of the West, where vitamin D supplements are easily available in a variety of foods. This is not true in many underdeveloped countries, however. (*Photo by UNICEF*)

amount of melanin present in the skin controls the amount of ultra-violet light reaching the photochemical zone, and hence the quantity of vitamin D produced from the precursor.

Because of the high light intensity in the tropical and subtropical regions where man probably evolved, early man may well have had a darkly pigmented skin, allowing just enough ultraviolet penetration to produce vitamin D in the safe range (0.01 to 2.5 milligrams a day). But as early man moved out of the tropics into the more cloudy and stormy temperate and arctic regions, heavy pigmentation prevented adequate transmission of ultraviolet light through the skin, and too little vitamin D was produced. When this resulted in rickets, the most heavily pigmented individuals never reached reproductive age. With the environment selecting in favor of less heavily pigmented skin, it was just a matter of time before most of the inhabitants of northern Europe became quite light complexioned.

When the sun's rays fall at an angle of less than 35°, much of the ultraviolet is filtered out by the atmosphere. When, in addition, smoke and airborne dirt are present in the atmosphere, as is char-acteristic of most industrial cities, there is little ultraviolet light that can penetrate to the level of the inhabitants. In European cities dur-ing the eighteenth and nineteenth centuries there was a high in-cidence of rickets, probably because of the combination of factors. In Jamaica, where the sun's rays never fall at an angle of less than 50° and there is little industrialization, rickets is unknown.

What about Eskimos, who live farther north than most Euro-peans, and yet are darkly pigmented? Eskimos eat large quantities of fish, seal, and whale liver, all of which contain generous amounts of vitamin D. Thus, isolated from dependence upon sunlight for vitamin D production, there was no selection for lighter pigmentation.

Discovery of the relationship between vitamin D and rickets led to the supplemental use of cod liver oil and, more recently, the addition of irradiated ergosterol to milk. Today no child, regardless of his geographical location, need develop rickets from the lack of vitamin D. If, that is, he can get enriched milk in his diet. However, not all people are adapted to drinking milk. Many Asians, for ex-ample, lose the ability to digest milk as they mature (see Chapter 22).

Assuming that vitamin D synthesis was the main cause of pig-ment variability in man, the present level of culturation has usually removed this characteristic from active environmental control. There are, however, other aspects of skin pigmentation besides its role in vitamin D synthesis. Pigments also protect the skin from sunburn. Even pale skin becomes more heavily pigmented, or tanned, upon exposure to sunlight. This reversible mechanism allows maximum absorption of ultraviolet in the winter when especially needed, yet protects the skin in the summer, when people are less heavily clothed and vitamin D is abundantly produced. Apparently, in the past, the selective advantage of freedom from rickets has outweighed greater susceptibility to sunburn or, more rarely, skin cancer in lightly pigmented people.

Sickle Cell Anemia

The normal red blood cell in man has the shape of a biconcave disc. Occasionally hemoglobin, a protein that readily combines with oxygen, is imperfectly formed in certain red blood cells. Such cells have a caved-in or sickle-like appearance (Figure 1.13). These sickle cells are less efficient carriers of oxygen, resulting in a general oxygen deficiency called sickle cell anemia. This condition is found in a broad belt stretching from tropical Africa, Sicily, Greece, and southern Turkey to India. In some places up to 20 percent of the population carries this trait. If the inheritance of the condition is complete (or homozygous), the individual usually dies long before the reproductive age; but if the inheritance is incomplete (or heterozygous), survival is not only possible, but ironically, the individual has a selective advantage. Recently scientists noticed that although the homozygous individuals didn't live to reproduce, the frequency of sickle cell carriers remained stable in the population. Further investigation found that sickle cell anemia somehow gave the sufferer a resistance to a common form of malaria, hence the sickle cell mutant was selected for in tropical areas.

With improved sanitation and mosquito control programs in sickle cell areas, leading to a lower incidence of malaria, the selective advantage would be removed; in time one would expect the sickle cell carrier to be selected against and the incidence of sickle cell anemia reduced in the population. In environments where malaria is not common, sickle cell anemia may be debilitating enough to be selected against by the environment. Occasionally a sickle cell carrier dies or is severely incapacitated when traveling in a jet plane. Although jets have pressurized cabins, they maintain a pressure equivalent not of sea level but 5000 feet; a sickle cell carrier is not able to maintain adequate oxygenation of his body at this pressure, while a normal person experiences no problem at all.

Fig. 1.13 • Normal red blood cells look like dimpled dimes; sickle cells are collapsed into crescents. (*Anthony Cerami*)

Many biological adaptations have taken place in man since his culturation began. More recently, the use of technology in place of physical adaptation has made it possible for man to insulate himself from the effects of the natural world by creating artificial environments. Today, man can occupy an incredible variety of ecological niches without the necessity of any structural change in his body—the upper atmosphere without growing wings, the bottom of the sea without gills or fins, and most recently, outer space. We might conclude from this that man's technology has been superbly successful and that the environment is conquered.

But the environment has not been conquered. To the contrary;

our independence of it is a illusion, engendered by our remoteness from a world we see through safety-glass windshields and thermo-paned picture windows. We cannot elude the environment, for even our insulations have had their effects. Consider, for example, the effect on our health of mechanical locomotion and the resulting lack of exercise; consider the incidence of respiratory disease caused by overheated and underhumidified houses. The very technology we invoke to insulate ourselves from the natural world creates massive perturbations in that world. We respond by increasing our isolation from the natural environment to protect ourselves from the hazards we have loosed, when we should be directing our technological resources to the amelioration of these man-made disturbances. With each turn of this vicious cycle we become more remote from the natural world outside our artificial environments, and less willing to make the massive economic commitment that the restoration of environmental quality will require.

In the following chapters we shall try to understand the paradox of a technology that seems to protect man from his surroundings but has produced an environment more hostile than the ancestral cave.

FURTHER READING

Ardrey, R., 1966. *The territorial imperative.* Atheneum, New York. A lively book based on Lorenz's notion that aggression in man is instinctive.

Blum, H. F., 1961. "Does the melanin pigment of human skin have adaptive value?" *Quart. Rev. Biol.* **36,** pp. 50–63. Thorough review of the pro and con arguments about the role of pigmentation in man's evolution.

Bresler, J. B. (ed.), 1966. *Human ecology.* Addison-Wesley, Reading, Massachusetts. A collection of papers skillfully culled from the extensive literature dealing mostly with man's physiological adaptation to environment.

Cohen, Y. A., 1968. *Man in adaptation, the biosocial background.* Aldine, Chicago. Intriguing view of man's evolution from an anthropological point of view.

Dubos, R., 1965. *Man adapting.* Yale University Press, New Haven, Connecticut. A bacteriologist-humanist looks at man's evolution. The perspective is broadly medical.

Harrison, R. J. and W. Montagna, 1969. *Man.* Appleton-Century-Crofts, New York. Honestly illustrated, no-nonsense approach to man's evolution from the anatomical point of view.

Howells, W. W., 1967. *Mankind in the making.* Rev. ed. Doubleday, New York. Beautifully written account of the evolution of man, intended for the layman, but fitting the pieces of the story together with great care.

Loomis, W. F., 1967. "Skin-pigment regulation of vitamin-D bio-synthesis in man." *Science* **157,** pp. 501–506. Logically presented case for the relationship between skin pigmentation and vitamin D production.

Lorenz, K., 1966. *On aggression.* Harcourt, Brace & World, New York. The classic work by a pioneer animal behaviorist, documenting the evidence for man's innate aggression.

Montagu, A. (ed.), 1968. *Man and aggression.* Oxford University Press, New York. A reasonable defense of the view (disagreeing with Lorenz, Ardrey, and Morris) that aggression is learned, not instinctive, in man.

Morris, D., 1967. *The naked ape.* McGraw-Hill, New York. Imaginative explanation of many of man's idiosyncrasies, not all of which is acceptable to anthropologists, psychologists, and animal behaviorists.

Pfeiffer, J. E., 1972. *The emergence of man.* Rev. ed. Harper & Row, New York. The broad scope of this book offers something for everyone.

Schaller, G. B., 1972. *The Serengeti lion, a study of predator-prey relations.* University of Chicago Press, Chicago. Some extraordinary observations on the behavior of lions suggesting that lions do kill each other and their prey wantonly if given an opportunity. A good antidote for some of Lorenz's theories.

2

MAN "MASTERS" HIS ENVIRONMENT

MAN'S ENVIRONMENTAL MANIPULATIONS have continued in greater or lesser degree for thousands of years, but only recently has an increased awareness of the repercussions of these manipulations surfaced in the popular consciousness. As so often happens when national or international concern has been aroused over some issue, the search for a scapegoat begins. Hitler blamed the "international Jewish conspiracy" for economic problems in Germany; Senator Joe McCarthy made much of a Communist threat in the 1950s; but Pogo was much closer to home when he said, "We have met the enemy and he is us." However, "us" is *too* close to home for comfort, so the search goes on. For a while it looked as though the Judeo-Christian ethic might be the problem. The historian Lynn White in a much-discussed paper suggested that the overthrow of animism—which replaced an infinite number of spirits in all objects with a finite body of saints living collectively in an abstract heaven—began the process of disregard for natural environments. The scripture (Genesis 1:26) that gave man dominance over nature and permitted great environmental license evolved into a dualistic concept of man *and* nature rather than man *in* nature.

There is a natural tendency to romanticize what is past or remote, masochistically supposing that we of today are uniquely to blame for environmental problems. But the Romans were the Americans of their day, masters of the superb technology that ran arrow-straight "interstates" all over Europe and the Mediterranean world (Figure 2.1) with a modicum of help from a Judeo-Christian ethic.

If ethical considerations had a truly significant impact on man-environment relationships, then China with its long tradition of Buddhism and Taoism would be, despite its population density, a shining example of man's sympathetic handling of nature. But China is one of the most ravaged countries in the world. Obviously there is a vast gulf between people's ideals and their day-to-day activities, what-

Fig. 2.1 • Besides roads, the Romans built walls—in this instance, Hadrian's Wall in Northumberland, England, which runs up and down hill with as little concern for the terrain as is shown by today's interstate highways. (*British Tourist Authority*)

ever their ethical allegiances. More critical factors have been suggested: democratization, making more wealth available to more people; the Industrial Revolution, which greatly increased the productivity of the individual worker; and the outcome of these, the growth of affluence that leads to proportionally greater urbanization and environmental waste.

The search for a scapegoat, then, is a dodge to avoid facing the real issues. Few people have a personal environmental ethic beyond the aesthetic of a lovely view or a nice day. Our institutions have been very slow to respond to either old or new problems. And we have a childlike faith that technology will somehow always come to the rescue like the cavalry in the last reel of a Western. These are the roots of our ecologic crisis. But now, some details.

DOMESTICATION OF THE LAND

Somewhere in the tangles of his cultural evolution, man learned that by taming certain animals he could not only provide himself with meat, hides, and wool, but also milk, butter, and cheese. It was much less work to follow a herd of domesticated cattle, sheep, or goats than to chase down elusive wild game.

In contrast to their wild predecessors, most domesticated animals are not adapted to strenuous wandering and are dependent upon readily available water and food supplies. This means that unless the grazing range has particularly good soil and abundant rain, the stock may graze beyond the ability of the grass to recover, ultimately forcing the herd to move to another pasture. The repercussions of overgrazed range are seen in the Middle East and along the Mediterranean. This land originally supported thin, open forests with a good growth of grass on the mountain slopes and shrubs and grass on the plain. Homer, writing around 900 B.C., mentioned wooded Samothrace and the tall pines and oaks of Sicily. However, incessant grazing by growing herds of sheep and goats gradually eliminated such forests by destroying the seedlings that replaced old trees. Because the trees were not replaced, a park-like forest developed with widely spaced trees and grass covering the ground (the origin, perhaps, of the western concept of park?). Finally there were no trees left at all. As grazing pressure increased, even the grass disappeared. Without trees or grass to keep the soil in place on steep slopes, extensive erosion occurred. Today much of the soil that once covered the mountain slopes, making possible the growth of trees and grass, lies in deep layers on the plains or in the heavily silted rivers and their deltas. Many of the seaports of the ancient Mediterranean world are now miles from the sea because of this greatly accelerated silting process. The steep mountain slopes are stripped to bedrock in many parts of Turkey, Greece, Yugoslavia, Italy, Spain, and North Africa. Although the lower slopes are often densely covered with a growth of evergreen shrubs, called chaparral, garrigue, or maquis, this vegetation is of little economic value compared with the grass and forest that have been lost.

From Forest to Field

The forests of central Europe were far more resilient than the Mediterranean forests. With more plentiful rainfall distributed evenly over the year, central Europe was covered with a luxuriant forest of beech and oak, so dense and forbidding that permanent settlement was delayed until culturation enabled man to clear trees and plant crops. This was apparently done by chopping down or girdling trees, burning the brush, and then planting grain in the ashes—a practice that produced a good crop. As the forest began to grow back, stock could be pastured on the grasses and weeds that persisted under the thickening cover of trees and shrubs. Then the farmers moved on and the forest returned.

These intermittent agricultural disturbances, while relatively minor, did create conditions that stimulated the evolution of weeds. Plants found new environments in the campsites adjacent to the clearings. Hardpacked soil eliminated competition from plants indigenous to the forest, while trash heaps, enriched with organic remains, gave those plants able to tolerate the bright sun an excellent place to grow and reproduce (Figure 2.2). To this day most garden

Fig. 2.2 • Shepherd's purse, here beginning to go to seed, is one of hundreds of common weeds that have thrived in man-disturbed environments. (*Grant Heilman*)

plants and weeds depend on man to provide direct sunlight for their best growth, and quickly disappear when man allows the regrowth of forest trees or shrubs, which shade them. Moreover, man has encouraged the development of weed species by intentionally moving plants or seeds from one place to another, thus bringing together plants from widely separated areas as crop seed contaminants, and allowing them to hybridize and evolve into weeds. Then, by regular agricultural practices, the life cycles of weeds became closely synchronized with the crop plants, assuring the contamination of man's crops with certain weed seeds.

Clearing the forest affected man's agriculture in still another way. When the forest vegetation and its animal population were disturbed or reduced, insects, such as the Colorado potato beetle, that lived on natural vegetation transferred their attention to man's crops. Also, some of the parasites of the vanishing animals were able to make the transference to man, for instance, the malaria-carrying mosquito. As farming replaced hunting as the principal means of getting food, human populations grew much larger and more stable, creating centers for maintaining and spreading parasites and disease. Indeed, the larger the population of a village, town, or city, the greater became the danger of epidemic or plague. Therefore, many diseases, pests, and weeds date only from the beginning of agriculture, which allowed permanent settlements.

Forest clearance in Europe reached an early peak during the Pax Romana, but after the Roman Empire fell to pieces and during the period of civil and religious wars, plagues, and other misfortunes which followed, the forest once again advanced, bringing about what was literally a "dark age." With the Renaissance and the rise of nation-states, clearing for farmland was resumed and the great European forest was again reduced to isolated patches. Because of the tradition of long-term ownership of land by one family, often for hundreds of years, the farmland was generally well cared for and today remains reasonably fertile and productive.

The European forests made this transition from forest to field over a period of at least a thousand years, but the North American deciduous forest fell in a far shorter time. To describe how dense and extensive this forest once was, it has been suggested that in 1620 a squirrel with an inclination for travel could have journeyed tree to tree from the Hudson to the Mississippi without once touching ground. Forgotten in the romance of this claim are the woodland Indians, who, unlike the buffalo hunters of the plains, burned clearings in the forest and raised pumpkins, squash, and corn. Indeed, the crops grown in those Indian clearings kept the Pilgrims alive that first winter in Massachusetts. Soon, however, the new colonists were clearing their own fields, and by 1820 much of New England was crop or pasture land. It took but a short time to discover that upland New England farms often produced more stones than pumpkins, and so farmers began moving west to New York State, then to Ohio, Indiana, and Illinois, abandoning the poor farmlands back East to the forest that covers much of New England today. Over two-thirds of Connecticut—a very small state, approximately 50 by 100 miles, with a population of 2.5 million—is covered by forest, much of it second or third growth. Worthless from a forester's point of view, this suburban forest is of incalculable value as an aesthetic and recreational resource. Some idea of the extent of former clearance can be seen from the number of stone walls running through forests over a hundred years old (Figure 2.3). Hardly a woodlot in New England lacks its crumbling stone wall covered with poison ivy and inhabited by blacksnakes, skunks, and chipmunks.

Fig. 2.3 • At one time this stone wall separated fields or pastures in the rolling countryside of Westchester County, New York.

Abandonment of poor or worn-out soil became a common pattern in the settlement of North America, especially in the South, as the boyhood wanderings of Abraham Lincoln attest. But today people no longer move from poor farms to better farms; the movement is usually from the farm, poor or rich, to the city. Hence, a second wave of land abandonment is now returning land that should never have been farmed back to forest.

INDUSTRIALIZATION

The effects on the environment of clearing the forests have been evidenced in many ways, not all of them obvious. To begin with, goats and sheep, insatiable as they have been for the past two thousand years, were not the only consumers of trees and vegetation in the Mediterranean world. The sole source of fuel throughout the classical and medieval periods in this region was wood. Forests were also cut to build ships for use in wars as well as for trading, because the settlements around the periphery of the Mediterranean depended on ships for communication and trade. Responding to the stimulus of this overseas trade, industries such as iron-working, glassmaking, and pottery began using large quantities of charcoal. Initially, their needs were met near the cities, but as the forests were cleared and overgrazing by sheep and goats prevented their regrowth, charcoal had to be hauled from farther and farther away, opening even more areas to grazing. Finally, the supply of charcoal, a relatively clean-burning fuel, could not keep up with the growth of industry. The subsequent discovery of coal led to increased water and air pollution problems (Chapters 6, 8). But what concerns us in this chapter are the environmental repercussions of removing coal and ore from the ground.

Stripping the Earth

Digging large holes to extract ores causes little disturbance in the environment of an area when the surrounding countryside is already barren, dry, and starkly eroded, as it is in the ore-producing states of the Rocky Mountains and Southwest (Figure 2.4). But in the humid, forested East, the disturbance is noticeably destructive. With the exception of the Mesabi Range in Minnesota, ore mines are rather small-scale and infrequent in the midwestern and eastern United States. Of far greater impact is the removal of coal in stripping operations. The desirable coal seam often lies below layers of soil and rock, which must first be removed.

Not too many years ago, strip-mined land was abandoned; when the coal was gone, there remained the desolation of steep piles of discarded earth or overburden alternating with the trenches from which the coal had been removed. Because the soil was often acidic from minerals leached from the piles of overburden, very few plants could grow, and the area became a wasteland.

Today most states have strict laws requiring backfilling of the trenches, recontouring of the ground surface to some semblance of its original state, and the planting of trees. Red pine or hybrid poplars can usually survive and even do well if the acidity of the soil is not too great. Such reclaimed areas, called spoil banks, may be used for

Fig. 2.4 • An open pit copper mine in Utah does not seem nearly as out of place in the cold desert of the Great Basin as it might in the lushly vegetated East. The rows of black dots are railroad cars which emphasize the enormous size of the mine crater.

recreation, particularly when they have collected water to form ponds for swimming or boating. Two problem areas remain, however: those areas strip-mined before reclamation laws came into effect, and certain parts of Kentucky where coal seams occur in horizontal strata near the ridge tops. When roads are built to mine these veins or mountain tops are removed to expose them, the refuse is dumped down the sides of the mountain to collect in the valleys. When wet, these unstable slopes constantly slip and cannot be planted with vegetation. Once the coal is gone, mountainsides, entire valleys, and streams are so disrupted that any economic or recreational use is impossible for many years (Figure 2.5).

The problem in Kentucky continues to intensify, in part because of economic factors. Strip mining is, without question, profitable—return on investment can exceed 100 percent in some instances. Because of the speed of strip mining, significantly lower labor costs ($.50 per ton versus $2.75 per ton for shaft-mined coal), and close to 100 percent recovery (only 50 percent in shaft mines where columns of coal must be left to support the shaft ceiling), stripped coal undersells mined coal by about $1.00 per ton. Small wonder that the proportion of coal obtained by stripping has jumped from 35 percent in 1965 to 50 percent in 1973. Fully 75 percent of the stripped-mined coal produced in 1970 was burned to produce 34 percent of the electricity generated by steam plants. But despite much talk about an impending energy crisis in this country (see Chapter 10), a fair

Fig. 2.5 • Contour strip mining for coal in West Virginia has destroyed the natural environment of the area. The ridge tops have been denuded and tailings slide into the river valleys below. (*USDA—Soil Conservation Service*)

amount of stripped coal is exported to coal-hungry nations in Europe.

Because of the rising demand for cheap coal, especially if it is low in sulfur, the huge reserves of low-sulfur coal in the western states are just beginning to be stripped. Several enormous power plants are in operation or under construction in the southwest where Arizona, New Mexico, Utah, and Colorado meet. Plans are under consideration for another strip mine–power plant complex in eastern Montana. All of this interest in strippable coal of course depends on its cheapness. But strip-mined coal is cheap only if the true costs of its removal are ignored or borne by the public at large. When the production costs are calculated in terms of water pollution abatement programs, the social welfare of people whose lives have been disrupted, watershed damage through siltation, tax losses from debased real estate, and loss of recreational values, the inflated profits shrink dramatically. The Corps of Engineers has estimated that the cost of restoring the Coal River watershed in West Virginia would exceed $26 million, approximately the value of the coal removed. The cost of environmental rehabilitation is high now and is likely to become higher in the future. But the price is willingly being paid in West Germany. Rich deposits of lignite in the state of North-Rhine Westphalia are being efficiently mined and the land restored to productive use at a cost of $3000–4000 per acre. However the state, through a land planning commission, exercises far more control over the mining process than does any American state at present.

TECHNOLOGY

Technology is the application of the laws of science to practical problems, frequently involving extremely clever and often prodigious engineering skills. Unfortunately these skills were rather narrowly applied as man attempted to harness nature more economically. We will look at three examples of the way man has thereby altered the face of the earth: the construction of both canals and dams depended on the growth of technology; stream channelization has depended more on economic considerations; but all have had broad ecological implications.

Connecting the Waters

More than one hundred years ago, the Welland Canal was opened, allowing ships moving from Lake Ontario to Lake Erie to bypass Niagara Falls. Then in 1932 the canal was deepened to thirty-six feet. With this deepening, lampreys began to move from Lake Ontario, where they were part of the natural fauna, into the western Great Lakes, where they had never been before (Figure 2.6). Within a few years the valuable catch of whitefish and trout in the western Great Lakes was sharply reduced, with severe economic repercussions in the fishing towns around the Great Lakes.

The lamprey feeds on the blood of large fish by secreting a chemical to prevent the blood from coagulating. In this fashion the lamprey slowly kills the fish. The lamprey spawns in small streams that empty into the lakes and the larvae that develop live in the streams for a few years, then move back into the lakes to complete their development, feeding on the valuable lake fish.

At first it seemed that the lamprey was here to stay. But in the late 1950s a chemical was found that would selectively kill the lamprey larvae without affecting other organisms. By the immensely laborious and expensive method of treating all tributary streams, some measure of lamprey control has been attained. Without question, it was the deepening of the Welland Canal that allowed the lamprey to enter new territory with such disastrous results. Fuller knowledge of the lamprey's ecological adaptability might have forewarned us of the possible introduction of the lamprey into the western Great Lakes, and precautions could then have been taken to avoid this in the construction of the Welland Canal.

Two other canals, the Suez Canal and the Panama Canal, both connect widely differing bodies of water, but have until recently had features that effectively prevented faunal exchange between the con-

Welland Canal

Fig. 2.6 • The conical structures of the oral disk of the lamprey are rasps which rub a hole in the side of the fish causing bleeding. The lamprey feeds on the blood and the fish eventually dies. (*U.S. Dept. of Commerce, Nat'l Marine Fisheries Service, Nat'l Oceanic and Atmospheric Admin.*)

29

Suez Canal

nected water masses. The Suez Canal connects the Mediterranean Sea
with the Red Sea. Before the last glaciation in Europe, the two seas
were connected naturally and shared a rich tropical flora and fauna.
Then a fall in sea level, through the removal of water to form glacial
ice, separated them, and the subsequent chilling of the Mediterranean
eliminated many of its tropical species. Today there are many open
niches in the Mediterranean, while the Red Sea has most of its biological niches filled. When the canal was dug across the Isthmus of
Suez, it intersected a small lake that was highly saline because its
bottom was covered with a thick layer of salt. Any organisms passing
through the canal had to be able to tolerate a change of salinity from
45 parts per thousand in the Red Sea to 80 to 100 parts per thousand
in the Great Bitter Lake section of the canal (Figure 2.7). Since few
organisms could tolerate such a change, this built-in salt barrier prevented most Red Sea organisms from passing into the Mediterranean.
In 1956 the canal was deepened to thirty-six feet; the greater flow so
diluted the hypersaline water of the Great Bitter Lake that today the
salinity of the entire canal is approximately that of the Red Sea.

In the years since dredging, over 100 species of invertebrates
have entered the Mediterranean from the Red Sea. Significantly, one
of the twenty-four species of fish that have made the transit, the
lizard-fish has become an important element in the fishing industry of
Israel, a rather ironic sidelight to Egypt's deepening of the canal.

Fig. 2.7 • A view of the Nile
Delta and the Suez Canal from
Gemini 4 shows the Suez
Canal running from the Mediterranean (left) to the Red Sea
(right). The Great Bitter Lake
is the small body of water between them, and the delta
lakes Manzala, Burullus, and
Idku are labeled A, B, and C
respectively. (NASA)

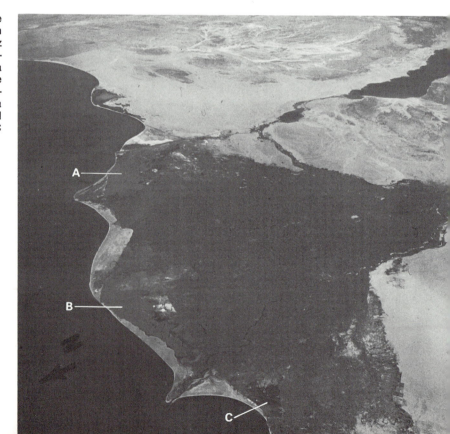

Because of available niches in the Mediterranean, introduction of new fauna from the Red Sea may very well increase the productivity of the eastern Mediterranean, without grossly upsetting the ecological balance. Some native Mediterranean species may be lost, but in this instance, the genetic and economic potential of the new species may be worth the risk.

Such is not the case with the Panama Canal. At the time of its construction, the spine of mountains running through the isthmus made the digging of a sea-level canal like the Suez impractical. The center of the canal is by contrast a freshwater lake to which ships passing between the Atlantic and the Pacific ascend and descend by means of locks. This lake serves as an effective barrier to the passage of most forms of marine life, but there is a possibility that because of the outmoded size of the Panama Canal locks, a new sea-level canal may be constructed or the present canal may be extensively rebuilt. A sea-level canal seems economically feasible only with the use of nuclear explosives to dig the ditch and expel the debris. When the costs of liability for property damage, relocation of nearby populations, and secondary excavation of rubble slipping back into the canal are added, the total expense may preclude the building of a sea-level canal.

Panama Canal

But if, despite the exorbitant cost, the canal is constructed, we must make use of previous experience from the Welland and Suez canals to project what might happen if animals from the Atlantic and Pacific can move freely from ocean to ocean. The oceans are quite different in many respects. The Pacific is colder, water temperature fluctuates more during the year, and the tidal range is greater— almost twenty feet, compared with one and a half feet in the Atlantic.

The two oceans have been separated by Central America for only three to four million years, so the fauna are certainly similar. Yet in this time there has been some divergence. According to Ira Rubinoff, a marine biologist at the Smithsonian Institution, several possibilities exist if the species were to mix in one or the other ocean:

1. If two populations were able to interbreed, the hybrids might eliminate either or both parent forms.
2. If the populations were farther apart genetically, the imperfect hybrids might eliminate the parents, but be incapable of long-term survival themselves.
3. If the populations did not interbreed but occupied similar niches, one might outcompete and replace the other.

Before construction of a sea-level canal, we should investigate these possibilities in a series of controlled experiments, looking at those species most likely to make the move from one ocean to another. Few would argue that a sea-level canal should not be built simply because a few species of fish might become extinct, but the implications are broader than this. The fisheries and sea life of both coasts of Central

America are of great importance to the economy of the region. Until we can predict with some assurance which species will dramatically increase their populations, or, conversely, which will become extinct, we cannot afford still another economically engineered, ecologically blind experiment.

Damming the Rivers and Flooding the Land

Dams are the opposite of canals in that they impede rather than facilitate access of organisms to an area. The most obvious feature of a dam is the water it impounds, but of equal importance is the environment flooded and the land that now surrounds the new body of water. Many dams have caused problems in one or more of these environments, but one stupendous project, the Aswan High Dam, has had some unique effects extending far beyond the dam site or its lake.

Constructed on the Nile in the United Arab Republic, the Aswan High Dam contains seventeen times the volume of the pyramid of Cheops at Giza and creates behind it Lake Nasser (Figure 2.8), six miles wide, ninety-three miles long, and 250 feet deep. One unusual aspect of this project is its effect on the Nile River. Historically,

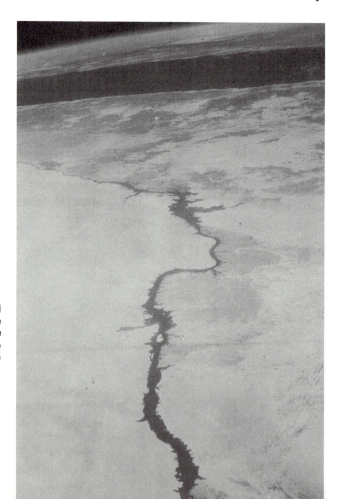

Fig. 2.8 • Lake Nasser is unusually long and narrow because of the configuration of the Nile floodplain that has been flooded. The Red Sea lies toward the top of the photo with the Arabian Peninsula beyond. (*NASA*)

the Nile has flooded every year, spreading fertile silt, derived from
the nutrient-rich basaltic lavas of Ethiopia, over the Nile valley and
delta. Over thousands of years this has created a fertile flood plain
in the Nile Valley that has nurtured civilization almost from its beginning (Chapter 3). As the fertile Nile flood water spread out into
the Mediterranean, it dramatically increased the productivity of the
coastal water from the Nile delta northeastward to Lebanon. This
fertilizing effect was measured by a count of the primary producers,
tiny algae cells that form the basic unit of the food chain in the sea.
Before the Nile reached flood stage, there were 35,000 cells per liter;
after flood stage, 2,400,000 cells per liter. One of the major harvesters of this primary food source has been the sardine; over 18,000
tons of sardines per year were caught in the UAR before the Aswan
Dam was built.

Since the dam, the Nile silt has been sedimented out on the bottom of Lake Nasser. Without the annual floods and silt deposition,
the Nile valley and delta will have to be fertilized artificially. Fortunately, the UAR is in a good position to do this: there are abundant
deposits of phosphate available, and power from the Aswan Dam is
already being used to produce nitrate. However, it would cost over
$100 million per year to manufacture enough fertilizer just to replace
the nutrients in the silt now lost annually to the bottom of Lake
Nasser.

The effect on the sardine industry has been more serious. After
closure, the annual catch fell from 18,000 tons to 500 tons, cutting
the overall production of Egyptian fisheries by almost 50 percent.
However, much of this loss may be recouped by fish harvested from
Lake Nasser. But there now seems to be some doubt that the lake
will ever completely fill because of the porous rock in the lake basin,
which allows water to seep away into underground aquifers. Of
course with the river running clear below the dam the delta of the
Nile has ceased to grow. In fact it has begun to erode, and it is feared
that the brackish Lakes Idku, Burullus, and Manzala (Figure 2.7),
which now buffer much of the intensively cultivated delta from the
Mediterranean, will become saline extensions of the sea and that cultivation will be adversely affected for several miles inland.

The creation of Lake Nasser has had another far-reaching effect.
Like Lake Kariba in Rhodesia, Lake Nasser has greatly increased the
breeding sites of the snail which carries a parasitic blood fluke causing schistosomiasis, malaria-carrying mosquitoes, and the black fly
that carries trachoma (an eye disease). Schistosomiasis, which has a
debilitating effect on the population, can be expected to increase ex-
plosively as land surrounding Lake Nasser is brought under cultivation. The only sure way to eliminate the infection is by increasing
sanitation facilities and by education, neither of which is widely available to the fellahin of Egypt.

33

Channeling the Streams

Although the Army Corps of Engineers has been associated with maintenance of navigable river channels in the United States since the late nineteenth century, channelization of smaller streams did not seriously begin until 1954, when Public Law 566 (the Watershed Protection and Flood Prevention Act) was passed by Congress. Under this law the Soil Conservation Service was charged with stream channelization to aid in the drainage of potentially farmable wetlands and to reduce flood damage. Thus we witness the spectacle of the Department of the Interior trying to save wetlands, the Department of Agriculture trying to reclaim them, while a third agency tries to reduce agricultural surpluses by land bank schemes!

Like the sorcerer's apprentice, a good bureaucracy strives greatly to carry out its tasks until directed not to. Unfortunately, until environmental awareness penetrates the decision makers, environmental degradation will continue. Some 8000 miles of streams in the United States have been "improved" since 1954 and another 13,000 miles of work is planned. Whatever the economic benefits, the environmental liabilities are many: destruction of waterfowl and game habitats, pollution of downstream lakes and reservoirs, increased upstream erosion and downstream flooding, interference with ground water recharge, and encouragement of floodplain development.

A case in point is the Blackwater River in northwestern Missouri near Kansas City (Figure 2.9). In 1910 a portion of the river was dredged, shortening it from 33.5 miles to 18 miles and sharpening the gradient from 8.8 feet per mile to 16.4 feet per mile. Both the main stem and its tributaries have greatly increased their cross sectional area, by 1173 percent for the former and 2000 percent for the latter. Because of this erosion most bridges crossing the river have been either replaced or lengthened, some several times. Beyond the terminus of the dredged section flooding has become more frequent, with floodplain deposits of almost 6 feet over the last 50 years. In addition, fish production in the dredged segment was estimated as less than one-fifth that of the undredged portion of the Blackwater.

The economic and material needs of people will continue to demand the development and completion of projects such as the construction of canals and dams, or the channelization of streams. We can continue to build on a purely economic and technological basis, picking up the ecological pieces as we go along; or we can become emotional and bemoan every lost earthworm or crushed ant, while turning a deaf ear to the crying of a hungry child. The only rational approach is to anticipate the massive technology-oriented

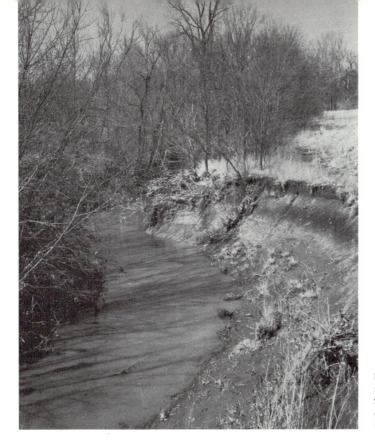

Fig. 2.9 • The Blackwater River in Missouri is typical of the many streams that have been transformed from productive aquatic systems to mere drainage ditches.

development projects of the future and weigh the prospects very carefully. In this way we may preserve some semblance of the natural world. On the basis of what we have learned from past projects, we should have some idea of the potential ecological problems previously unthought of by engineers or economists. In the time interval between a project's conception and realization—often several decades—much descriptive and experimental environmental information can be obtained. Mathematical models might be developed to make it possible to predict potential problems years in advance; at the very least, this would give sufficient research and development time to meet these problems.

There is a need to broaden the current approach to cost analysis, now largely based on what is economically and technologically feasible, to include other values, especially the recognition that projects exist in an environment that will be affected. Our first obligation is to other people, but we cannot ignore the plants, animals, and soil of the world as we scheme and dream, earthmovers and nuclear explosives ready. Our continued existence depends on the continued well-being of *all* plants and animals, not just corn and cows. Our disregard for the other organisms on earth is not just a manifestation of myopia; it is based on our abysmal ignorance of the interrelatedness of plants, animals, and their environments.

FURTHER READING

Aron, W. I. and S. H. Smith, 1971. "Ship canals and aquatic ecosystems." *Science* **174,** pp. 13–20. Several case histories are discussed.

Caudill, H. M., 1963. *Night comes to the Cumberlands: a biography of a depressed area.* Little, Brown, Boston. A graphic, moving account of man's rape of the Cumberland Plateau.

Emerson, J. W., 1971. "Channelization: a case study." *Science* **173,** pp. 325–26.

Malin, H. M., 1972. "Feds eye regulations for strippers." *Env. Sci. & Tech.* **6**(1), pp. 27–29. Reasonably current statistics on availability of strippable coal.

Marsh, G. P., 1885. *The earth as modified by human action.* Scribner's, New York. The first synoptic recognition of man's impact on his environment. Contains a fascinating wealth of detail.

Martin, P. S., 1973. "The discovery of America." *Science* **179,** pp. 969–974. An absorbing and plausible rationale for man's first major environmental effect, extinction.

Moncrief, L. W., 1970. "The cultural basis for our environmental crisis." *Science* **170,** pp. 508–512. An effective debunking of White's paper (see below).

Nephew, E. A., 1972. "Healing wounds." *Env.* **14**(1), pp. 12–21. Detailed examination of West German restoration techniques in the lignite fields near Cologne.

Rubinoff, I., 1968. "Central American sea-level canal: possible biological effects." *Science* **161,** pp. 857–861. Some background on what might happen if a sea-level canal comes about.

Sterling, C., 1971. "The Aswan disaster." *Nat'l Parks and Conser. Mag.,* August, pp. 10–13.

Veen, J. Van, 1955. *Dredge, drain, reclaim.* 4th ed. Nijhoff, The Hague, Netherlands. A thorough, well-illustrated treatment of the saga of dikes, polders, and land reclamation in the Netherlands over the last thousand years. But some voices are being raised in concern over this destruction of tidal habitats and their wildlife.

White, L., 1967. "The historical roots of our ecologic crisis." *Science* **155,** pp. 1205–1207. Lays the blame on ethical considerations.

3

SOIL, SAND, AND CYCLES

Despite tremendous technological advances since the agricultural revolution, man continues to depend on the soil for food. A complex mixture of compounds and elements necessary to man's existence on earth, soil plays a major role in the ecology of any particular area. Both plants and animals depend on it, for nutrition, either directly or indirectly. For this reason, it is important to understand something about its formation and nature.

SOIL FORMATION

Soil has a developmental history and is in some respects as alive as the plants that live in it. It contains not only organic components, but so many interdependent organisms that the resulting complex behaves as though it were one large organism. Long ago, farmers noticed the association between parent rock and soil type and assumed that a simple relationship existed everywhere. Chalky soils were found on chalk, sandy soils on sandstone, and clay soils on limestone. Later, however, when large areas of Asia, Africa, South and North America were explored, it was discovered that rather uniform soils were found in regions with varying types of parent rock, which suggested a more subtle relationship between bedrock and soil.

Weathering

Solid rock is continuously being broken down into small particles, a process referred to as weathering, which may involve either physical or chemical factors. Alternate heating and cooling of the rock itself or freezing and thawing of water in its cracks can, in time, fracture the hardest rock, as evidenced by the pile of rock fragments or talus found at the foot of most cliffs (Figure 3.1). This is the

37

Fig. 3.1 • Physical and chemical weathering on the cliff face formed this talus slope in Wales.

action of physical weathering. Other agents of physical weathering, running water and ice, can grind stones, pebbles, and sand into silt and clay, the finest components of soil.

The more the rock fractures by physical weathering, the faster chemical weathering takes place. A piece of granite, for example, may be broken down into its component minerals up to ten times more rapidly by the combined action of physical and chemical forces than it would be by chemical weathering alone. Since rocks are made of mineral crystals or older rock fragments cemented together, the minerals and the cementing substances are all soluble to a degree, and thus can be dissolved by rainwater. Pure rainwater would not make very great headway on an outcropping of rather dense and hard granite or basalt, but rainwater is rarely pure. Carbon dioxide dissolves in rainwater as it falls through the atmosphere, forming carbonic acid, which, although it is a weaker acid than vinegar, still attacks rocks more vigorously than pure water does.

Any parent rock can contribute only those minerals that it contains. Once released, however, the proportions of these minerals in the soil may be greatly affected by climate and the type of vegetational cover. A simple granite, for example, might contain three minerals: feldspar, mica, and quartz. Feldspar, a complex of potassium, aluminum, and silicon dioxide, breaks down in the weathering process into soluble potassium carbonate and kaolinite, a clay. Mica, containing magnesium as well as the oxides found in feldspar, decomposes into soluble carbonates, the sesquioxides of iron and aluminum (sesqui means 1.5 times, referring to the 3:2 ratio of oxygen to metal), and more clay. Quartz (silicon dioxide) simply weathers from large pieces to smaller pieces.

A soil weathered from granite, then, would be expected to contain carbonates, the source of many plant nutrients, sand, clays, and sesquioxides. But the climate strongly affects the rate at which these soil components are either retained or washed away, leached, by water percolating through the soil. In a cool, moist climate, the sesquioxides and clays tend to be leached away by rainfall, leaving behind porous, sandy soil called a podzol (from the Russian word meaning ash). Many of the light, greyish or brownish sandy soils of the temperate zones of the world are podzols. In the tropics and areas with a warm moist climate, the silica or sand is leached out, leaving behind the sesquioxides and clays. Because of the quantities of iron and aluminum in sesquioxide form, the soil is brightly colored in shades of red, orange, and yellow, and because of the clay content it is heavy and plastic in consistency. These lateritic soils (from the Latin for brick, an apt description of their dry state) or soils resembling them are found in the piedmont of the southern United States from Virginia to Texas, and in many other regions with warm moist climates.

SOIL STRUCTURE: A POROUS TRAP

If rock were to weather into tiny cubes like salt crystals, they would pack tightly together, leaving little room for either water or air. However, weathering produces rounded particles, and when these are packed together the interstices are filled with either air or water. In addition, the activities of plants and animals make large channels and holes that allow even more air and water to enter.

Immediately after rain most of the spaces between the soil particles are filled with water, much of which is lost to the water table by the action of gravity, or to the atmosphere by evaporation. The remaining moisture is gradually absorbed by plant roots and returned to the air via the plant's leaves. This transpiration—loss of water from the plant—is an integral part of the cycling of water in the environment (see Chapter 5).

Obviously, plants cannot grow without water, but what is less obvious is that the air content of soil is important too. Plant roots contain living tissues that are as vital to the plant as its leaves, stems, and flowers. While water can move into the plant root by simple diffusion, minerals dissolved in that water may require an output of energy by the root to be absorbed. The source of this energy is photosynthesis. Plant leaves are able to use sunlight to split water molecules into hydrogen and oxygen ions (charged particles). The oxygen ions combine to form gaseous oxygen, but the hydrogen ions are fed into a complex series of reactions along with carbon dioxide, and ultimately yield a sugar.

Sugars can be burned at once in an energy-yielding process called respiration, or transported to the roots, where the sugars supply the energy needed to absorb certain minerals from the soil. But oxygen must also be available in the soil. In waterlogged soil all the spaces between particles are filled with water and oxygen is in very low concentration. Plant roots therefore die from being unable to respire properly, and because of their critical function in supplying water and minerals, the plant also dies.

The feeder roots of trees, those roots active in absorbing water and minerals, are quite close to the soil surface, and are killed if the soil in which they are growing is covered with an additional six inches or more of new soil or fill. Such filling prevents air from reaching the level where the roots are growing. The reason many trees die when a house is built on a wooded lot is that the feeder roots of the trees are killed when soil removed from the cellar hole is spread over the roots of adjacent trees to a depth of several feet. Within a year the leaves yellow, and the tree dies. This need not happen if care is taken to build dry wells around the trees where soil will be placed. By backfilling up to the dry well, a zone of aerated roots is left adjacent to the tree, which permits the tree to survive until its roots can occupy the new soil (Figure 3.2).

Fig. 3.2 • By constructing a dry well around the base of a tree sufficient feeder roots survive to keep the tree alive until it can develop new roots in the new soil backfilled against the wall.

SOIL: THE ORGANIC COMPONENT

Soil with its inorganic components, sand, clay, air, and water, is still incomplete. The organic material that must also be present in fertile soil is called humus and is made up of partially decomposed vegetable material, leaves, rotting wood, and animal remains—their excrement as they live, their bodies when they die. Humus is the rich brownish-black crumbly top layer of leaf mold in temperate zone forests which you see when you brush away the dry leaves on the forest floor. Large animals, such as elephants, deer, or rabbits, make some contribution to humus; mice, insects, and worms contribute even more. But, because of their numbers, bacteria, algae, fungi, and other microorganisms contribute more humus to the soil than all the big trees and animals combined. These organisms are also of critical importance in the decomposition of organic material and the cycling of nutrients.

Humus, which gets mixed with the mineral components of the soil by the burrowing activities of animals and the dying and rotting of plant roots, forms a fairly stable association with the clay in the soil, a clay-humus complex (Figure 3.3). This complex then forms aggregations, or clumps, of various sizes that determine the soil texture, which in turn helps to determine the kinds of plants that will grow in a particular kind of soil.

Fig. 3.3 • A schematic drawing of a clay-humus particle shows the relationship between mineral nutrients and soil fertility. Carbonic acid combines chemically with the mineral nutrient ions in the particle, leaving hydrogen ions behind.

As organic compounds and minerals in the soil gradually decompose, they dissociate into simpler and simpler compounds, finally forming ions that percolate through the soil and are attracted and held by the clay-humus particles. These ions comprise a reservoir that determines soil fertility. If there is an abundant supply of clay-humus particles well saturated with minerals, the soil is said to be quite fertile.

Carbonate ions from rainfall can extract minerals from the pool, leaving hydrogen ions in their place. If positively charged ions continue to be added to the soil from leaf fall, their loss to the carbonate ions that pass through is balanced, and the soil remains fertile. But if the replenishment of ions is interrupted or diminished, they will be leached away as carbonates and the soil will be less fertile.

This is what has happened since the settlement of Virginia in the seventeenth century. When rich forest was cleared and the land plowed and planted in tobacco, minerals were not replaced; in fact they were removed both by the leaching action of rainwater and the harvesting of the tobacco leaves. Without replenishment, the fertility dropped until the land was worn out. Large areas of the South were farmed in this fashion; much of this land was subsequently abandoned because of its low fertility. The economic disruption of the Civil War caused further areas to be abandoned. After a few years pines invaded the land and the extensive pine forests covering so much of the present-day South were formed. If you enter one of these forests in the late afternoon when the sun is at a low angle you can still see furrow marks from the last plowing, years ago.

Soil Acidity

As the clay-humus particles lose their mineral nutrients the soil becomes less fertile. When these minerals are replaced with hydrogen ions, however, the soil becomes more acidic. Acidity is defined as the concentration of hydrogen ions, which is usually expressed by the symbol pH. A pH of 1 means that one in 10,000 parts by weight of the soil is composed of hydrogen ions. A pH of 2 means 1 in 100,000 parts; 3, 1 in 1,000,000 parts, and so forth. The lower the pH, the greater the acidity. The pH scale runs from 0 to 14. A pH that decreases from 7 to 0 is becoming increasingly acid; a pH that increases from 7 to 14 is becoming increasingly alkaline. A pH of 7 represents neutrality.

If most of the minerals normally held by the clay-humus particles are replaced by hydrogen ions, the pH drops to about 4. On the other hand, if the clay-humus particles are saturated with minerals, the pH rises to about 7, close to neutral. If the soil is supplied with organic acids or lime, the pH may be lower or higher than these two values. For most soils the pH is between 5 and 7. A low soil pH may affect the solubility of various minerals and, hence, adversely affect the activities of vital microorganisms.

We cannot, however, infer the fertility of the soil from its pH alone. If we had a soil with a large quantity of clay-humus particles but they were only half saturated with minerals, the pH would probably be below 7. Another soil with a very much smaller quantity of clay-humus particles that were fully saturated with minerals would probably have a pH close to 7. An assumption that the latter was more fertile than the former on the basis of pH might be quite mistaken. A more accurate way to measure the fertility of soils is to find the number of clay-humus particles present, determine from that the nutrient exchange capacity of the soil, and from that its fertility.

Man has had his greatest impact upon the water, air, and organic components of soil. This impact is especially obvious in the beaches and dunes characteristic of many coastlines. At first glance, sand hardly seems a soil at all compared with that in our gardens or under our lawns. But if moisture is available, coastal sand qualifies as a soil by supporting vegetation, mostly simple but vigorous growth of a few grasses and shrubs. Beaches and dunes when thus viewed as a soil-vegetation complex demonstrate graphically how man may interfere with normal cycles that produce and maintain these fragile features of our landscape.

CYCLING OF BEACHES AND DUNES

The east coast of North America from Long Island south to Yucatán, Mexico is characterized by a chain of barrier islands isolated from each other by inlets and from the mainland by lagoons or bays. Until recently these barrier islands have retained a wild beauty that is characteristic of the interface between land and sea, just as the Great Plains embody the beauty of the land-sky interface. Because of this charm these island beaches have been popular recreation areas for many years. Visitors first made day trips by excursion steamers, then built crude summer cottages, and finally plush, year-round homes.

Barrier islands and their beaches, formed largely of sand, are quite unstable, despite their characteristically heavy cover of grasses, shrubs, or even low trees. Both the barrier island and its beaches have an erosion cycle: that of the island is hundreds or perhaps thousands of years long, that of the beach is annual. These cycles cause barrier islands to grow and enlarge or erode and shrink, unlike rocky islands, which usually just get smaller as they are eroded away.

Most sandy coastlines are fringed with a row of dunes parallel to the beach. During times of rapid sea rise following glacial periods, the lower areas to the lee of the dunes are flooded and become shallow lagoons. The former dune ridges become an offshore barrier island (Figure 3.4a). The results of this process are almost diagrammatically shown along the Georgia coast (Figure 3.4b), where a retreating sea has stranded six different skeins of barrier islands, including the present Sea Islands.

The movement of longshore currents parallel to the mainland, and the force of storm waves coming in at various angles to the beach line, keep island barriers moving in two directions—along the coast

MAN IN
THE LANDSCAPE

Fig. 3.4 • (a) One of the most plausible theories about the origin of barrier islands holds that at any given sea level dunes form along the coast. When the sea level rises the low areas behind the dunes are flooded, separating the dunes from the mainland and forming the barrier islands. (*From Hoyt, J., 1969. "Barrier island formation,"* Geol. Soc. of America Bull. *78, pp. 1125–1136, fig. 2*)

Fig. 3.4 • (b) A series of barrier island chains marks former sea levels along the Georgia coast. (*From Hoyt, J., 1968. "Geology of the Golden Isles and lower Georgia coastal plain." In Maney, D. S., F. C. Marland, and C. B. West, eds.,* The future of the marshlands and Sea Islands of Georgia. *A record of a conference convened by the Georgia Natural Areas Council and the Coastal Area Planning and Development Commission. CAPDC, Brunswick, Ga. 31520 or Univ. of Ga. Marine Inst.*)

and toward the land. Longshore currents move in a fairly constant direction and tend to remove sand from the up-current end of the islands and to deposit it on the down-current end. Structures built on the eroding end eventually topple into the water; those on the growing end will soon be hundreds of yards from the water's edge.

In the winter large storm waves stir up the sand on the beach face and pull it offshore, eroding the beach by as much as two or three yards. Anyone visiting the beach at this season might well wonder what became of last summer's broad expanse of sand. During the succeeding spring and summer, gentle wave action pushes the sand back onto the beach from the offshore bars where it was deposited during the winter storms. By Memorial Day or the Fourth of July weekend, the beaches on the eastern coast of the United States are restored. This annual cycle is even more obvious along the California beaches. Although these beaches are not usually associated with barrier islands but are found at the foot of cliffs, they have the same rhythm of winter removal and summer replacement of sand. This can be strikingly seen in the educational film, *Beach, A River of Sand.*

The Disappearing Coastline

Barrier island beaches, then, are constantly on the move, and part of their charm as natural phenomena lies in this movement. But how many times has an eager summer resident built a house in the dunes or on the beach only to find to his dismay that his house and property are moving into the sea.

The fact is, beaches, barrier islands, and sand bars are ephemeral features of the environment. They will persist if sea level, climate, sand supply, and offshore currents remain constant, but not in exactly the same position from year to year. Problems arise when man wants to stabilize the unstable.

Cape Hatteras National Seashore Area in North Carolina provides a classical statement of some of these problems. The trouble began with the establishment of a permanent road in the 1930s. To protect the road from the sea it was decided to stabilize the low dunes running along the upper beach. Snow fencing was set up, grasses planted, and the dunes began to grow to impressive height, effectively shielding the road from storm waves. But as the dunes grew in height the width of the beach narrowed from 450 feet to less than 90 feet. In addition, the foreslope of the high dunes was attacked by every storm, which severely eroded and occasionally breached the dunes. The reason for this is straightforward: the profile of an unaltered natural barrier island is low, with few dunes of any size (Figure 3.5). The barrier island accommodates the energy of storm waves by presenting no serious obstacle to their attack. The sea simply washes over the barrier island, leaving a delta of sediments

Fig. 3.5 • Core Banks, part of Cape Lookout National Seashore in North Carolina, has been little influenced by man. Storms are free to wash over the barrier island, leaving deltas on the lagoon side that are soon vegetated by marsh plants. The dunes are low, so the wave energy is expended in overwash rather than beach erosion; thus the beach is quite wide compared with that in Fig. 3.7. (*National Park Service photo by Paul J. Godfrey*)

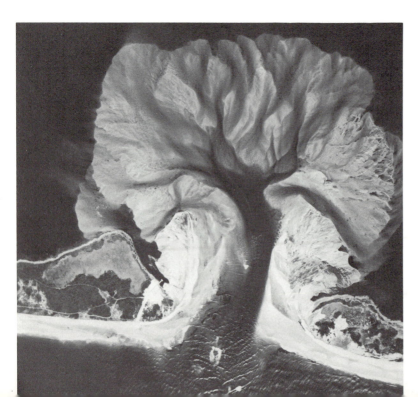

Fig. 3.6 • The Moriches Inlet on Fire Island, New York, just after it was opened by a storm. Sand washed into the bay by the storm waves will be vegetated by salt marsh species after the inlet silts up. (*U.S. Air Force*)

Fig. 3.7 • After the road was constructed, fencing was used to stabilize and increase the height of the dunes which run parallel to the beach. While the road was protected, the beach began to erode away. (*National Park Service photo by Paul J. Godfrey*)

(Figure 3.6) in the lagoon. The delta is soon colonized by marsh plants and the inlet or channel silts up. When the dunes have been stabilized, preventing easy overwash, the wave energy is concentrated on the beach and foreslope of the high dunes and these features erode (Figure 3.7). When the dunes are breached the road is either washed out or covered with sand.

In nonpark areas where the road has stimulated extensive development the extremely expensive answer has been to restabilize breached dunes, close inlets, and build groins or stone piers to prevent sand from being washed away, defying the inevitable like King Canute. On publicly owned barrier islands, the best solution would be to remove existing roads and prohibit new ones. Once the sea is free to wash over the barrier island, equilibrium will be restored. Transportation can be supplied by mainland-based ferries and by small buses able to run along the hard sand of the beach leaving nothing but tread marks that are smoothed by the next tide.

A simple but revolutionary solution would be to put all beaches and barrier islands under public ownership, and allow only the most temporary and replaceable recreational facilities to be built. Then we could let these ephemeral features come and go as they please. For

Solutions

47

those who fear that all beaches would soon slip away into the sea, it can be argued that there will always be sand beaches for recreation somewhere because sand eroded from one beach is not usually dumped into the middle of the ocean, but is piled up on another beach somewhere.

If such a prescribed dose of public ownership is unpalatable, then some form of land use control might be established. This could be administered as a zoning regulation which would prohibit the construction of houses on the most unstable sites. The ultimate alternative to such measures is, of course, complete freedom to build a house wherever one wishes—on the top of a sand dune, the high tide line, or even a quarter-mile out to sea. But in exchange for this right the builder should not expect public subsidy when his investment is threatened by shifting sand or storm-whipped waves.

Man also attempts to "improve" beaches. Because of strong longshore currents and lack of sand, many stretches of coastline are covered with cobbles and boulders. By pumping sand from offshore deposits, an artificial beach can be constructed over the rocks, but no one should be surprised when after a few years the sand has been returned by the waves to its former location. Such a man-made beach can be maintained only with constant effort and expense.

Artificial beaches can be stabilized to some extent by constructing groins out into the sea at right angles to the shore. These intercept the longshore currents that sweep the sand away and allow some sand to be retained (Figure 3.8). Even so, sand must be pumped in every few years to maintain a continuous beach, and the groins must be kept in good repair.

Fig. 3.8 • Despite the construction of these groins and the pumping in of sand from offshore, currents moving from the foreground to the background have begun to remove the sand, exposing the original surface of cobbles that formerly comprised the beach.

Beaches and dunes represent one type of cycling. But whatever the climate or type of soil, the vegetation found on most soils plays a critical role in another type of cycling. As we saw in our discussion of the weathering process, nutrients are captured by the roots of plants and ultimately made available to other elements of the biological community. Without this cycling mechanism, most soils would be so lacking in nutrient content that their cultivation would be impossible.

SOIL CYCLES

An ecosystem is any group of organisms interacting with each other and their environment; thus a drop of pond water, the entire pond itself, or the region in which the pond is located can all be considered ecosystems. The ultimate source of the energy that powers the ecosystem is sunlight. But this energy is dissipated as it flows through the system from the photosynthesizing primary producers, algae or grass, through a series of consumers, hunters, and harvesters. Without continued recharging from sunlight, ecosystems would soon run down and their component organisms die. Water, foods, and minerals, however, recycle indefinitely in a regular, often subtle way, and it is precisely this cycling of materials through the ecosystem that can so easily be disrupted by man. The continued well-being of ecosystems —and of man, who is a part of nature—depends upon what man does about keeping these natural recycling mechanisms intact.

Almost all minerals, nutrients, and gases have distinct cycles and cycle periods. The most obvious cycles are those of oxygen, carbon dioxide, and water, because photosynthesis, respiration, and transpiration are interwoven and complementary (Figure 3.9). In the

Ecosystems

Fig. 3.9 • Three cycles, oxygen, carbon dioxide, and water, are intimately interlocked through the action of biological, chemical, and geological processes.

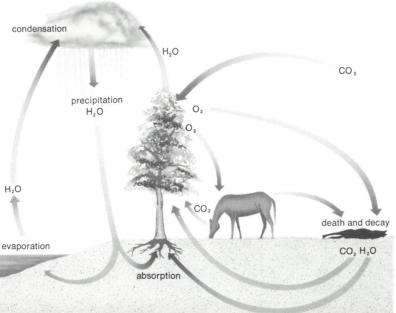

condensation

H_2O

CO_2

precipitation
H_2O

O_2

O_2

H_2O

CO_2

death and decay

evaporation

CO_2 H_2O

absorption

production of sugar by photosynthesis, green plants use carbon dioxide and water, releasing oxygen as waste. When these same plants respire, they use oxygen and release carbon dioxide and water, some of which is reused. Animals also need oxygen to produce energy from food, releasing carbon dioxide and water as waste. Much of the oxygen needed by animals comes from plants, and much of the carbon dioxide required by plants comes from animals. Thus oxygen, carbon dioxide, and water, interrelated in various physiological processes, cycle in the environment. Because cycles such as these involve living organisms as well as geological and chemical processes, they are called biogeochemical cycles.

Mineral Cycling in Temperate Forests

Plants, along with the animals that consume them, perform important roles in the maintenance of the environment as a whole, roles that are not always obvious to man. Gases are by no means the only materials cycled in ecosystems. The roots of many hardwood forest plants readily absorb minerals from deep in the soil. Some species are selective, absorbing greater quantities of some mineral ions than others. Dogwood, for example, selectively absorbs calcium, much of which goes into the developing dogwood leaves in the spring. When the leaves fall at the end of the summer, calcium is added to the nutrient pool held by the clay-humus particles, making it available to plants with shallower root systems. Dogwood thus acts as a pump, bringing calcium from deep in the soil to the surface.

When the leaves of mineral-rich deciduous plants decay after their fall, they decompose and release minerals to the clay-humus particles, resulting in a rich soil with a pH around 7. Let us suppose that because white pine is considered to be more valuable than most hardwoods, we decide to cut down a grove of hardwoods and plant pine. Pine roots are less efficient than hardwood roots at absorbing minerals, so pine needles have a lower mineral content. In a few years the supply of rich hardwood leaves would be replaced with pine needles. More minerals would be lost from the clay-humus particles than would be replaced from the decaying pine needles. As a result, the fertility would decline, the hydrogen ions would accumulate, and the soil would become acidic; that is, the pH would drop. The fall of both fertility and pH would be accelerated by the resin content of the pine needles, since resin forms organic acids in the soil which lower pH and retard decomposition, further decreasing the availability of minerals. Thus, man by his manipulation of vegetation affects the nature of the soil.

Mineral Cycling in Tropical Forests

In a temperate forest, most of the minerals made available by decomposition of organic litter or disintegration of the parent rock are quickly absorbed by plant roots and incorporated into the vegetation. If you were to stand in an oak-hickory forest in midsummer you would find several inches of slowly rotting leaves covering the rich topsoil, itself black with incorporated humus. Conversely, a tropical rain forest has such a continuing high rate of organic litter decomposition that no mineral pool has time to accumulate. Directly beneath the most recently fallen debris is a heavy, clay-containing, mineral soil. As a result of this tie-up of all available minerals in the standing vegetation, the cycling of minerals is rapid and direct. As soon as a leaf falls, it is decomposed and its minerals are absorbed by plant roots and channeled into the growth of another leaf. So tight is this cycling process that those few ions not absorbed by plant roots but leached through the soil into the water table, and then out of the system, are replaced by ions picked up by the tree roots from the slowly disintegrating bedrock below or from rainwater.

When a tropical forest is cut, minerals are suddenly released faster than crop plants or the remaining trees are able to use them. They leach out of the system and fertility drops sharply. If the disturbance covers only a few acres, weeds and short-lived successional species quickly invade the area, shield the soil, and begin to restore the balanced mineral cycle. But when very large areas are cleared, this kind of recovery may be impossible. The lateritic nature of the soil also becomes part of the problem. When the forest is cleared, the heavily leached sesquioxides are exposed to high temperatures, and they bake into pavement-hard laterite. Once formed, laterite is almost impossible to break up, and areas that once supported lush forest quickly become scrubland at best.

Ignorance of this aspect of forest ecology has had disastrous consequences in some tropical forests. A few years ago a large tract of rain forest in the Amazon Basin was cleared and cultivated. In five years the fields were virtually paved with hard-packed laterite. The ancient Mayan and Khmer civilizations of Mexico and Cambodia probably fell into decay in part because of the destruction of their soil by laterite formation following rain forest clearance.

In view of the abundant evidence that clearing tropical regions of their protective vegetation can often lead to ecological disaster, it it disturbing to read of short-sighted proposals to reclaim and develop large areas of "unproductive" forest. Schemes like the Mekong River Project, which would convert the Mekong River in Southeast Asia into a series of pools like the Tennessee River in the United States, are long on engineering skills and short on ecological knowledge. Much of the fertility of the Mekong Basin comes from the silt left by annual flooding. Deposition of this silt behind a series of dams would benefit no one. The large-scale forest clearance that would have to accompany such an extensive environmental change would probably accelerate the laterization of the soil and result in double-edged destruction under the guise of development: the resources of the forest would be destroyed, and the soil would be ruined for any future growth of desirable vegetation.

THE NITROGEN CYCLE

Approximately 78 percent of the air we breathe is nitrogen. In combination with carbon, hydrogen, and oxygen, nitrogen is an important part of the proteins vital to life. Because of its role in protein synthesis, nitrogen is also one of the most important elements to plants. Absorbed as nitrate or ammonium ions, nitrogen contributes to rapid and luxuriant growth. We know that we can easily supply nitrate and ammonium ions to plants artificially, but how do plants get their supply naturally? Since over 90 percent of the world's supply of nitrogen exists as a gas in the atmosphere, let us examine a series of conversions within the ecosystem that began in the atmosphere. Lightning strikes contribute a few pounds of nitrates per acre per year. In the soil certain bacteria are able to "fix" nitrogen into ammonia (NH_3), using as their energy source the foods produced by photosynthesis of a host plant. These organisms often live in swellings or nodules on the roots of plants such as beans and clover. These microorganisms are capable of fixing 200 pounds of nitrogen per acre per year into ammonia.

But many plants are unable to use ammonia directly and some are poisoned if ammonia accumulates. Fortunately another group of soil bacteria is able to oxidize ammonia into nitrite (NO_2), and still another oxidizes nitrites to nitrate (NO_3). From these sources, nitrogen is made available for the synthesis of protein. But what is the fate of nitrogen contained in proteins when plants or animals die? These proteins are broken down by bacteria into their component amino acids, and the amino acids are in turn broken down into ammonium (NH_4) ions. Bacteria convert the ammonium into nitrite, then either degrade this to nitrogen gas or oxidize it back to nitrate, where it is absorbed by plant roots and the cycle begins again (Figure 3.10).

Perturbations and Their Repercussions

In late 1966 a water company in California's San Joaquin Valley notified its customers that its water was considered unsafe for infants, for it contained more than 45 parts per million (ppm) of nitrate, close to the level that may cause the occasionally fatal disease methemoglobinemia. Methemoglobinemia is the term used to describe a depletion of oxygen in the blood of infants. Because of the almost exclusive milk diet of babies, their stomach and small intestines are less acid than those of adults. This sometimes allows bacteria, which convert nitrate into nitrite and which are native to the alkaline large intestine, to survive in the upper portions of the digestive tract. If water with high levels of nitrate is fed to such infants, these bacteria convert the nitrate to nitrite. Although nitrate is relatively innocuous, nitrite is able to bind itself to hemoglobin in the red blood cells. Normally hemoglobin carries oxygen as oxyhemoglobin all over the body, supplying the respiring cells with oxygen. Nitrite, however, unites more readily than oxygen with hemoglobin-forming methemoglobin. The more nitrite produced, the more methemoglobin formed and the less oxygen available to the body's tissues, resulting in an

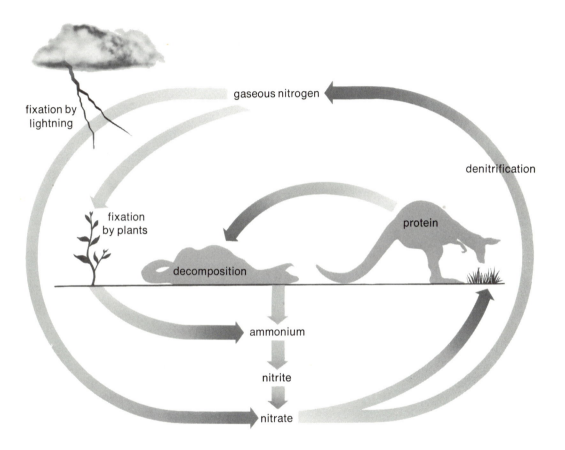

Fig. 3.10 • The nitrogen cycle is of critical importance to the growth of plants that supply the amino acids necessary to build animal proteins.

anemia. This condition has been observed when the nitrate ranges from 66 to 1100 ppm in the water supply of infants.

The problem occurred in the San Joaquin Valley in California because of certain irrigation practices. Early settlers found a fertile valley but only six inches of rain a year. Wells soon provided irrigation water, but after a number of years of intensive pumping, the well water level fell. At the same time, leaching of both natural and artificially added sources of nitrates produced an enriched zone of nitrates deep in the soil. Then in 1951 the Friant-Kern Canal bringing water from the Sierra Nevada was opened. The water table began to rise again because of the plentiful surface of water, the reduced demands from wells, and drainage from the irrigated fields. When the water table reached the enriched nitrate zone, the nitrates dissolved and began to contaminate the wells used for drinking water.

Another case of nitrate pollution of a water supply was discovered in Decatur, Illinois, in the 1960s. Routine analysis of tap water showed a nitrate concentration above the 45 ppm recommended by the Public Health Service. The source of the water was Lake Decatur, an impoundment of the Sangemon River. Both the river and lake had a nitrate content of 45 ppm. Since the concentration of nitrate varied with the season, an agricultural source was suspected.

The dominant crop plant in Illinois, corn, requires either a very fertile natural soil or one heavily fertilized with nitrates to produce well. As the natural fertility of the soil declines when crops are removed, breaking the natural nitrogen cycle, crop yields decline as well unless fertilizer is added. In 1945 the average yield of corn in Illinois was 50 bushels per acre with relatively small amounts of fertilizer being used. By 1958, 100,000 tons of fertilizer were used statewide to produce a yield of 70 bushels per acre. In 1965 it required 500,000 tons of fertilizer to produce 90 bushels per acre. Apparently while the yield increases arithmetically, the fertilizer requirement increases logarithmically.

Over this period of increasing fertilizer use the nitrate level in the Illinois River increased by a factor of 5. But was this "new" nitrate from heavily fertilized fields or a result of population growth? While most nitrogen is in the typical configuration N_{14}, about 0.4 percent exists in a slightly different form, N_{15}. The nitrogen in soil, manure, and sewage has higher levels of N_{15} than does the fertilizer nitrogen, which is manufactured from the nitrogen in air. The ratio of these two forms of nitrogen can be determined on a mass spectrometer. When such an analysis was performed by researchers at Washington University, the nitrate was low in N_{15}, suggesting that as much as 60 percent of the nitrates in Lake Decatur came from fertilizer used on the surrounding farmlands.

Water is not the only source of excess nitrate. Leaf crops that are heavily fertilized with nitrate to encourage vigorous growth may contain greater amounts of nitrate than are desirable. An infant given strained spinach that contains excess nitrate might risk conversion of enough nitrate to nitrite to cause methemoglobinemia. As a result of this danger, the Food and Drug Administration has established as maximum 40 ppm of nitrate in canned food. Thus the consumer is now better protected, but producers are faced with a new challenge. Because of leaching, only about 20 percent of the fertilizer applied to a crop is actually taken up by that crop. To counteract this, the tendency has been to fertilize more heavily, leading occasionally to dangerous accumulations of various minerals in the harvested crop. Yet if the land is not fertilized with nitrates the yield of leaf crops constantly falls. This presents the farmer with a rather serious problem. He must somehow balance meeting the FDA standards with maximum fertilization. If productivity of farms can be maintained only by overfertilization, then less soluble fertilizers that prorate their nutrient effect over an entire growing season must be developed. Other solutions have not as yet been sought because of the relative ease of fertilization. The alternative is pollution of underground water tables that recharge very slowly, some requiring thousands of years.

Logging and Nitrate Runoff

Overfertilization of cropland is not the only source of excess nutrients in the soil. A fascinating experiment set up by F. H. Bormann and G. E. Likens on a series of watersheds in a New Hampshire forest carefully monitored the minerals being added to the systems by rain and snow, and those being removed via stream flow, allowing a balance sheet to be drawn up. After a few years of undisturbed operation one watershed was logged with as little disturbance as possible. The trees were cut and delimbed but left on the forest floor, and a herbicide was used to prevent sprouts and weeds from growing. After two years the nitrate ions increased dramatically from 0.9 to 53 ppm. An unexpected result of this experiment was that the overfertilization of a sparkling brook produced a dense growth of algae (see Chapter 6).

Normally nitrate is highest in winter when plants are dormant, lowest in summer when plants are actively growing and using nitrate. As was apparent from the experiment just described, clear cutting, or felling all the trees in a given area, increases runoff by as much as 40 percent over the year (four hundred eighteen percent in normally low-water months when the vegetation loses much water from its surface by transpiration). Increased runoff increases the rate of leaching, but the reduced root absorption decreases the removal of minerals from the material leached out, and results in the loss of nitrate from the system. If watersheds are to be properly managed for high-quality water, logging must be recognized as a potential source of excess minerals as well as of silt from erosion.

Obviously we still need to know more about the details of the nitrogen cycle, in particular how it is affected by land use patterns.

NUTRIENT CYCLING

When we examine the nutritional needs of plants, we find that there are two groups of nutrients: those that are required in large quantities, macronutrients, such as nitrate, phosphate, calcium, potassium, and others; and those needed in very small quantities, micronutrients or trace elements, such as cobalt, boron, copper, vanadium, and so on. The need for macronutrients was recognized long ago and was supplied first by natural manures, then by synthetic fertilizers. The need for the micronutrients was a more recent discovery. In 1906 Sir Frederick Hopkins first broached the idea of accessory food factors, which were later recognized as micronutrients. Humans as well as plants need these micronutrients as components of vitamins, enzyme systems, hormones, and their precursors.

The process of harvesting, taking crop plants from the field for processing or consumption, removes many minerals incorporated in plant tissues. Unless these minerals are replaced, the fertility of the soil declines, gradually in temperate areas, precipitously in the tropics. Routine fertilization replaces lost macronutrients—sometimes too efficiently, as we have seen. But the micronutrients may be depleted without symptoms of conspicuous deficiency developing in the plants, so their loss is more difficult to detect. Trace element deficiency is not confined to plants, however; it can occur in animals and in humans as well. In man, its subtle, debilitating effects have sometimes been mistaken for mental deficiency.

In Australia a certain chronic disease of cattle was found to result not from bacteria or viruses, but from a deficiency of cobalt in the soil. When this trace element was added to the soil in the proper amount and passed on to the stock through the grass, the "disease" disappeared.

Certain trace elements are now added to fertilizers, but we still lack knowledge of all such elements that play a vital role in our metabolic well-being. Progress is being made, however, in determining the roles played by trace elements in normal and abnormal metabolism. A deficiency of copper may cause defects in the connective tissue forming the walls of blood vessels, leading to hemorrhages; chromium seems to play a role in the sugar metabolism of the body; and vanadium may inhibit formation of cholesterol (the fatty substance deposited on blood vessel walls).

We cannot assume that all the vitamins and trace elements for which we have a physiological need are automatically provided by food from our supermarkets. Yet as the population continues to expand, demanding more food from less land, it is absurd to insist that only organically grown food is fit for human consumption, to imply that super phosphate from a bag is somehow unnatural, if not sub-

versive. But in this age of additives (see Chapter 12), it is too easy
to cut corners by adding something to take the place of something
milled, processed, boiled, or frozen out. The pluses and minuses
may balance, but no one seems to be certain that they do. To assure a
reasonable intake of vitamins and nutrients one need not become a
health food faddist and subsist on carrot juice and wheat germ. But a
widely varied diet, including green and yellow vegetables, fruit, and
grain, as well as protein and carbohydrates, *is* essential. The problem
is nicely summed up by Dr. Walter Mertz of the Walter Reed Army
Medical Center:

Nobody can state for certain that all the elements with an essential function are known, and nobody can take for granted that our diet contains
sufficient quantities of these. Careful animal experimentation must attempt
to define deficiency symptoms and to clarify the biochemical and nutritional role of trace elements. Only then, can we hope to apply this
knowledge to the detection and prevention of deficiency states in man
and to their possible correlation with the many slow, long-term degenerative disease processes that concern us.[1]

Perhaps our general lack of concern with soil results from our
increasing remoteness through urbanization. Man is rarely able to
exact a yield directly from soil. Only through the medium of vegetation are we able to stabilize soil and control its yield of water and
minerals. While we are more or less aware of what happens to vegetation as a result of insect attack, industrial or automotive fumes (see
Chapter 8), fire, or drought, the soil is still too often regarded as a
kind of incinerator capable of absorbing and decomposing all and
every material developed by man. For most of man's history this has
been true. The fungi, bacteria, algae, protozoa, and arthropods have
had an enormous capacity to cope with organic materials. But in the
past few decades, the production of materials unknown in nature has
overwhelmed the resources of the soil. Plastics and biocides persist in
the soil because so few organisms can break these materials down.

If soil is to be maintained as a resource, we must give much
greater priority to understanding its cycles, to untangling its structure
and function, and to understanding its fauna and flora and their relationship both to natural environmental factors and to those imposed
by man's use. Until then, we run the grave risk of destroying a resource and permanently upsetting a balanced system before its full
potential or basic importance to man is fully appreciated.

---

[1] Conference on trace substances in environmental health. 2nd. University of Missouri. 1968. D. D. Hemphill (ed.), 1969. *Trace substances in
environmental health, II.*

FURTHER READING

Bormann, F. H. and G. E. Likens, 1967. "Nutrient cycling." *Science* **155,** pp. 424–429. Explores the movement of nutrients in the ecosystem.

Committee on Nitrate Accumulation, 1972. *Accumulation of nitrate.* National Academy of Sciences, Washington, D.C. A review of the sources and fates of nitrate in the environment.

Dale, T. and V. G. Carter, 1955. *Topsoil and civilization.* University of Oklahoma Press, Norman. Fascinating historical review of man's impact on the soil.

Dolan, R., P. J. Godfrey, and W. E. Odum, 1973. "Man's impact on the barrier islands of North Carolina." *American Scientist* **61,** pp. 152–162. Well illustrated account of why the Outer Banks seem to be washing away.

Eyre, S. R., 1968. *Vegetation and soils, a world picture.* 2nd ed. Aldine, Chicago. A comprehensive coverage of soils—their nature, origin, and interaction with climate and vegetation.

Hyams, E., 1952. *Soil and civilization.* Thames and Hudson, London. Intriguing account of the effects of soil destruction upon history, with examples from the classical world, Far East, and United States.

McNeil, M., 1964. "Lateritic soils." *Scientific American* **211**(5), p. 96. Explores the results of laterite development in tropical ecosystems.

Kohl, D. H., G. B. Shearer, and B. Commoner, 1971. "Fertilizer nitrogen: contribution to nitrate in surface water in a corn belt watershed." *Science* **174,** pp. 1331–1334. Detailing of the elegant technique that differentiates between various sources of nitrogen.

Richards, R. W., 1952. *The tropical rain forest.* Cambridge University Press, Cambridge, England. The most comprehensive treatment of the rain forests of the world.

Wagner, R. H., 1964. "The ecology of *Uniola paniculata* in the dunestrand habitat of North Carolina." *Ecol. Mon.* **34,** pp. 79–96. Discusses the important reciprocal relationship between a dune grass and its sand dune habitat.

4

NATURE IN CAPTIVITY

THE FOREST and sand dune ecosystems have been especially vulnerable to man, who has, first by necessity and then by habit, viewed nature as an adversary against which he must constantly struggle. Today, ironically, it is the environment that must constantly struggle, by means of its self-repair mechanisms, to defend itself; not against those who would manipulate or destroy it but against those who would quite literally love it to death. Population growth coupled with renewed interest in outdoor activities is beginning to have a devastating impact on natural areas, from traditionally battered urban parks to formerly pristine alpine meadows. With such pressure likely to increase in the future it becomes all the more important to preserve carefully and restore present natural and recreational areas; and to acquire, soon, a broad spectrum of new ones.

PARKS

While much wilderness has been destroyed, some has been transformed into parks. This transformation is most clearly seen in densely populated England. The thick oak forest that once covered England was methodically cut for ship timbers and lumber and replaced by cultivated land. By the seventeenth century, forests were limited to those selected as crown lands to be reserved for royal hunting parties and the forests that frequently surrounded the country manor houses of the aristocracy. By the eighteenth century, these woodlands, managed for both timber and game, had lost much of their wildness. Often too, they were grazed, which tended to open the forest by curbing tree reproduction. Over the years these hunting parks became fixed in the public mind as representing natural vegetation (see Chapter 16). When the wealthy members of the rising middle class began to construct country estates, landscaping often reflected the influence of the hunting parks of the older estates.

In *Pride and Prejudice,* Jane Austen describes a walk in such a park:

> They entered the woods, and bidding adieu to the river for a while, ascended some of the higher grounds; whence, in spots where the opening of the trees gave the eye power to wander, were many charming views of the valley, the opposite hills, with the long range of woods overspreading many, and occasionally part of the stream. Mr. Gardiner expressed a wish of going round the whole Park, but feared it might be beyond a walk. With a triumphant smile, they were told, that it was ten miles round.

In the American colonies, New England towns and villages had their green or common, but these were obviously intended as places of public ownership where sheep and cattle could be conveniently pastured. The idea that parks were for public pleasure grew with surprising slowness. Besides, pleasure in an outdoor setting, or in any setting for that matter, was anathema to the established church of the late eighteenth and early nineteenth century in New England. This attitude began to change when orthodoxy gave way to the more liberal Congregational and Unitarian Churches in the nineteenth century. Indeed, moral fiber became so relaxed that one liberal man of some wealth donated a large tract of land to the City of Worcester, Massachusetts, as a public park of the *enjoyment* of the people of that

Fig. 4.1 • A typical town square in Pennsylvania includes a Civil War monument, flagpole, cannons, and flower beds.

city. Shortly thereafter, New York City became the second American city to establish a public park. Elaborately landscaped by Frederick Law Olmsted, Central Park set the tone for the creation of public parks from coast to coast, until few towns were without some patch of green with statue, fountain, or flagpole (Figure 4.1).

The National Park

The first explorers of the West, hardy trappers and Indian traders, brought back tales of fantastic mountains and valleys, geysers of steam, gleaming caves, and huge trees. Subsequent expeditions corroborated even the wildest of these tales. Because of the Victorian glorification of the romantic, there developed a demand that some of these natural marvels be preserved for the wonderment and enjoyment of future generations (Figure 4.2). Surprisingly, many such monuments of nature *were* preserved; and this at the height of an era

Fig. 4.2 • Thomas Cole and William Cullen Bryant view the sublimities of nature in a painting by Asher B. Durand. (*Collections of The New York Public Library, Astor, Lenox, and Tilden Foundations*)

of westward expansion when nature was considered by railroad, mining, lumbering, and cattle interests to have been created for man's exploitation. Between 1870 and 1900, four of the nation's best-known national parks were established: Yellowstone (1872), Yosemite (1890), Sequoia (1890), and Mount Rainier (1899). The National Park Service, which inherited these natural wonders when it was founded in 1916, had as its prime duty their protection and preservation. Similarly, the Forest Service inherited upon its establishment huge tracts of land formerly in the public domain to protect and manage for a sustained yield of forest products.

FORESTRY AND FIRE

Attempts have been made many times and by many cultures, to manage and regulate forests for water supply, grazing, and the cutting of timber. However, formal education in forestry did not develop until 1825, when schools were established in Germany and France. The concept of forest management came late to North America's seemingly inexhaustible forests. It was not until the superb white pine forests of New England and then the Great Lakes area were gone that people began to consider the possibility of managing the remnants and second growth for sustained yield. Since there were no schools in the United States, professional training had to be obtained abroad in the highly respected schools of Europe, particularly in Germany.

The major emphasis of European forestry in the nineteenth century was protection of the forest. Considering the population density of Europe even then, the demands made on forests for kindling were great, and although gleaning rights to this forest by-product were often passed from father to son for generations, forests had to be protected from poaching. But European forests also had to be protected from fire. Fire was considered an enemy in the carefully tended conifer forest plantations, and any role that it might have played in the ecology of natural vegetation was thereby blurred.

However, what was useful in the forest plantations of Europe was not necessarily useful in the incredibly large and ecologically diversified forests of North America. Because fire was considered intolerable in European forests, it was not surprising that this view was extended by European trained foresters to *all* forest types in North America. Fire exclusion came to be regarded as a basic article of forest management, first by private foresters, and later by the government's Forest Service, set up to manage the huge tracts of public land still covered by timber.

Fire exclusion

Longleaf Pine and Fire

After much of the best white pine had been cut in the Northeast and the Great Lakes area, attention passed to the hitherto neglected longleaf pine forests distributed in a band along the coastal plain from Virginia to Texas. The trees were well spaced, with a light cover of grasses on the forest floor. After the initial logging of longleaf stands, it was noticed that reproduction was poor, leading to replacement by much less valuable hardwoods or other species of pines. It was supposed that ground fires, quite prevalent in the region, and the foraging of hogs destroyed the seedlings, and that if both of these factors were controlled and enough seed trees were left to provide a seed source, the forest would soon regenerate itself.

A young forester by the name of Chapman took a close look at the ecology of longleaf pine and realized that the species, far from being eliminated by ground fires, actually seemed to have many features suggesting long association with fire as a part of its environment. He noticed that the young seedling at the ground surface develops a fat bud, surrounded by a tuft of needles. Then, for the next three to seven years, the seedling puts most of its energy into developing a deep taproot. Because of its close resemblance to a clump of grass, this phase of longleaf pine development is called the grass stage. By the end of seven years the seedling suddenly begins active vertical growth, and in a few years' time grows ten to fifteen feet. Thereafter the sapling grows at a more conventional rate but develops thick fire-resistant bark. Chapman concluded that periodic ground fires may singe the tuft of needles but leave the terminal bud and taproot intact. By using energy stored in the taproot over the three to seven years spent in the grass stage, the seedling is able to make exceptional growth in height over a period of two or three years, time enough to get the more sensitive new branches up and away from the danger of fire. Then development of thick fire-resistant bark protects the young tree from further ground fires.

Drawing on his observations as early as 1909, Chapman noted that "fire always has and always will be an element in longleaf forests, and the problem is not how fire can be eliminated, but how it can be controlled so as, first, to secure reproduction; second, to prevent the accumulation of litter and reduce the danger of a really disastrous blaze."[1] This sound appraisal of the situation was not taken seriously at first.

Data versus Dogma

Because he was not affiliated with the Forest Service, Chapman was free to speak out. The Forest Service seemed to have already made up its mind about the role of fire in all forests, contending that southern forests were no different than any other in their need to be completely protected from fire. An experiment was set up in Louisiana in 1915 to demonstrate the effect, both combined and singly, of

[1] Schiff, A. L., 1962. *Fire and water: scientific heresy in the Forest Service*. Harvard University Press, Cambridge.

foraging pigs and ground fires in eliminating longleaf reproduction.

A acre of young seedlings was divided into quarter-acre plots and treated as follows: one plot was burned and grazed, a second was unburned and grazed, a third was burned and not grazed, and a fourth was neither burned nor grazed. By 1918 the two grazed plots were completely cleared of longleaf pine seedlings. By 1924 the Forest Service reported that in the third plot some seedlings were killed and were not growing as well as the larger number of seedlings in the fourth plot, so fire was condemned for damaging longleaf reproduction and growth. Its fire policy thus affirmed experimentally, the Forest Service terminated the experiment and landowners were strongly urged to eliminate fire from their pine stands.

Chapman, however, continued to follow the now officially abandoned experiment and found that by the late 1920s the trees in the third or fire-thinned plot began making much faster growth than the unthinned trees in the fourth plot, where the crowded trees began to stunt each other's growth. The Forest Service, however, though informed of these results, never got around to publishing them and continued in its well-established dogma that fire must be eliminated from longleaf stands.

Too frequent burning of the Southeast reinforced the now rigid attitude of the Forest Service toward fire. Seeing no possible value in woods burning, the Forest Service in the late 1930s hired a young psychologist named Shea to explore the psychological reasons behind the stubborn tendency of the local people to burn the woods. As Shea put it, "it was hoped that here might be found a point of vaccination that with an improved educational serum would reach the germs of the woods-burning desires."[2] Reinforcing the dogma that no good could be found in a destructive practice, Shea concluded that:

the sight and sound and odor of burning woods provide excitement for a people who dwell in an environment of low stimulation and who quite naturally crave excitement. Fire gives them distinct emotional satisfactions which they strive to explain away by pseudo-economic reasons that spring from defensive beliefs. Their explanations that woods fires kill off snakes, boll weevil and serve other economic ends are something more than ignorance. They are the defensive beliefs of a disadvantaged culture group.[3]

In a final effort to modify these inherited beliefs, the Forest Service invented a talking bear to symbolize the grave danger that fire held for all forests and their inhabitants. To this day a solemn, reproving, or imploring Smokey the Bear asks the customers to "please help prevent forest fires" in national parks and forests from coast to coast.

[2] Shea, J. P., 1940. "Our pappies burned the woods." *Amer. For.* **46**, p. 159.
[3] Ibid.

Then in the late 1930s a children's novel, *Bambi,* was made into a feature-length cartoon by Walt Disney. Who can forget the climatic scene where Bambi, Thumper, and all their friends flee from that worst of all evils, the forest fire. By now the man in the street had it firmly in mind that there was no place for fire in the forest.

But by the early 1940s, clearly something was wrong in the great longleaf pine forests in the Southeast. Both private and public forests, which had been with great labor protected from fire, were afflicted. There was no reproduction. There was a very damaging brown spot fungus on the needles of the grass stage seedling. There was massive invasion of worthless hardwoods, which shaded out the longleaf seedlings. And worst of all, there were catastrophic fires fueled by twenty- to forty-year accumulations of combustible trash. In some forests more wood was lost from a single fire than was supposed to have accrued from all those years of fire protection. Beginning to doubt the advice of the Forest Service, private owners set up controlled burning to reduce the danger of wholesale conflagration in their forests. By the late 1940s the Forest Service was finally forced to admit that controlled burning, at the right time and under the right conditions, not only was effective in reducing dangerous litter accumulation, but was absolutely necessary to the survival of longleaf pine stands. Controlled burning would assure that the longleaf pine seeds had a mineral soil seedbed free of litter, that the brown spot disease would be reduced by occasional singeing of the grass stage leaves, and that hardwood invasion would be reduced (Figure 4.3).

Controlled burning

Today the Forest Service, as well as most foresters, realizes that the belief that fire in the forest environment is *always* bad is far from correct. No one suggests that indiscriminate use of fire by anyone, anytime, is appropriate; rather, that in the right time and place, with the proper conditions, for certain species, fire can be an essential management tool. Where evidence exists that fire has always been a regular part of the environment, as in the ponderosa pine forests of the Southwest, in the longleaf pine forests of the Southeast, or Douglas fir in the Northwest, it makes little sense to exclude fire because it is damaging to spruce forests in Bavaria or white pine forests in Wisconsin. But the Forest Service is still left with a dilemma: on the one hand it must ask people to follow the admonition of Smokey the Bear not to set forest fires; yet on the other hand, it must deliberately burn many of those same forests as a necessary management practice upon which the very existence of many western forests and national parks depends.

Problems in the Parks

Preservation of major geologic features like the Zion, Bryce, or Grand Canyons is relatively easy. They change so slowly that we have the illusion in our short lifespans of no change at all. But lesser

65

Fig. 4.3 • (a) This Texas pine stand is crowded with hardwood invaders that shade out pine seedlings. (b) Periodic controlled burning removes the fire-sensitive hardwoods. (c) This pine forest in Mississippi is kept "clean" by periodic prescribed burning. (*Photographs from U.S. Forest Service*)

environmental features such as plants and animals are constantly changing, at a rapid rate. One such feature common to our national parks was the stately groves of trees—Douglas fir, ponderosa pine, redwood—their massive trunks suggesting the columns of great cathedrals (Figure 4.4). Beneath, like a primordial garden, the ground was carpeted with grasses and flowers. The coincidental resemblance of these forests to the European concept of a well-tended park surely

Fig. 4.4 • An open grove of ponderosa pine in Washington. Maintenance of the park-like aspect of this grove depends upon periodic ground fires to remove competing species, thin ponderosa pine seedlings, and prepare the soil for the germination and growth of the next generation of trees. (*Photo by H. Weaver, courtesy of H. H. Biswell*)

gave early impetus to their preservation as parks, but ironically, it very nearly brought about their ultimate destruction.

Park officials seemed to ignore the fact that trees grow old, die, and are replaced by other trees, all in a dynamic balance with various environmental factors. Instead they excluded fire, removed diseased trees, and even tried to control natural predation to protect the "good" animals from the "bad." All efforts were made to stop time—to preserve the scene just as the early settlers saw it.

But in a few years things began to go awry. With the elimination of fire, twigs, needles, branches, and fallen trees began to pile up

(Figure 4.5). Decomposition of these fallen materials proceeded slowly in the hot, dry western forests. Without fire to remove litter, seedlings of the desirable species were often unable either to germinate or, if they did, to survive the shade.

Under the regime of fire elimination, the desirable trees were unable to reproduce in national parks, and seedlings of shade-tolerant species, such as conifers, began to invade, forming a dense undergrowth that obscured the beautifully spaced trees the parks were set up to preserve (Figure 4.6). This not only impaired the view, but because of the deep layer of highly combustible litter and the equally combustible understory of conifers, parts of some national parks became veritable tinderboxes. No longer could a fire be localized near the ground in needles and dry grass. Once the litter was ignited the fire would quickly jump into the understory and then into the crowns of the giant ponderosas, redwoods, or Douglas firs. A crown fire is the end of any forest, and if many of them occurred, the recreational, aesthetic, and scientific qualities of parks would disappear.

That this is not altogether a recent problem can be seen in an 1887–88 report of the commission charged with the responsibility of

Fig. 4.5 • Accumulated litter representing years of protection from fire. Should this ignite, the entire forest would be lost. (*Photo by H. Weaver, courtesy of H. H. Biswell*)

Fig. 4.6 • Protection from fire allowed these weed trees to obscure the more scenic sequoias and increase the danger from wildfire. (*Photo by H. Weaver, courtesy of H. H. Biswell*)

Several times during the period of my labors on your behalf, it required suddenly almost the entire force of twenty or thirty men to divert the all consuming course of forest fires on the floor of the valley. Since the annual practice of the Indians in burning off the dried grasses and leaves has been discontinued, and even forbidden by law, the accumulation of vegetable matter beneath the trees has been practically undisturbed, until a growth of young pines has sprung up all over the valley, and destroyed much fine meadow land. A campfire carelessly left, or a match thrown among the leaves, has caused several fires within the past two or three years that could not be extinguished. They burned until the walls, the roadway, and streams defined and determined their course.[4]

This was true of Yosemite in the 1880s and it is now true of Lassen, Sequoia, and King's Canyon. The expense of hand-clearing trash species and removing forest floor litter from huge areas of rough country is enormous. To attempt this with bulldozers would severely injure tree roots, expose the forest floor to erosion, and be impossible altogether in inaccessible areas. Fire is too long overdue to be used now without grave danger. What can the Park Service do? The only practical solution, regardless of expense, seems to be hand-clearing, acre by acre, and removal of combustibles, followed by controlled burning every few years to keep litter to a minimum and to eliminate trash trees when they are still small seedlings. But because this hand labor is so terribly expensive, the task will take many years—years filled with the risk that large tracts of forests will suddenly explode in a grand conflagration, should a fire ever get out of control some hot, windy day in late summer.

The preservation of forests that began ten, fifty, or one hundred years ago is clearly impossible if we try to extend their existence by merely erecting a protective "fence." A forest *can* be maintained in the way that a hedge can be kept at a desired height—with active management requiring constant care. If we are to manage effectively, we must through ecological research devise management techniques and then apply them unrelentingly.

Wilderness Areas

The idea of preservation in a time-stopping sense dies hard. Many who have been disillusioned by the reality of changes brought about by increasing use of national parks have pressed for wilderness areas—remote, inviolate, unspoiled. But the idea of a brooding, mysterious, and awesome wilderness is finished. Man now views the wilderness in an entirely different light. In an age of crowded, dirty cities, the wilderness has come to symbolize a refuge, the last place where man can breathe clean air, drink freely from streams, and get

[4] Gibbens, R. P. and H. F. Heady, 1964. "The influence of modern man on the vegetation of Yosemite Valley." *Calif. Agric. Exp. Sta. Manual* 36.

away from other people. Unfortunately, wilderness has become a symbol viewed in romantic anthropocentric terms, its cycles of destruction and renewal ignored because they extend beyond the life-span of man, an attitude that will surely lead to the same difficulties encountered in the national parks. A true wilderness should be viewed biocentrically; its forests must be free to burn, free to be attacked by insects, free to be blown down by storms, and free to be carried away by floods, all because these are natural events to which the forest is adapted to respond. When left alone, forests regenerate themselves. The new forests may be different from the old, but that is the way things change in a natural ecosystem.

The idealized wilderness is a myth. We must have the wilderness, but not for selfish gratification. At stake is the survival of ecosystems, free to change in accordance with natural variations in the environment, without interference from man with his notions of good and evil, a last refuge of absolute freedom in a world of increasing technological control. To allow some measure of this ecological freedom, a wilderness must consist of at least 100,000 acres in a fairly compact unit, without usable public roads or conspicuous signs of human activity. A block of land of such size is often free from the intrusive effects of the surrounding man-dominated countryside. In 1965 there were sixty-four tracts meeting these standards in the United States, and some have since been designated as reserves following the passage of the Wilderness Act in 1966 (Figure 4.7). Despite their size, however, wilderness areas share the plight of the national parks.

Fig. 4.7 • The areas in black are over five miles from the nearest road, railroad, or navigable waterway. This is America's remaining wilderness. (*From C. Tunnard and B. Pushkarev,* Man-Made America, *Yale University Press, 1963.*)

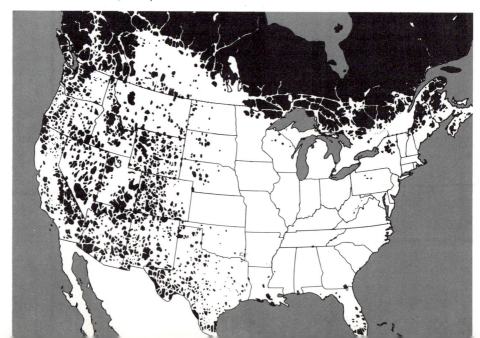

People, Parks, and Wildernesses

If people lived within national parks and wilderness areas by hunting and gathering, their populations limited by the availability of food, these areas could last indefinitely. But they will not last, because people, sustained from outside sources, are pouring into these areas in increasing numbers, destroying by their very presence the natural landscape that the parks were set up to preserve. On one holiday weekend in 1970, the Yosemite Valley, seven miles long and two miles wide, contained over 50,000 people, packed bumper to bumper in their cars. As a result of such traumas, in 1971 motor vehicles were prohibited in the eastern end of the valley; the remaining roads were made one-way; shuttle bus service was inaugurated, carrying 1,600,000 people the first year; and plans were made to eliminate automobiles from the valley floor altogether in a few years.

Another solution might be to build just outside each national park a large, comfortable reception center that would display the wonders of the park, using lifelike dioramas, slide shows, films, and perhaps closed-circuit color television. A wag has suggested that if a three-color bumper sticker were available at such a center, proclaiming that a visit had been made, many visitors might be satisfied by that alone. For those who insist upon venturing beyond the safety and comfort of their automobiles, a system of monorails or cable cars could be installed to allow close observation of plants, animals, and earth formations, affording spectacular views unavailable from the ground, while protecting the ground from human traffic. For the minority (probably less than 5 percent) who want direct physical interaction with the park, a system of backpack campgrounds, connected by trails, rotated periodically to avoid overuse, might be arranged. Such a program would relegate roads, buildings, and parking lots to the park perimeter and at the same time would satisfy the three major types of park users.

Although it is unlikely that such a program will soon be implemented, the deluge of park visitors may be handled in the interim by computers. Beginning in 1973, half the campsites in six national parks—Acadia, Everglades, Yellowstone, Grand Teton, Grand Canyon, and Yosemite—were reserved for a computerized reservation. The prospective camper now has the option of making a reservation at a particular park for a certain day at a computer outlet provided by American Express. By paying the campsite fee plus a service charge he receives a printed confirmation, which he presents to the forest ranger. While the remainder of the campsites in these parks are still available on a first-come, first-served basis, the system may be applied to the rest of the 30,000 campsites maintained by the National Park Service across the country if it works satisfactorily.

THE HIERARCHY OF NATURAL AREAS

Many types of natural ecosystems—indeed the most scenic ones—are found in national parks and wilderness areas. But there are also immense portions of the west held by various governmental agencies that are used for timber, watershed protection, grazing, and recreation, where a fine variety of natural ecosystems is found. While the scenery is not usually as spectacular as that in the parks, there is far more opportunity to enjoy the outdoors free from crowds of people. In the future, as the population of the United States increases, more and more of this public land will probably have to be managed strictly for recreation. If this can be done, the pressure on existing parks may be reduced. Public lands of all types will also serve in future years as the last source of national parks, for most suitable private land will soon be either developed or priced beyond reach.

State parks fill an important role in making the natural world available to people who cannot travel to the great western national parks. Quite generously scattered across the country, state parks are within an hour's drive for most people. Finally, local parks provide contact with some semblance of nature, if only in the form of grass, trees, and squirrels.

As one descends the hierachy of reserved natural environments from national park to city park, their size and scenic qualities decrease, but their *use* increases dramatically. Central Park, although huge by city park standards, is literally being worn to the bedrock by the enormous pressure exerted upon its facilities by New Yorkers (Figure 4.8). Unless alternate recreational facilities are provided in

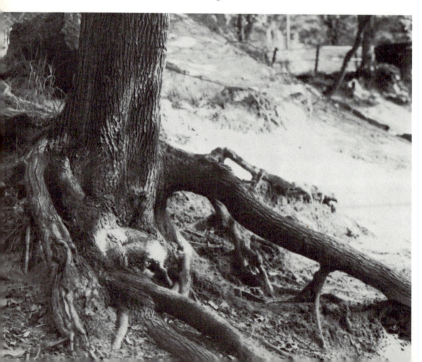

Fig. 4.8 • When tree roots are exposed the bark thickens. More than one foot of soil has eroded from the base of this tree in Central Park since its roots were first exposed.

the form of smaller vest-pocket parks, or far more maintenance is provided for Central Park than is currently available, or both, the park will succumb like an isolated field of wheat to a swarm of locusts.

Scenic Easements

We conventionally think of natural reserves as separate, specified areas. There are many other possibilities. One of the most useful of these is the device of *scenic easement*. Easement allows a property holder to retain ownership of his land, yet he can sell or lease the right-of-way, air rights, or the right to develop. This means that land of great scenic value could remain in private hands, but the right to change the landscape by development might be sold or leased to a governmental body. The government, of course, would not exercise this right, but would leave the land in its natural state. The property owner would get some financial return for his sacrifice in not developing his land, and the public would gain by not having to buy the land outright to preserve its beauty.

A classic example of the use of scenic easement was the securing of easements on the development of land across the Potomac River from Mount Vernon, thereby preserving a view that Washington himself must have seen. Without such an arrangement the opposite bank would doubtless have sprouted high-rise apartments, marinas, and shopping centers, destroying the illusion of stepping back into history that is so carefully and beautifully fostered at Mount Vernon. While we have preserved a view, we have also saved a stand of irreplaceable riverbank forest.

Parkways

Easements are especially useful in the construction of parkways and the preservation of riverscapes. Their scenic value is a tremendously important aspect seldom considered in plans to acquire and reserve open space. Driving along the Blue Ridge Parkway in Virginia and North Carolina gives the feeling of intimate contact with nature. Although the Park Service bought most of the land bordering the Parkway, much has been leased to local farmers (Figure 4.9) to preserve the interplay of woods and fields. Thus the monotony that might otherwise have resulted from long tunnels through the trees has been relieved.

Rivers

A combination of federal, state, and local parks together with scenic easements has been proposed to retain the relatively unspoiled beauty of the Connecticut River in New England, the Hudson in New

Fig. 4.9 • This landscape along the Blue Ridge Parkway in the Southern Appalachians is purposely being managed to provide an interplay of forests and cultivated fields that is aesthetically pleasing. (*National Park Service*)

York (Figure 4.10), and the Potomac in the mid-Atlantic states. Ironically, the pollution of these rivers has prevented the commercial development that has already spoiled many cleaner streams. But time is running out: as the pollution in these three rivers is reduced, steps must be taken to forestall the usual haphazard development projects with ecologically sound master plans that will assure the continued beauty of the rivers commensurate with recreational and economic uses.

Fig. 4.10 • A river valley does not have to be a virgin wilderness to be beautiful, as this view of the Hudson River Valley from Bear Mountain State Park in New York demonstrates. (*Photo by Harry Thayer, N.Y. State Dept. of Commerce*)

Trails

In 1937 the Appalachian Trail was opened, running 2,000 miles from Mount Katahdin in Maine to Springer Mountain, Georgia (Figure 4.11). Like the proposed scenic rivers, the trail is a composite of federal, state, local, and private ownership, which has worked well over the years, allowing public access to some superb natural areas. Other shorter foot trails have been established around the United States and many are in the process of development. In the planning stage, for example, is a north-south trail through Florida. Since trails can be rapidly destroyed by overuse, there must be adequate alternatives to spread the traffic more equitably.

Fig. 4.11 • The Appalachian Trail runs for 2000 miles along the crest of the Appalachian Mountains from Mt. Katahdin in Maine to Springer Mountain, Georgia.

Nature Preserves

Among the smallest natural reserves are so-called nature preserves. Usually they are small-scale, under 1000 acres, often under 100 acres. They are composed of ponds, salt marshes, woods, prairies

—features that are in danger of disappearing altogether. Mostly privately owned by schools, clubs, and conservation organizations, they serve a vital purpose in preserving natural areas too small to be considered for national or state parks. The Nature Conservancy, a private organization, has been instrumental in saving many of these small areas from bulldozers or sanitary landfill operations and holding them for future enjoyment and study.

Other areas capable of multiple use are golf courses, rights-of-way of various types, and even cemeteries. All preserve open space that can be used for alternative purposes.

Nature in an Urban Setting

All such categories represent occasional use, at best, by man. The nature many of us live with is in our backyards, or if we are apartment dwellers, the trees in the street, the ivy growing on the wall, or the geraniums growing on the windowsill. The connection of man and nature, however tenuous, can be maintained in the dreariest apartment by a window full of house plants as carefully nurtured as the oldest sequoia tree in Sequoia National Park. To a confirmed urbanite, the sight of an ailanthus tree growing behind a tenement, or pushing its way through the soil at the edge of a parking lot, is a treasured thing indeed (Figure 4.12).

Fig. 4.12 • Nature in the city is often limited to a window full of house plants, and scraggly ailanthus trees in unlikely places.

The greatest problem in the effort to save some fragments of a once-natural world has been the failure to recognize the hierarchy of natural systems from parking-lot ailanthus to redwood; each community has its place in the world and each is equally important. We must have as broad a spectrum of natural environments as can be assembled, from entire forests, deserts, mountains, and lakes down to ponds, bogs, woodlots, and single trees. Our spectrum should also include a wide range of uses, from inviolate wilderness available only for scientific study, to open space so heavily used that its vegetation must be continually renewed. Man is, of course, a part of the natural environment just as the natural world is part of man. While the opportunities still exist, land must be earmarked for *all* categories of natural areas. Once a salt marsh has been filled with waste, a sand dune leveled for houses, or a valley flooded by a reservoir, the options for preservation and continued renewal are forever lost, for there are limits to environment reconstruction (Figure 4.13). Failure to act now can only impoverish future generations, which will have problems enough to bear. Let us give them at least a glimpse of how planet earth once looked.

Fig. 4.13 • There will be no renewal for this salt marsh being "reclaimed" along the Connecticut coast.

FURTHER READING

Abbey, E., 1968. *Desert solitaire: a season in the wilderness.* McGraw-Hill, New York. A beautifully written account of Abbey's experience as a park naturalist in Utah.

Darling, F. and J. P. Milton (eds.), 1966. *Future environments of North America.* Natural History Press, Garden City, New York. Papers and superb discussions about the future of the natural world in North America.

Darling, F. and N. D. Eichhorn, 1967. *Man and nature in the national parks.* The Conservation Foundation, Washington, D.C. A brief survey of the problems arising from overuse of the National Park system.

Huth, H., 1957. *Nature and the American.* University of California Press, Berkeley. A leisurely look at our changing attitudes toward nature.

Johnson, W. A., 1972. *Public parks on private land in England and Wales.* Johns Hopkins Press, Baltimore. An interesting contrast to the problems faced in preserving natural areas in the United States.

Leopold, A. S., 1963. "Study of wildlife problems in national parks." Transactions of the Twenty-eighth North American Wildlife Conference, pp. 28–45. A very perceptive study that points out the basic ecology behind the problems.

McHarg, I. L., 1969. *Design with nature.* Natural History Press, Garden City, New York. A very personal view of the landscape by a landscape architect with exceptional ecological awareness.

Nash, R., 1967. *Wilderness and the American mind.* Yale University Press, New Haven, Connecticut. Traces the wilderness concept through American history.

Shepard, P., 1967. *Man in the landscape.* Random House, New York. A literary review of man's past and present relationship with nature.

Webb, W. P., 1931. *The Great Plains.* Grosset & Dunlap, New York. Still a classic statement of how a landscape has influenced not only its own development, but the course of the American nation.

Whyte, W. H., 1968. *The last landscape.* Doubleday, New York. A fine sourcebook on what to do about the disappearance of open space.

U.S. Department of Commerce, 1966. *A proposed program for scenic roads and parkways.* Washington, D.C. Handsomely illustrated guide to the proper integration of the highway into a scenic landscape.

U.S. Department of Interior, 1966. *Potomac valley, a model of scenic and recreational values.* Washington, D.C. Outlines the opportunities of preserving the scenic and recreational values of the Potomac.

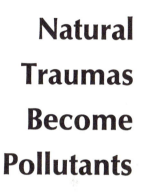

2

Natural
Traumas
Become
Pollutants

This steel mill on Chesapeake Bay is an ongoing catalog of environmental ills resulting from the obsolete economic dictum of externalizing environmental costs. (*Grant Heilman*)

5

"WATER, WATER, EVERYWHERE,..."

THE SUMMER OF 1965 found New York City as closely described by Coleridge's verse, "Water, water, everywhere,/Nor any drop to drink" as its inhabitants would ever want it to be. Picture the irony of a city surrounded by water, yet plagued by low water pressure. In New York car washing or lawn watering was prohibited, whereas Los Angeles—in a desert, comparatively speaking—had plenty of good water for every possible use. The 1960–1967 drought in the Northeast underscored the seeming paradox of a well-watered East suffering recurrent shortages of water while a dry West luxuriated in its abundance.

Once people become accustomed to drawing water from a tap they quickly forget that their faucets are connected by long miles of pipe to a supply somewhere. A continuous flow of water is assumed to be an inalienable right, like the "right" to toss an empty beer can into a lake or a woods. This seemingly inexhaustible supply of water must be replenished ultimately by precipitation falling somewhere.

THE HYDROLOGIC CYCLE

All water is locked into a recycling process called the hydrologic cycle. Solar energy, especially in the tropics, evaporates water from the ocean surface, filling the air mass above with large quantities of water vapor. When these warm, moist maritime masses conflict with cool, dry air over large land areas, some of the water vapor precipitates out. The reason temperate latitudes (contrary to what the name implies) produce frequent storms is that contrasting warm and cool air masses mix between the high and low latitudes. Although it may seem, when caught in a convectional thunderstorm triggered by this meeting of air masses, that *all* the water vapor is quite abruptly being dumped from above, no more than 20 percent of the water held by maritime air masses is lost over land.

Once it was felt that most of the rainfall over land was the result of evapotranspiration, a term that describes the combination of evaporation from the land surface together with transpiration from vegetation. If this were true, land use might have a profound effect upon precipitation both in quantity and distribution. So it was hoped by some and feared by others that by manipulating evapotranspiration, precipitation could be brought under control. It was later determined that almost 90 percent of the precipitation falling over land areas is oceanic in origin, with perhaps 10 percent derived from land sources. This is not to say that moisture released by a magnolia tree in Louisiana into a moist air mass from the Gulf of Mexico will not fall on a corn plant in Ohio, but the amount would be insignificant. The principal effect of ocean-generated precipitation is to moderate the effect that man through his land use practices can have on continental precipitation.

Water Management

Precipitation that has fallen can be controlled to some extent by various management practices. J. E. Church, one of the early pioneers in hydrology, once observed that "forests catch the falling snow directly in proportion to their openness, but conserve it, after it has fallen, directly in proportion to their density." Entrapment of snow by fences or sheltering belts of trees can also increase the water content of adjacent soil significantly.

In dry areas, treatment of the watershed itself can affect the yield of water. By replacing chaparral or forest with grassland, evapotranspiration can be sharply reduced and more runoff water made available for man's use. This has been successful in some watersheds. Going a step further, small areas of desert have been experimentally blacktopped to obtain close to 100 percent runoff from summer storms. This works well for a while, but the paving is difficult to maintain under the extreme conditions. Another less drastic approach is to remove trees and shrubs that line the banks of many southwestern streams. These plants—salt cedar, cottonwoods, willows, and mesquite—absorb close to 25 million acre-feet of water per year (an acre-foot is one acre flooded to a depth of one foot) and transpire it into the dry air. The expense of removing these plants and the impact of removal on wildlife, however, have slowed implementation of such management schemes.

Water Demand

The chief fact to be learned by examination of the hydrologic cycle is that the water supply of the earth is fixed. Approximately

97.3 percent of the earth's water is in the oceans; a very tiny amount of this, 0.007 percent, is distributed annually as precipitation over the land areas. Much of the remainder is locked up for the present in ice and snow; glaciers and ice caps cover 11 percent of the land area and pack ice and ice floes cover 25 percent of the ocean. The amount of fresh water immediately available to people is a small fraction of the total water found on earth. Of the earth's total water supply, 3.59×10^{20} gallons,[1] the United States receives a rather generous 4.3×10^{12} gallons a year in rainfall and its rivers and lakes contain another 6.5×10^{11} gallons. Since this is about fifty times the average yearly demand of 1.12×10^{11} gallons, there should be an ample supply of water for years and years of unrestricted growth.

As population increases, however, the supply slowly becomes critical—not because it is used up, but because of the tendency for abundant water sources (rivers or lakes) to be used for waste disposal and the unrealistic "need" for absolutely pure water for *every* human use. The problem, put quite simply, is not so much short supply as it is thoughtless and unnecessary contamination of the supply coupled with irrational demands for purity.

Metabolic Need

Most terrestrial animals have a limited water need; many satisfy this need from their food with a very small supplement of water. Some desert animals, such as the gerbil, fulfill all their water needs from their food supply and are able to exist without an independent supply of water. Man's needs are quite modest. Although composed of about 65 percent water, man requires only a few quarts a day for maintenance, and, since much of this is supplied by food, often a pint or two is adequate. This innate ability to conserve water evolved over a very long time and probably had certain survival value where water was scarce, as in desert areas inhabited by primitive man.

Cultural Use

Culturation changed this irrevocably, however, and although today's civilized man *needs* little more water than his ancestors of 20,000 years ago, he *uses* incredible quantities in his daily activities. Flushing a toilet requires three gallons, while a single shower uses thirty gallons. Food preparation, laundering, and house maintenance add further to a total of about 180 gallons per typical residence a day.

---

[1] Expotentials are a handy way of expressing huge numbers. For example, 10^2 is shorthand for 100; 10^3 for 1000. So 10^{20} would be 100,000,-000,000,000,000,000.

Public Water Supply

Supplying this quantity of water to huge urban areas has required long-range planning by all communities. Founded in 1781, Los Angeles managed to supply a small population with water derived from the annual precipitation of about fifteen inches. But a drought in 1862–1864 put an end to the huge cattle ranches that surrounded the town and the inevitable subdivision process began. Further population growth accompanied the arrival of the Southern Pacific Railroad in 1877 and the Santa Fe in 1885. By 1900 it was clear that future growth of the Los Angeles basin would require a large, dependable water source, available only in the Sierra Nevada some 200 miles away. The most convenient and economically feasible Sierran source was the Owens Valley on the Nevada border, but Owens Valley was already occupied by ranches and small settlements that were naturally unwilling to become ghost towns for the benefit of distant Los Angeles. The ensuing struggle ended with the construction in the 1920s of a 200-mile-long aqueduct which assured Los Angeles' continued growth and the abortion of similar hopes for Owens Valley. The Colorado River was tapped in the 1930s after a long, contentious struggle by the watershed states over water rights to the Colorado's flow. By the 1960s even the combined water resources of the Colorado River and Owens Valley seemed limited, so the rivers in underpopulated northern California are now being tapped (Figure 5.1). This tremendous project, costing well over $3 billion, will assure southern Californians of adequate water until they finally begin to question the desirability of endless growth.

Encouraged by the enormous demand of arid regions for water, a plan of continental dimensions, called the North American Water and Power Alliance (NAWAPA), was prepared by an engineering firm to "utilize the excess water of the northwestern part of the North American continent and distribute it to the water deficient areas of Canada, the United States, and Mexico." Making use of the unusual geographical, geological, climatological, and hydrological features of North America, this plan would take water now being poured into the Arctic Ocean and store it in an interconnected system of reservoirs at high elevation (Figure 5.2). This water could then be redistributed by means of a reservoir-canal-river system throughout the continent, not only producing a network of navigable waterways but generating power as the water flows to the sea. The cost has been estimated at $200 billion and it would require twenty years to construct.

Such mammoth projects involving interbasin transfer of water are now being scrutinized more closely than in the past, particularly by the western states that have always assumed that water was the sole factor limiting growth, and must be supplied regardless of cost. But both assumptions—inevitable growth and the role of water in

Fig. 5.1 • The California Water Plan is designed to supply water-poor southern California with water from water-rich northern California. The completed sections of the system are shown here in black. (© *1971 by The New York Times Company. Reprinted by permission*)

that growth—are finally being challenged. At least two states, Colorado and Oregon, have stopped encouraging growth by withdrawing support from some of the mindless boosterism that can be so destructive of environmental values—small step, perhaps, but the beginning of an important change in attitude. It suggests that people have begun to recognize that every ecological system has a carrying capacity, and that when this is exceeded, the total economic and environmental costs soar well beyond the immediate return provided by increased growth.

Another factor that must be scrutinized much more closely in the future is cost. Most municipalities condone excessive waste of water; directly, by poorly constructed and maintained mains that leak up to 50 percent of the water they were designed to carry; and indirectly, by unrealistically low charges for water consumed by the public. Federal agencies, the Army Corp of Engineers, the Bureau of Reclamation, and the Soil Conservation Service persist in carrying out water-oriented projects—damming, irrigation, channelization, draining—whose utility is becoming increasingly marginal and which often work at cross purposes (see Chapter 19). Irrigation in many western states is so heavily subsidized that if the beneficiary of the projects was required to pay the costs, such agricultural development schemes would cease. Arizona uses 90 percent of its water supply for agriculture, which contributes only 10 percent of the state's income. Yet to support that 10 percent the billion-dollar Central Arizona Project has been ballyhooed for years.

Fig. 5.2 • The projected North American Water and Power Alliance (NAWAPA): the implications of technology on the environment are enormous in this scheme; the ecological implications are unknown. (*Courtesy Ralph M. Parsons Co.*)

In view of such expensive struggling for water it is more than a little surprising to realize that *there is no shortage of water in the United States, nor is there likely to be in the future,* no matter how many people live in the country and make demands on its water supply (see Table 5.1). Except for irrigation, no major water use consumes more than a fraction of the total water used.

Table 5.1

United States Intake and Net Consumption of Water
(10^9 gal/day 1965)*

	INTAKE	NET CONSUMPTION
Irrigation	142	116
Steam generated power	111	1
Industry	74	12
Public water	23	1
Rural domestic	6	4

* Dykes, D. R., T. S. Bry, and C. H. Kliner, 1967. "Water management: a fashionable topic." *Env. Sci. & Tech.* **1** (10), pp. 780–784.

All of the water now "in captivity" is to be found in reservoirs, water supply systems, and industrial and municipal waste systems. If all of this water were efficiently recycled, the annual increase in demand for new water would be exceedingly small. Loss to the atmosphere through evaporation from soil, rivers, reservoirs, plants, and animals, as well as metabolic demands by population increases, would easily be replaced by annual precipitation.

A Domestic Recycing Scheme

To be sure, a society that completely recycles its water will be a long time in coming. But such a plan is not as visionary as it might seem. At the present time, all water piped into our homes is grade A drinking water regardless of the use we make of it. It makes little sense to throw away three gallons of grade A water just to dispose of a small volume of wastes in a toilet, or thirty gallons to get rid of a little detergent with its dirt and lint from clothes. A priority system of water use could easily be established, reserving grade A water for drinking, cooking, washing, and bathing. Grade B water reprocessed from grade A uses would be quite adequate for outdoor uses: car washing, lawn watering, filling swimming pools (these would still be cleaner than most public bathing beaches), and flushing toilets.

Grade A water would be supplied from a well or reservoir direct to the faucet. After grade A uses, the waste water would be filtered and stored in a basement tank for grade B uses (Figure 5.3). In a

Fig. 5.3 • With a few relatively simple and inexpensive modifications in the plumbing of a typical house, a considerable amount of water can be saved. Drainwater from the bathtub and sink flows by gravity to a collecting tank in the basement and then is pumped back upstairs to provide flush water for the toilet. (*After Mark Bigatel*)

recent pilot experiment, water from filtered laundry and bath water was simply stored in a basement tank and subsequently pumped into toilets for flushing, reducing water consumption by 39 percent in a typical suburban home. Even under prototype conditions, which are far more expensive than mass produced versions, the capital cost was only $500 and the savings in water and sewer rental costs $20 a year, which over the lifetime of the house (twenty-five years) would pay for the cost of installation. Such a simple recycling device could cut domestic water consumption by almost 50 percent if widely adopted.

A more comprehensive system would be to recycle carefully treated and purified sewage effluent into the water supply, drastically curtailing the need for new water. Doubtless this strikes a repugnant note to most of us who feel that the water that comes out of our tap is or should be as pure as distilled water. On the other hand, communities that draw their water supplies from rivers are already simply recycling the effluent from the town above, particularly when under drought conditions many rivers continue to flow only because of the effluent contributed by towns along the banks.

Drinking Our Effluent

A classic case of intentional, but desperate, reuse of effluent for water supply took place in Chanute, Kansas in 1956–1957. The drought of 1952–1957 was the most severe in Kansas history. By the fall of 1956 the Neosho River, which was Chanute's water supply, simply stopped flowing. Shutting off the water supply to industry was considered but rejected, because so many people would lose their jobs. Wells were then suggested, but well water was too saline for use. Hauling was too expensive and logistically complex. Finally recycling was adopted as the best solution. As luck would have it, a new sewage disposal plant constructed in 1953 produced an effluent lower in its bacterial count than the Neosho itself. The effluent was pumped into a holding pond for a number of days, then prechlorinated and recycled. The resulting water, carefully monitored by the state board of health, met minimum standards, although it had a musky taste and odor, a pale yellow color, and produced foam when agitated. Some people were concerned about the effectiveness of chlorination in eliminating viruses, particularly the polio virus (this was before the various vaccines) and infectious hepatitis. And their concern seemed justified, since 20,000 to 40,000 cases of hepatitis were suspected to have resulted from similarly treated water distributed by the city of New Delhi, India a few years before.

After a few months the drought was broken by rains, the river resumed flowing, and Chanute returned to its previous water source. A doctors' survey conducted after resumption of normal water supply showed there were no illnesses that could be related directly to the use of recycled water. Apparently the major problem was the increasing levels of various salts. If recycled water is to be used in the future, it will be necessary to have rigorous filtration systems to remove odor, taste, and color, and precipitation processes to remove salts. The virus problem will be more difficult to solve.

Recent evidence suggests that Down's syndrome (mongolism) may be related to cases of infectious hepatitis contracted by mothers in early pregnancy, in the same way that German measles infections during early pregnancy have been associated with birth defects. Unless a fail-safe means of removing or killing viruses in sewage effluent can be perfected, full recycling of effluent cannot take place. It is only a matter of time, however, before population pressure requires the extensive recycling of *all* available water resources, including those which have been traditionally regarded as unfit for human use.

Industrial Recycling

Water reuse is not limited to the individual user. Industry must also make a greater effort to recycle and reuse some of the enormous quantities of water it uses in processing various products and services. As an example of what can be done, Kaiser Steel in Fontana, California has employed various recycling methods to cut the water requirements for manufacturing steel from the industry average of 65,000 gallons per ton to 1400 gallons per ton.

With the added incentive of an even tighter water supply and more stringent state and federal pollution standards, even such gains as these can be exceeded. Finally a point will be reached when, beyond initial charging of a system, very little additional water will be needed. This means not only more but cleaner water will be available in the environment for other uses.

DESALINIZATION

Many articles have been published about the coming wonders of desalinization, complete with glowing promises about severing our dependence upon natural precipitation, making deserts bloom, changing climate, and other age-old fantasies of mankind. Any desalinization process, and there are many, requires energy, but only one process, electrodialysis, removes the salt directly. While desalinization is usually thought of as a sea-edge process, any brackish water is potentially desalinizable; the lower the concentration of salt, the more efficient and less expensive the separation. However, the need is often most critical in dry climates near an ocean, which provides the nearest source of water.

There are many methods of desalinization, some rather complex, and all ingenious (Figure 5.4). The older processes, stills, use heat energy; the more recent processes, electrical energy. With the development of nuclear power plants it has been suggested that the two be combined, producing both electric power and fresh water. Such a plant projected for an artificial island off the coast of southern California was planned to produce 150 million gallons of water a day.

Fig. 5.4 • Desalinization of water is now a technically feasible possibility. Different methods of desalinization are shown schematically: on this page, (a) a multi-stage flash distillation process; on the next page, (b) an electrodialysis cell, and (c) an electromagnetic separation. (Modified from Chemical and Engineering News, June 8, 1970)

(b)

(c)

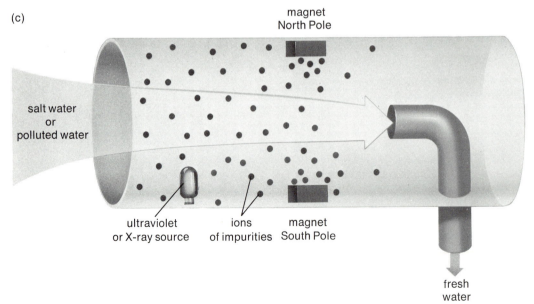

Perhaps one of the most ingenious schemes for augmenting fresh water supplies was suggested by Gerard and Worzel a few years ago (Figure 5.5). Many small coastal islands in the tropics have grossly inadequate fresh water supplies because of their tiny catchment area, but by exploiting their exposure to constant prevailing trade winds of high moisture content and their proximity to cold,

Fig. 5.5 • Proposed water recovery plant. (1) Large-diameter pipe to deep water. (2) Pump. (3) Connecting pipe. (4) Condenser. (5) Freshwater reservoir. (6) Windmill electric generator. (7) Baffles to direct wind. (8) Small turbine to recover water power. (9) Lagoon receiving nutrient-rich water for agriculture. (10) Community enjoying cooled dehumidified air. (*Modified from R. D. Gerard and J. L. Worzel, Science 157, 1967. Copyright 1967 by the American Association for the Advancement of Science.*)

deep, nutrient-rich water, a variety of benefits could be obtained. Cold sea water pumped through an insulated pipe to a baffle would provide the cool condensing surface for condensation of moisture from the warm, moist trade wind. The condensed water could be stored in a cistern below, and the cool, dehumidified air could be used to air condition nearby buildings. The sea water coolant, rich in nutrients, could be released into a lagoon whose productivity would be increased, thereby supplementing the protein resources of the island. At least a part of the power needed to pump the water from offshore depths could be provided by a windmill powered by the same trade winds.

The biggest problem with desalinization is the cost. Although costs are being reduced constantly by building larger and more efficient plants, desalinization still compares unfavorably with competing sources of fresh water. Even if desalinization costs were reduced to those comparable with other water sources, there remains the problem of transport from the coast to the inland valleys and deserts. Moving anything against gravity is extremely expensive (all of the existing far-flung water supply systems work for the most part *with* gravity flow). It has been said that if the quantity of water needed is large enough, it is cheaper to bring water from Baffin Bay to Baja California than to desalinate and transport sea water to the place where it is needed. The very existence of the California water scheme or the projected NAWAPA lends credence to that statement.

Desalinization will surely come to play a much larger role in the water supplies of major coastal communities in the future, but unless we develop some low-cost energy source to push water uphill, desalinized water will not quickly transform the arid regions of the earth into lush, tropical gardens. Indeed, to do so might have major climatological implications. Perhaps by the time this is feasible we will have developed the ecological insight to predict the consequences of such drastic changes.

WEATHER CONTROL

Besides the management of rainwater after it has fallen, there is the possibility of weather control. Ever since artificial snow was produced in a deep freeze some years ago, a great deal of interest has been generated in seeding clouds to generate rain and break up hurricanes. The simple fact of the matter is that until recently there were no weather data available for over two-thirds of the world's land area and nearly all of the water area. This lack of information made weather forecasting inexact and widescale manipulation impossible. While reconnaissance satellites are becoming extremely useful in helping to cover weather information gaps—for example, in various polar, tropical, and oceanic areas—the ability to use satellite information to project and predict weather is making slow headway. Weather modification of large areas is therefore not yet possible. Indeed, at our present level of knowledge any broad-scale experiment in the atmosphere aimed at making permanent changes would be irresponsible in the extreme (see Chapter 17). However, a great deal of experimentation has been carried out on a local basis, and the results, which range from inconclusive to promising, bear further examination.

Seeding certain types of clouds with silver iodide crystals provides water-attracting nuclei for condensation of the tiny supercooled

Seeding droplets of water that make up the cloud. As these droplets condense on the silver iodide nuclei, drops of water are formed that ultimately may grow to precipitable size. Evaluation of seeding attempts is extremely difficult because of the great variability of cloud types. While some experiments have increased rainfall around 10 percent, the increase is rarely more than 20 percent, and, in some instances, a decrease in rainfall has probably occurred. Other problems include the accurate calculation of lag time (the time from seeding to the time rain falls), for clouds cannot be seeded directly over the desired rainfall area. Seeding must take place as much as 50–100 miles upwind to allow for raindrops to form. Their trajectory from cloud to target area is especially important if the precipitation is to be snow. Also, most successful seeding attempts have used cumulus clouds or orographic storms, that is, unstable clouds formed by air being forced up over mountains. Efforts to seed large cyclonic storms, that is, those associated with low pressure areas, have not been very successful.

Hail suppression has been reported by the Russians, who have fired silver iodide directly into potential hail-producing clouds using antiaircraft shells or rockets. But because of the extreme variability in the formation and fall of hail compared with rain, proof of success is elusive.

Lightning suppression is being attempted by introducing silver iodide into clouds, in the hope that the large number of ice crystals formed would relieve the electrical potential before enough charge had accumulated to result in a lightning stroke. Metallic chaff dipoles —small pieces of foil—dropped into thunderheads might have the same effect.

Fig. 5.6 • Satellite photos of Hurricane Debbie (1969). The photo on the left was taken three hours after the hurricane was seeded. The photo on the right was taken the next day. Seeding apparently accomplished some temporary reduction in the storm's wind velocity. (*Environmental Science Services Administration*)

95

Hurricane and tornado modification is making slow progress. Recent experiments with a hurricane in the Gulf of Mexico have succeeded in lowering the wind velocity around the eye by seeding with silver iodide (Figure 5.6), but reducing the force of tornadoes by direct action has not yet been achieved. Progress has been slow because there is no precise way to predict conditions likely to spawn tornadoes. Until tornado modification becomes feasible, a more accurate warning system is our only defense.

The action necessary to change the weather or climate over large areas may never be within man's grasp, but there are probably many instances where the various controlling factors are delicately balanced and man may already have the ability to effect radical change. The question is, as always: does man have the insight to predict the consequences of his manipulations, and then can he exercise restraint where he must?

FURTHER READING

Anonymous, 1968. "Reusing storm runoff." *Env. Sci. & Tech.* **2**(11), 1001–1005. Ingenious scheme for using a presently wasted resource.

Dykes, D. R., T. S. Bry, and C. H. Kliner, 1967. "Water management: a fashionable topic." *Env. Sci. & Tech.* **1**(10), pp. 780–784. Overview of the water supply-demand problem.

Gerard, R. D. and J. L. Worzel, 1967. "Condensation of atmospheric moisture from tropical maritime air masses as a fresh water resource." *Science* **157,** pp. 1300–1302. A very clever scheme to utilize the environmental features of a tropical island for water and energy supply.

Hickman, K., 1966. "Oases for the future." *Science* **154,** pp. 612–617. A review of many of the problems in increasing the supply of fresh water.

Hodge, C. (ed.), 1963. *Aridity and man.* AAAS Publ. 74, Washington, D.C. A fine collection of papers dealing with various aspects of water supply.

Maass, A., 1951. *Muddy waters: the Army Engineers and the nation's rivers.* Harvard University Press, Cambridge, Massachusetts. A critical examination of the role of the Corps of Engineers in man-handling the environment.

Nace, R. L., 1967. "Water resources: a global problem with local roots." *Env. Sci. & Tech.* **1**(7), pp. 550–560. An excellent summing up of the water supply problem.

Okun, D. A., 1968. "The hierarchy of water quality." *Env. Sci. &*

Tech. **2**(9), pp. 672–675. Suggests an approach to a meaningful reuse of water.

U.S. Department of Interior, 1968. *The a-b-seas of desalting.* Office of Saline Water, Sup. Doc., Washington, D.C. A concise, well-illustrated discussion of the various techniques of desalinization.

6

"...NOR ANY DROP
TO DRINK."

TRADITIONALLY SPEAKING, pollution, like prostitution, is any departure from purity. But in the environment it has come to mean departure from a normal rather than a pure state; otherwise we would have to say that water is polluted with algae, fish, or fowl.

Pollution is, of course, a subjective and often emotionally charged term, one that man has used to describe the environmental traumas of fire, overproductive or deoxygenated water, smoke- or dust-filled air, and background radiation that have always existed. Natural traumas become pollution when man's activities overwhelm the capacity of an ecosystem to handle them, thereby causing an imbalance in the system. This imbalance requires an enormous input of energy to correct. Further, because of man, environmental traumas are now regional instead of local, continuous rather than episodic. Water and, as we shall see in Chapter 8, air, are the two most polluted areas in our environment.

Dilution of water pollutants

Because of its unique bipolar properties, water readily dissolves a great variety of substances; those substances that remain insoluble are at least dispersed. Where human populations are small and widely scattered, dilution *is* a reasonable solution to waste disposal. Rivers and streams are, in a few miles, able to carry out the work of breakdown, absorption, oxygenation, or consumption, and small amounts of pollutants are effectively dispersed.

In primitive times, by the time waste water from a village of a few hundred people had been carried a few miles downstream, the water had been purified naturally—at least, enough to be drinkable in the next village. When villages became towns, and towns cities, rivers could no longer break down the quantity of pollutants poured into them, and the effluents from each population center added to the water supply of the next. Running water almost everywhere became sewage water, seemingly putting an end to the concept of dilution as a solution to pollution. But when the true costs of proper pollution control at the source became widely recognized, the idea of dilution was revived. One of its current incarnations is in a proposal to build two dams on the New River in Virginia to generate electric power. The proposal provides that 400,000 acre-feet of the reservoir's total capacity of 3,261,000 acre-feet will be released periodically to flush chemical wastes from the Kanawha River at Charleston down the Kanawha into the Ohio, the Mississippi, and finally to the Gulf of Mexico. If this project is constructed a precedent will be set and the concept of water quality storage (a euphemism for dilution of pollution) will be applied elsewhere. One certain result will be the relaxation or postponement of industrial pollution control programs at the source of the pollution and a continuing or increased level of pollution in this country's rivers.

CATEGORIES OF WATER POLLUTANTS

Saline Wastes

Occasionally, salt brines from mines or oil wells are released into normally fresh water. Whereas some organisms can tolerate a certain range of salt concentration, many disappear as fresh water becomes brackish. If the salt level remains constant, salt-tolerant species often appear and reproduce successfully. A marine species of the green seaweed *Enteromorpha* became quite abundant on the bottom of a New York stream that was badly polluted with salt brine. However, because of constantly fluctuating levels of most pollutants, organisms that might otherwise adjust to the new conditions rarely have the opportunity to become naturalized, so neither native nor newly introduced organisms can survive.

A far more serious type of salt pollution has developed as the use of sodium chloride–calcium chloride mixtures on snowy roads in winter has increased to over 6 million tons per year. Although limited to states with a winter snow problem, the quantities of salt used in these states is staggering: 20 tons per mile or 4 pounds per linear foot along each side of a two-lane road. The salt runoff changes the osmotic balance of the soil bordering the highway and makes plants much more susceptible to drought. Sugar maples seem especially vulnerable to salt in the soil, and ancient specimens are dying along rural roads all over New England. Massive intrusions of salt into the water table from carelessly placed and unprotected storage piles have contaminated many wells, and runoff from highways into storm sewers and natural watercourses is beginning to affect aquatic life in some places. Apparently because the chloride ion complexes strongly with mercury, contamination of fresh water with deicing salt can release mercury from polluted bottom sediments. The problems this excessive use of salt can cause seem a high price to pay for snow and ice melting. One cannot help but wonder how many accidents are caused by poor visibility through salt-smeared windshields, or by the slippery slush that quickly follows application of salt to highways. Perhaps salt is indispensable to maintaining safe highways in winter, but far more salt is being applied than the job requires. Faulty equipment, poor supervision, and bad judgment all contribute to the problem. But now that the environmental fate of highway salt is being traced (Figure 6.1), steps can be taken to assure at least its proper use, and at most, its replacement by safer alternatives.

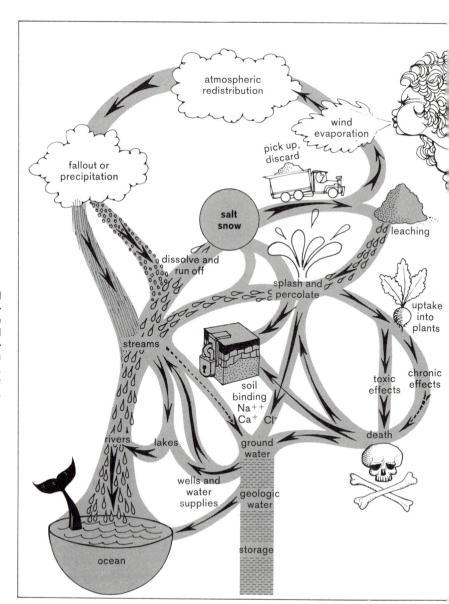

Fig. 6.1 • The environmental fate of salt applied to highways in winter is rather more involved than was supposed in the past. The possible effects go well beyond the death of plants along the roadside. (*Drawing by Jim McClure in* Farm Economics, *February, 1973*)

Acid Wastes

In 1803, T. M. Morris made one of the first references to acidic drainage in America when he noted, "The spring water issuing through fissures in the hills, which are only masses of coal, is so impregnated with bituminous and sulphurous particles as to be frequently nauseous to the taste and prejudicial to the health."[1] Morris was referring to natural seepage, which has apparently always been a minor problem. But since the opening of the nation's first coal mine in Pennsylvania in 1761 the problem has become far more widespread. Of course, acidic water from mines increases with production of coal, but it also continues long after mines are abandoned.

Joshua Gilpin visited a coal mine in 1809 and stated that "above the coal is several feet of a mixed kind of bad coal and iron abounding in sulphur and in vitriol efflorescing in white and yellow crystals."[2] The materials Gilpin was describing were iron pyrites, which enter streams following strip mining. Certain bacteria are able to obtain energy by changing iron from one oxidation state to another (ferrous to ferric). They use the energy to respire and reproduce. With an unlimited source of food they multiply greatly; the by-products are the orange-red precipitate of ferric hydroxide that stains riverbanks and channels, and the acidity caused by the release of sulfuric acid.

This set of reactions not only drastically lowers the pH of the stream and contributes huge quantities of iron and sulfate to the water, but it reduces the availability of oxygen in the stream as well. In the low-water season of summer and early fall, the pH may rise to 4.5, but with the high water of winter and spring, flushing of wastes from strip mines and mine shafts by tributary streams may lower the pH to 2.5. At pH levels below 4.0 all vertebrates, most invertebrates, and many microorganisms are eliminated. Most higher plants are absent, leaving only a few algae and bacteria to save the stream from total sterility. One species of higher plant, the needlerush (*Eleocharis acicularis*), is quite resistant to low pH and shows great promise as a possible means of rehabilitating acid-damaged streams.

Acid mine drainage is a serious problem wherever coal is mined. In the United States, the most severely affected areas are parts of Pennsylvania, Ohio, West Virginia, Kentucky, Illinois, Missouri, and Tennessee; in Europe, parts of the British midlands, France, Germany, and Poland. Most coal mining areas now have strict regulations which have greatly reduced pollution from contemporary sources. Regulations, however, do nothing for those shafts or stripped areas already abandoned which continue to discharge acid wastes into the

[1] Braley, S. A., 1954. "Acid mine drainage: I The problem." *Mechanization* **18**, pp. 87–89.
[2] Ibid.

nearest drainage. Control of these abandoned workings will have to come bit by bit, adding enormously to the ultimate cost of acid pollution control.

Although coal mining operations add the greatest amount of acid to streams, other industries produce wastes which affect pH too. Continuous casting and hot rolling mills use sulfuric acid (pickling liquor) to clean oxides and grease from the processed metal. When the pickling solution becomes diluted with iron salts and grease, it must be replaced with fresh solution. Since over 50 percent of all steel products are treated with pickling liquor at the rate of fifteen gallons of liquor per ton of steel, a considerable volume of spent liquor is disposed of every year. After pickling, the metal is rinsed to stop the pickling action; every ton of steel requires another eighty gallons of water, which picks up about 15 percent of the acid. Because of their acidity these spent pickling and rinse solutions are a most difficult waste water problem.

Organic Wastes

The most common pollutants are organic. These materials are not poisonous to stream life, nor do they necessarily affect pH. Their effect is more subtle. Most organic materials are attacked by bacteria and broken down into simpler compounds. To do this, bacteria require oxygen. The greater the supply of organic food, the larger the population of bacteria that can be supported, and the greater the demand on the oxygen supply in the water. This demand for oxygen by BOD the bacteria is called biological oxygen demand, BOD. The BOD is a useful index of pollution, especially that related to the organic load of the water. Because all stream animals are dependent upon the oxygen supply in water, the BOD is of particular importance in determining which forms of life a polluted stream is capable of supporting. Fish have the highest oxygen need; usually cold-water fish require more oxygen than warm-water fish. Invertebrates can tolerate lower concentrations of oxygen, and bacteria still lower.

Two invertebrates are able to tolerate such a low oxygen level that they have become recognized as a kind of biological index of oxygen depletion. One is a small worm called *Tubifex,* and the other is a bright red larva of a tiny midge called *Chironomus.* Both live in the bottom mud in numbers that are inversely proportional to the oxygen content of the water. In badly polluted water *Tubifex* populations have been estimated at twenty thousand individuals per square foot. In unpolluted water *Tubifex* may be absent altogether.

Sometimes when the organic load is especially great and the water warm, oxygen is so depleted that the usual oxygen-requiring or aerobic decomposers are replaced by species that don't require oxygen (anaerobic), with a quite different group of end products (Table 6.1).

Table 6.1

*Comparison of Decomposer End Products under Differing Conditions**

AEROBIC CONDITIONS	ANAEROBIC CONDITIONS
$C \rightarrow CO_2$	$C \rightarrow CH_4$
$N \rightarrow NH_3 + HNO_3$	$N \rightarrow NH_3 + amines$
$S \rightarrow H_2SO_4$	$S \rightarrow H_2S$
$P \rightarrow H_3PO_4$	$P \rightarrow PH_3$ and phosphorus compounds

* Klein, L., 1962. *River pollution Vol. II. Causes and effects.* Butterworth, London.

While methane, CH_4, is odorless, amines have a fishy smell, hydrogen sulfide, H_2S, smells like rotten eggs, and some phosphate compounds have a wormy smell. When added to the smell of decaying fish or algae, the shift to anaerobic conditions is not a pleasant one.

Now and then severe pollution from excess organic matter occurs in natural situations, as when heavy leaf fall temporarily overwhelms a small stream, creating a large BOD and possibly killing fish. But most organic wastes are related to man's misuse of his environment by the release of untreated human and animal wastes as well as those from industrial processing—blood, milk solids, grease, pulp liquor, washings from fruit and vegetables.

The Human Factor

Human waste disposal is not a recent problem, though it certainly has been made more difficult by the sharp population rise of the past century. The Old Testament met and solved, for its time, the question of human waste disposal: "and thou shalt have a paddle upon thy weapon; and it shall be when thou wilt ease thyself abroad, thou shalt dig therewith and shalt turn back and cover that which cometh from thee."[3] Millions of farmers in Asia still utilize a similar approach, fertilizing their fields and rice paddies with "night soil." The people of medieval towns and cities, being unable to return waste to the soil, dumped it from their windows into the street, requiring well-bred gentlemen of the day to defend their sensibilities by carrying an orange stuck with cloves to mask the foulness of streets. Urban people today prefer sewage treatment plants in which organic wastes can be broken down by bacteria, so that the effluent does not deplete the oxygen in rivers and lakes.

But with our increasing level of affluence, this ecologically sound pathway is being short-circuited. For example, people are spending

[3] Deuteronomy xxiii: 12–13.

more and more of their leisure hours aboard a great variety of pleasure craft (Figure 6.2), many with toilets but few with means to handle the waste properly. In most instances waste is dumped overboard with little or no treatment. In addition to the sanitary wastes there is litter (one pound of paper, cans, and bottles, and one-half pound of garbage per capita per day of boating); bilge water containing lubricating oil leakage; and in large ships, ballast. The average freighter carries 1000 tons of ballast (usually water taken from one polluted harbor and released in another) making ships pollution carriers par excellence. In 1964, of the 50,000 ships clearing United States customs, 16,000 released 16 million tons of polluted ballast water. Other sources of vessel waste include wash water from cleaning operations, usually containing high concentrations of detergents, and accidental cargo discharges through negligence, collision, grounding, or sinking. The Torrey Canyon disaster, which will be discussed in Chapter 7, is just one example of this.

Fig. 6.2 • Most of the boats in this marina are large enough to have onboard toilets, which dispose of their wastes directly into the water. (*Photo by Mario Marino*)

The United States Navy until quite recently disposed of its shipboard wastes over the side, causing an enormous sewage problem in large ports when the fleet was in. Consider how much waste is generated in a single day by a nuclear aircraft carrier with several thousand men aboard!

Some possibilities for aboard-ship sewage disposal include holding tanks, incinerators, biological treatment systems (BTS) and maceration-disinfection. Holding tanks are adaptable to any size vessel and minimize mechanical failure, but they are large and heavy. They are also suitable only for human wastes, and require shore support facilities that are nonexistent even in some large ports. Incinerators are lightweight, but have a high power requirement. Like holding tanks, they are limited to human waste, and, in addition, can be a fire hazard. A BTS can handle all kinds of sewage, since it is a miniature secondary sewage treatment system, but it is quite large and heavy and requires trained personnel for proper operation. Maceration-disinfection is an easily installed, lightweight system, occupying little space, but it assures no reduction in BOD, or mineral and organic content: its disinfectant efficiency is not always certain and it is subject to failure. Perhaps the most promising approach is a variant of the recirculating flush toilet. The airline model can handle 80–100 uses before servicing; larger models, capable of servicing fifty people for 7–10 days, have been developed for ships.

Animal Wastes

Humans are not the only waste producers, of course. Kansas, which is not overpopulated, has such a water pollution problem that no stream or river in the state is safe for drinking or swimming. The cause is the more than 200 feedlots scattered across the state (Figure 6.3). The 5.5 million cattle and 1.3 million hogs daily excrete the sewage equivalent of 70 million people. Most of this goes directly into streams and rivers with little attempt at control. Because of this pollution load some towns have had to use ten times the usual level of chlorination to purify their drinking water. New regulations will require feedlots, which process millions of livestock a year, to treat their wastes before releasing the effluent. Even so, it will take five to ten years before pollution from this source can be reasonably controlled.

Fig. 6.3 • The animal wastes from this midwestern feedlot enter the nearest surface drainage system without treatment. The waste from one cow equals the wastes of ten people. (*U.S. Dept. of Agriculture*)

Treatment of Human and Animal Wastes

Once the organic materials are broken down into their components by sewage treatment, the minerals must be removed to avoid overfertilizing the environment. Precipitation removes only some of the minerals; those like nitrates, which are highly water soluble, are more difficult to deal with. A useful approach, first used by a large food processor in New Jersey and later developed and refined by a research team at The Pennsylvania State University, uses the soil and its plants and animals as a living filter. Effluent is piped from a sewage treatment plant and sprayed on a field, pasture, or forest at a rate that is commensurate with soil absorption (Figure 6.4). If effluent is applied at a rate of two inches a week, 129 acres could filter a million gallons a day. The plant roots and soil microorganisms are especially efficient in removing phosphates and nitrates, the two nutrients that cause the most problems in aquatic systems.

Fig. 6.4 • Year-round spraying of treated sewage effluent onto forest and crop lands at Pennsylvania State University fertilizes the soil and filters the effluent, returning clean water to the water table. (*Photo by R. S. Beese*)

Gratifyingly, experimentally grown crops benefited from the fertilizing effect of effluents; hay yield increased 300 percent and the nitrogen content of wheat increased by 30 percent. But there are problems with this approach. It requires the lack of surface drainage, which would short-circuit the process, and deep soil to allow maximum filtration.

A recent extension of this technique has been its application to revegetation of strip mine spoil banks. Because of their dark surface and exposed position, spoil banks become excessively hot and dry in the summer, exceeding even the broad tolerance of weeds. In addition the soil is usually both infertile and acid, and hence establishment of seedlings is impossible.

Application of sewage effluent to these banks ameliorates all these problems. Moisture, which also cools the ground surface by evaporation, allows a broad range of plants to become established, and the mineral nutrients which abound in the effluent encourage rapid and luxuriant plant growth. Finally, the addition of two inches of water a week helps leach some of the acid into lower soil levels, which allows the establishment of species requiring less acid soil. Since transport of the effluent is expensive, the spoil banks and sewage disposal plants should be as close together as possible. The potential benefits of the scheme might even justify the location of sewage plants in the center of spoil areas. Once vegetation is established and the vicious cycle of excessive heat and inadequate moisture is broken, the pipes distributing the effluent could be shifted to another barren area. When the area surrounding the sewage plant is revegetated, the effluent could be used indefinitely to increase the productivity of the new forest.

Industrial Waste Treatment

The waste water from some industrial plants—food processing, textiles, rubber and plastics, machinery, and transportation equipment—can usually be accommodated by the plant's own sewage treatment equipment or by nearby municipal facilities for a fee. Wastes from paper, chemicals, petroleum, coal, and primary metals are not so easily handled because of the acidity, toxicity, or resistance of the wastes to bacterial breakdown.

Recycling and reuse of water in the pulp and paper industry has increased tremendously since 1950 and will continue as water supplies and pollution standards tighten. Lignosulfonates, a by-product of paper manufacturing, are being used because of their dispersal and sequestering characteristics as additives to drilling mud in oil wells and as stabilizers of unpaved road surfaces in the summer. Even though 425 million pounds of lignosulfonates are used yearly, these markets are small compared with the supply.

Reuse and by-products

There are other possible uses of pulp wastes. Over 40 percent of the domestic vanillin supply (artificial vanilla flavoring) is synthesized from lignosulfonate, as are glacial acetic and formic acids. The sugar xylose, which is quite abundant in pulping liquor, is very sweet but has few calories, making it a potential substitute for cyclamates. The complex sugars from hemicellulose can be converted to ethanol (grain alcohol), for which there is a steady industrial demand. Hemicellulose can also be eliminated from the effluent by culturing a yeast, *Torula,* which can be harvested as a valuable protein source. However, competition with established chemical industries already producing these materials is great, and these by-product activities are not as profitable as they may appear.

The economic elimination of the small processor and his replacement with much larger operations is already reducing pollution loads in industry. Food processing plants that are large and diverse enough to afford their own disposal plants can effectively reduce the organic content of their wastes and ultimately the BOD of the aquatic environment. This has been demonstrated with canning plants, meat packers, and dairy product processors.

Most industrial problems, then, can be handled by recycling, reuse, by-product utilization, more efficient and effective equipment, and biological breakdown of organic wastes. For those problems remaining, more radical methods have been proposed.

Deep injection wells

When oil wells are drilled, large quantities of salt brine are brought to the surface with the oil. One approach to disposing of the brine has been to pump it, under pressure, down dry wells or new wells drilled for this purpose. In the past forty years over 40,000 brine injection wells have been drilled or utilized. The chemical industry and others have followed this example, using deep injection wells for the disposal of highly toxic or difficult-to-handle wastes generated in various processes. As surface disposal becomes more and more restricted, other industries will doubtless use this technique as the cheapest and most convenient way to get rid of wastes. But deep injection wells cannot be put down anywhere; there must be permeable sedimentary rock layers such as limestone or sandstone, capable of transmitting fluids and porous enough to hold large quantities of liquid in the spaces between the grains of rock. To contain this liquid properly the porous layer must be bounded above and below by impermeable layers. The desired stratum must also be below the water table to prevent contamination (Figure 6.5). Although these requirements seem rather specialized, over half the land area of the United States appears to be suitable for injection, especially the southeastern coastal plain and the interior lowland of the Midwest. Wastes put into deep injection wells must have a low enough solid content to avoid filling the pores of the rock stratum, thereby reducing its perme-

Fig. 6.5 • A deep well injection must deposit the waste fluid in a permeable layer of rock bounded by impermeable layers. The water table lies much closer to the surface.

ability. Potentially dangerous is the generation of heat by radio-active wastes, which may produce explosive steam, and the precipita-tion of salts or heavy metals, which may contaminate water supplies.

Despite these restrictions, wells are used by petrochemical, chemical, steel, plastic, pulp, and photoprocessing industries. The effluents they dispose of include phenols, cyanides, phosphates, chlorides, nitrates, chromates, sulfates, alcohols, chlorinated hydro-carbons, acetates, and ketones. Although most of the wells are be-tween 1000 and 6000 feet deep, some are shallower than 1000 feet, and others as deep as 12,000 feet.

One such deep injection well near Denver, Colorado, has stirred up much controversy. This well was drilled through the flat strata into the crystalline bedrock 12,000 feet below. Fluids were both injected and deposited by gravity flow at rates varying from 2 to 6 million gallons a month, beginning in March 1962 and terminating in Febru-ary 1966. Since a number of minor earthquakes had been recorded in the Denver area during this period, it was thought that the deep in-jection well might have caused them. A comparison of earthquake fre-quency and waste injection (Figure 6.6) does indeed show an inter-esting correlation. The chances were calculated to be only one in 2.5 million that the earthquakes associated with the injection well in both time and space were solely due to coincidence. Apparently the fluid injected into the basement rock lubricated fracture lines, allow-ing tension built up over the years to be released through a series of slippages along these fault lines, which resulted in earthquakes. As the volume of fluid injected increased, either there was more slippage or a greater area was affected. The only solution beyond cessation of in-

Fig. 6.6 • When earthquake frequency is compared with the volume of fluid injected into a disposal well at the Rocky Mountain arsenal near Denver, Colorado, a cause and effect relationship is strongly suggested. *(Modified from J. H. Healy et al.,* Science 161, *1968. Copyright 1968 by the American Association for the Advancement of Science.)*

jection would be to remove some of the fluid already present, or, barring that, to wait for a natural decrease in the resistance to diffusion as the fluid percolates further and further from the well.

This example raises several questions about deep injection wells and liquid waste disposal in general. One wonders, in light of the Denver example, whether such wells should be used at all in areas of seismic activity. In addition, the qualifications outlined previously suggest that storage capacity is limited and perhaps should be reserved for wastes difficult or awkward to handle by conventional means. Injection wells certainly should not be used for mere convenience or economy. The presence of faults even in the most stable rock casts doubt on the ability of supposedly impermeable rock layers to contain wastes permanently and to prevent their introduction to the water table. Finally, we know little about confining the lateral spread of liquids in a stratum; this information would be of great significance should we wish to retrieve these wastes in the future. There are many instances of yesterday's garbage becoming today's food, and there may come a time when it would be most useful to recover as resources the wastes that we discard today.

The near future will doubtless see a sharp increase in by-product recovery and in the reuse and recycling of water. As this takes place, industry's role as a major polluter of water will steadily decrease.

EUTROPHICATION

No natural body of water is without life. Those with relatively small populations of plants and animals are called oligotrophic; their low level of fauna and flora results in very clear blue water. Lakes with high populations of aquatic life are termed eutrophic; because of the density of organisms the water is turbid and green. The causes of the eutrophication are many and varied but generally can be traced to the abundance of three nutrients: phosphorus, nitrogen, and carbon.

The concentration of phosphorus in aqueous environments is normally low, about 0.02 parts per million (ppm), because most phosphates are insoluble in water. If the input of phosphorus into aquatic systems is increased much beyond this figure, it is either organically fixed by organisms and ultimately deposited in sediments, or precipitated directly. Either way, excess phosphorus is incorporated into sediments eventually, but the pathway is of critical importance.

When a nutrient normally in short supply is added to a system it may trigger a rapid growth of organisms in that system (Figure 6.7). Phosphorus is just such a trigger factor in most aquatic systems. Microorganisms are often held in check by a lack of phosphorus, but when it is made available they use it immediately, and with their rapid reproductive rates they can quite suddenly become extremely

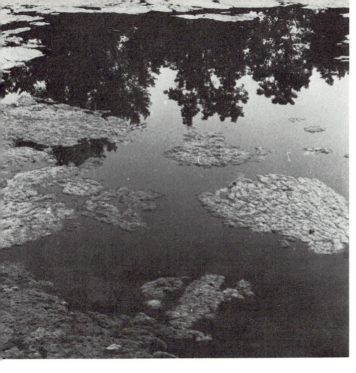

Fig. 6.7 • A mass of filamentous algae in the nutrient-rich water of a trout hatchery in Pennsylvania.

abundant. This population explosion, usually of one or more species of algae, is called a bloom. Most freshwater ecologists or limnologists call an aquatic population a bloom when there are 500 individuals of a species per milliliter of water. If a given amount of phosphorus caused a bloom of a certain intensity, we could easily set up maximum allowable concentrations of phosphorus and control the bloom. Unfortunately, the information available on phosphorus and algal blooms indicates that there is no one-to-one relationship between phosphorus concentration and algal blooms. In some Wisconsin lakes, algal blooms have been recorded with a phosphorus concentration of 0.01 ppm. In other lakes, blooms have been observed where the phosphorus concentration was 0.001 ppm and yet other lakes have not had blooms with a phosphorus level of 0.05 ppm, fifty times as great.

This has led to the assertion that organic matter is really the key to the eutrophication problem, that carbon is more often limiting than phosphorus. Adding large amounts of organic matter to aquatic systems stimulates a bloom of bacteria, which produce large quantities of free carbon dioxide, as much as 20 mg per liter. This in turn stimulates the rapid reproduction of algae that contribute to the overall eutrophication of the system. Those who believe that phosphorus is the limiting factor respond that blooms occur in water with very low levels of phosphorus because the phosphorus is in the algae and not the water. If both the phosphorus *and* organic material were removed from effluents before discharge into natural aquatic systems, as is done in tertiary treatment systems, the controversy would become quite academic.

What are the sources of phosphorus, both natural and man-made, in the environment? The data in Table 6.2 show that domestic waste and agricultural runoff are the largest contributors. The phosphate content of domestic sewage is about evenly divided between human waste and household detergents.

Table 6.2

Sources of Phosphorus in Water Supplies*

SOURCE	PRODUCTION, MILLION LBS/YR	PHOSPHORUS DISCHARGE, PPM
Domestic waste	200–500	3.5–9
Industrial waste	†	†
Agricultural runoff	120–1200	0.05–1.1
Nonagricultural runoff	150–750	0.04–0.2
Urban runoff	11–170	0.01–1.5
Rainfall	3–9	0.01–0.03

* Task Group Report, 1967. "Sources of N and P in Water Supplies." *JAWWA* **59**, pp. 344–366.
† No estimate available.

Detergents

The unusual cleaning powers of detergents are not the result of a single magic molecule engineered by far-seeing chemists. There are many components, each tailored for a specific cleansing function, yet working together so that the result is greater than the potential of any single part. Typically, a detergent contains a surfactant, a builder, a silicate, and carboxymethylcellulose.

The surfactant or sudsing agent (30 to 50 percent of the total) is an organic agent that is able to penetrate between the soil particle and cloth or fabric. This is because the surfactant molecule is bipolar; one end is attracted to the dirt particle and the other is attracted to water. This bipolar action combined with machine agitation quite effectively removes dirt and, by lowering surface tension, grease as well.

Once in the wash water, dirt must be kept in suspension; this is one function of the builder. Another function is to soften water, eliminating the scummy precipitate soap produces in hard water, water containing a large amount of calcium and magnesium salts. While varying greatly from brand to brand, most detergents contain up to 50 percent builder. The remaining detergent ingredients, though present in small amounts, complement and complete the cleaning action of the surfactant and builder. Sodium silicate is added as a corrosion inhibitor and carboxymethylcellulose is an anti-redeposition agent that helps keep removed dirt in suspension until the wash water is flushed away.

At one time the surfactant caused problems because of its tendency to keep sudsing in the environment beyond the washing machine. A change in the structure of this component in 1965 alleviated this problem. But an even greater problem has since been recognized. The builder, which comprises up to half of the packaged detergent, contains an impressive amount of phosphorus in the form of tripolyphosphate. Since this as well as the other detergent ingredients must be water soluble to perform its cleansing function, it is also available to microorganisms when it enters the environment in sewage effluent. Initially the potential fertilizing effect of tripolyphosphate was probably not considered important enough to worry about, but with the enthusiastic acceptance of detergents by the public, millions of pounds of phosphorus were soon being added to the environment each year. The result was algal blooms in increasing frequency across the country. Up to a certain point bloom organisms increase the useful productivity of streams and lakes, making available a more abundant food supply for higher links of the food chain. But in very large numbers these blooms cause problems.

Since many bloom organisms are photosynthetic, the oxygen level in the water during the daylight hours is sharply augmented as oxygen is released in the photosynthetic process. But during the night, when there is no photosynthesis to balance respiration which consumes oxygen, the oxygen content of the water may fall below the level necessary for respiration of higher forms of animal life; fish kills often result. This is a particular problem in the still water of lakes. In addition, windrows of algal mats washed up on lake shores decompose, producing hydrogen sulfide and other by-products, often making bodies of water unsuitable for use of any kind.

The connection between detergent builders and algal blooms was eventually recognized, and the industry is presently dealing with its second unanticipated problem. Unfortunately we cannot assume that once phosphates were somehow removed from detergents in manufacture or processed out of sewage effluent, algal blooms would quickly disappear, since detergents are only one source of man-generated phosphorus.

But it is an unnecessary source, since we are not obliged to use detergents. Nearly 60 percent of the population of the United States lives in softwater areas, where sodium stearate or soap-based cleansers would work as well as detergents. For the hard water regions, lime soap dispersing agents (LSDA) are effective. These compounds—anionic fatty alcohols, fatty amides, fatty acid esters, and nonionic compounds based on tallow alcohols—disperse the lime soaps or scum normally precipitated in hard water. They are easily available from tallow, and are biodegradable and nontoxic. Since they have been used for years in bar soap, it would not take an industrial revolution to transfer the LSDAs to laundry use. But unfortunately for the environment detergents are more profitable

Alternatives to detergents

113

than soap for the industry to manufacture and sell. Ostensibly because of dangers in caustic soda replacements for phosphates in some laundry products that appeared in response to the controversy over phosphates in detergents, the Environmental Protection Agency, the Food and Drug Administration, and the Council on Environmental Quality reversed their decision to support a phase-out of phosphates and at last hearing were urging the public to purchase phosphate-based detergents as the lesser evil. The government would evidently rather support removal of phosphates from the effluent than control their entrance in the form of detergents.

Agricultural Runoff

Despite the enormous quantity of phosphorus being added to the environment via detergents, we are adding several times this quantity in the fertilization of farmlands. As the world's population continues to increase, the need for greater productivity in agriculture becomes critical, and increased fertilization is one of the chief means of accomplishing this. But when highly soluble fertilizer is used, 10 to 25 percent is leached away into the surface runoff before the plants are able to use it. This represents a loss to the farmer and a burden to the aquatic ecosystems, which are being grossly over-fertilized. Another agricultural source of phosphorus is animal waste. Livestock, increasingly grouped in feedlots, contribute 70 million pounds of phosphorus per year, almost as much as detergents.

While the phosphorus problem will never be completely solved, there are several ways by which the various sources of man-generated phosphorus can be brought under control.

SOME SOLUTIONS TO THE PROBLEM

Where it exists at all, sewage treatment consists of one or two stages. Primary treatment simply removes suspended solids; after the filtering process, the remaining liquid is released into the environment. Secondary treatment goes a step further and utilizes bacteria to break down into inorganic compounds the sludge or organic wastes that were settled out in the primary process. The water is then filtered, chlorinated to remove the bacteria, and released. The BOD is greatly reduced because of the inplant decomposition of organic material, but the effluent still contains large amounts of phosphates, nitrates, and most other nutrients known to promote plant and animal growth. Primary and secondary treatment of sewage is only partly effective. Some kind of tertiary treatment to remove nutrients *must be included* if artificial enrichment of surface water is to be controlled.

There are several possible tertiary treatments using existing technology. The effluent can be prebloomed, that is, held in shallow tanks or pools for a time to allow algal growth to remove nutrients. Algae can then be filtered out, leaving the water considerably reduced in its nutrient content. Problems with this approach are the space necessary for the holding ponds and the need for supplying heat to keep algae growing at a high rate in the winter in most northern states. Perhaps heat generated from the activated sludge phase of secondary treatment or coolant water from a power plant could be used to heat the water to allow a longer season of algal growth. The algae filtered out of the final effluent could be processed and sold as a protein feed for livestock, generating some economic return. A more promising solution uses lime or alum to take phosphorus out in the form of insoluble precipitates that can then be treated to re-generate the precipitator and perhaps provide some form of phos-phorus that has agricultural or industrial use. Unfortunately nitrates in sewage are much more difficult to remove because, unlike phos-phates, most nitrates and ammoniates are water soluble.

As for the other man-generated sources of phosphorus, there is a growing tendency for livestock and poultry to be concentrated in feedlots or various controlled environments, raising the possibility of treating animal wastes in the same manner as domestic sewage. In contrast, fertilization of agricultural land, which is by far the greatest source of man-generated phosphorus in the environment, presents far more serious management problems. If highly soluble fertilizers, which are immediately available regardless of the plant's ability to make use of them, could be replaced by less soluble fertilizers, which release nutrients more slowly and keep pace with crop needs, there would be a substantial decline in nutrient leaching into the water table or surface water. The problem of industrial phosphorus sources and the phosphorus contained in bottom muds is still un-solved.

One thing is certain, however: if phosphorus-stimulated algal blooms are to be significantly reduced, ways must be found to reduce the release of man-generated phosphorus into the environment. If this cannot be accomplished effectively, then much greater emphasis must be placed on the control of organic matter, which in some situations acts as a trigger factor.

WHERE SOLUTIONS FAIL

Most problems of excessive aquatic productivity, however un-pleasant, have been limited in scope to streams, segments of rivers, and small lakes. But with the appropriate conditions, there is no upper limit on the size of overproductive waters: large lakes, seas, and ultimately even oceans can become overfertilized with cata-strophic results.

The Aging of Lakes

It was long thought that lakes aged gradually, progressing through a sequence of events from oligotrophic birth through slowly increasing eutrophy to a marsh, and, finally, a forest. But things are not quite so simple. Studies in the lake district of England seem to indicate that this simplistic picture of lake development should be re-examined. Apparently the thickness of bottom deposits in some lakes bears little relationship to the internal productivity; it is the drainage basin that is the major source of sediments. When erosion in the watershed of a lake is high, minerals are removed and sealed away in lake bottoms before nutrients locked in these minerals have a chance to be leached away in the process of soil formation. During periods of less intense erosion when more leaching of soil takes place, the nutrients pass into the lake and stimulate biological productivity. When the watershed lacks vegetation, erosion is rapid; when it is covered with trees, erosion is much reduced. The history of a lake and its watershed can often be read by analyzing the bottom mud and tracing the abundance of various nutrients layer by layer. When this was done at Esthwaite Water in England, sodium and potassium levels were found to be quite low soon after the lake was formed. As the watershed became covered with vegetation and erosion was checked, the levels of sodium and potassium rose. At the time of forest decline associated with man's activities about 5000 years ago, these nutrients decreased again and have continued declining up to the present. A similar picture was obtained from work done on Linsley Pond in Connecticut—one of the most studied bodies of water in the world.

The implication of this work is that lakes may be eutrophic or oligotrophic from the very beginning and may remain in their initial state for thousands of years. They may change rather suddenly when man intervenes; or they may even cycle from oligotrophy to eutrophy and back again to oligotrophy. So it now seems unlikely that all lakes are in the inexorable process of gradual change from oligotrophy to eutrophy. If the old pattern no longer represents the true situation, we can no longer predict what will happen. Detailed studies of just what is happening in grossly overfertilized lakes right now are necessary. From this, we may be able to make some predictions about the future of the lakes. No remedial action can be taken until the base lines are drawn.

Death of a Great Lake?

Of all the overfertilized lakes in the world, Lake Erie is one of the largest and best known. In fact, the western basin of this lake is so polluted that discussion has arisen on whether this portion of the lake is merely dying or already dead. On the other hand, the eastern basin is relatively unpolluted.

Lake Erie, like the four other Great Lakes, is a glacially scoured lowland that was filled by the melting continental glacier around twelve thousand years ago. Although Lake Erie is not the smallest Great Lake, because of its shallowness (mean depth of fifty-eight feet) it has the smallest volume. This shallowness is at the root of its subsequent problems.

Once clear and filled with valuable fish and game, western Lake Erie today is the embodiment of all that can go wrong in an aquatic system. The more desirable species of fish have disappeared; the once clear water is filled with excessive numbers of microorganisms; mats of filamentous algae at times cover whole square miles of lake surface; swimming is impossible in many places because of the quantity of untreated sewage in the water and the decaying vegetation covering the beaches; oil scums often cover harbors and coves, making boating and water sports unpleasant.

Of all the Great Lakes, Lake Erie is adjacent to the largest number of population centers, each having varied industries with effluents to be disposed of. Erie produces paper; Buffalo, flour and chemicals; Cleveland, petrochemicals and steel; Toledo, glass and steel; and Detroit, automobiles, steel, and paper (Figure 6.8). Naturally, large population centers contribute quantities of sewage, both treated and raw. And the agricultural land between the cities adds pesticides, herbicides, and fertilizers, giving no respite. Bear in mind this is (or was) a "great lake," 240 miles long and 50 miles wide, with a volume of 109 cubic miles!

The first symptom of lake-wide problems in Lake Erie was seen in the 1920s, when the fish crop—50 million pounds annually of cisco, whitefish, pike, and sturgeon—began to fall. By 1965 the catch of these commercially valuable fish had fallen to just 1000

Fish as a pollution indicator

Fig. 6.8 • When Great Lakes cities with a population of 100,000 or more are spotted on a map, the clustering around Lake Erie helps to explain the critical nature of that lake's pollution problem.

Ft. William •
Lake Superior
Duluth •
Lake Huron
Toronto •
Lake Ontario
• Rochester
Hamilton •
• Buffalo
Lake Michigan
Milwaukee •
Detroit •
Lake Erie
• Erie
Toledo •
• Cleveland
Chicago •
• cities over 100,000

pounds. This doesn't mean that Lake Erie is devoid of fish, for there are still about 50 million pounds of other types of fish caught every year; but the more valuable species have been replaced by catfish, carp, drum, and other less valuable types (called "rough" fish by commercial fishermen).

Why did this happen? During the warm season of the year all but the most shallow lakes stratify into three layers (Figure 6.9), a warm surface layer, epilimnion, a transition layer of rapidly changing temperature, thermocline, and a cold deep layer, hypolimnion. Under normal conditions whitefish, cisco, and other cold-water fish move to the deeper, colder hypolimnion in summer, and return to the surface in fall when storms and colder weather mix the layers and the lake again becomes thermally homogeneous.

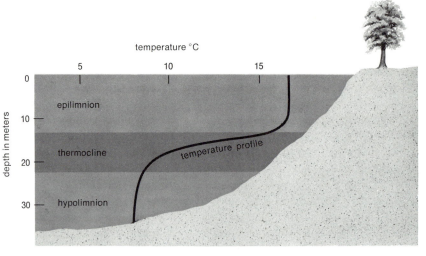

Fig. 6.9 • In the summer deep lakes stratify into a cold hypolimnion, a zone of rapid temperature change or thermocline, and an upper layer of relatively warm water. The dark line describes this temperature profile.

With increasing eutrophication, problems develop. Normally, a certain number of algae are found in the warm surface water during the summer. At the end of the growing season, they die and fall to the bottom where they are broken down by bottom bacteria into inorganic nutrients which are held in the bottom mud. By the next summer, enough nutrients have leached from the watershed into the lake to support another population of algae. No more algae are able to grow than accumulated nutrients are able to support, and so a balance is maintained. When nutrients pour in from man-generated sources, the balance is upset. Great blooms of algae released from their dependence on natural nutrients fill the epilimnion during the summer, sinking to the bottom during fall and winter. When the lake stratifies the following summer, extra thick organic layers at the bottom so stimulate the growth of bacteria that the oxygen level of the hypolimnion begins to fall, often falling below the minimum requirements of fish. Because cold water can hold more oxygen than warm water, cold-water fish such as trout or cisco have developed, through evolution, higher oxygen requirements than warm-water fish such as carp or catfish. Hence, when the oxygen level of the cold deep water begins to fall, the valuable cisco, whitefish, and pike are eliminated.

The Future of Lake Erie

The deoxygenation of the hypolimnion previously discussed is far more serious than the loss of valuable fish. Under normal oligotrophic conditions, when the organic sediment load is light enough to allow reasonable oxygenation of the hypolimnion, the bottom mud is covered by a thin layer of iron in the insoluble or ferric oxidation state, usually ferric hydroxide. This compound not only absorbs much phosphorus from the water, but because of its insolubility seals the bottom mud, preventing exchange of nutrients in the mud with the water above.

When oxygen is exhausted, the iron changes to the soluble ferrous state. This not only puts phosphorus into solution as ferrous phosphate but exposes the bottom mud, allowing free entrance of nutrients into the lake water. The problem in Lake Erie, then, is simple and overwhelming. Wherever pollution stimulates algal blooms, which later decompose and deoxygenate the bottom, nutrient-rich mud is exposed that is capable of contributing its nutrients to the lake water, stimulating still more algal growth, deoxygenating more lake bottom, thus closing a vicious cycle. Even if all man-generated pollutants or nutrients were prevented from entering the lake, there might still be quite enough nutrients already in the lake to continue algal blooms with concomitant problems far into the future.

Although the misuse of Lake Erie makes a rather depressing story, other, smaller, lakes have been snatched from the green slime of recreational death by active community concern and careful planning. Lake Washington, located east of Seattle, is 24 miles long and 2 to 4 miles wide (Figure 6.10). Fifty years ago Seattle was far to the west and the eastern shore was sprinkled with small towns, each with fewer than 10,000 inhabitants. The western shore was either parkland or large estates. Then in 1940 a floating bridge spanned the lake and population growth on the eastern shore began in earnest, rising from 33,000 in 1950 to over 120,000 in 1970. Although many of the larger communities built sewage treatment plants, the outfall of these and of the septic systems of communities without sewers flowed into the lake. By 1958 dissolved oxygen had disappeared from even the deepest part of the lake, driving out the salmon; water clarity had diminished to 2.5 feet, and algal growth made summer recreational use of the lake something less than a pleasing experience. The problem wasn't Seattle's, really, because most of its sewage was piped into Elliott Bay, an arm of Puget Sound. The difficulty lay in the typical hodge-podge of tax

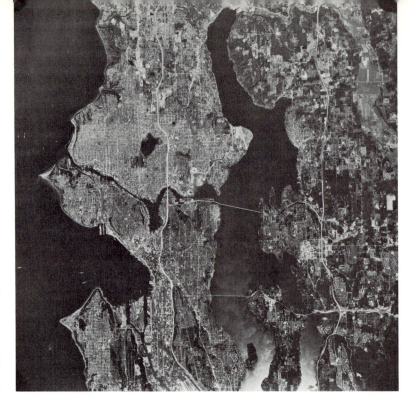

Fig. 6.10 • Considering its location, surrounded by Seattle and its suburbs, it is not surprising that Lake Washington became badly polluted as its eastern shore became urbanized. But when the people living in its drainage basin faced the problem squarely, effective measures were taken to restore the lake to the recreational uses enjoyed in the past. The body of water to the left is Puget Sound. (*NASA*)

districts, townships, and cities that surrounded the lake. Clearly, the answer was a metropolitan system to pull everything together.

The Metro concept was pioneered by a young lawyer, James Ellis, and a citizens' committee, which developed legislation supporting a metropolitan district that would handle sewage disposal, transportation, and land use planning. While the legislature passed the bill, a majority vote was required in both Seattle and the surrounding communities. In early 1958 voters in Seattle approved Metro, but the suburbs killed it. Since the entire Metro concept was apparently too progressive for 1958 it was watered down to include just sewage disposal. After a long smelly summer, which must have convinced many voters that enough was enough, the Metro package was resubmitted and passed in September 1958. By constructing interceptors all around the lake and building four large treatment plants on Puget Sound, all sewage discharges into Lake Washington were halted. Within a couple of years, even before the system was completed, the BOD in the lake dropped by 90 percent, fish returned to deep water, clarity increased to 15 feet, and phosphorus levels fell from 70 ppb to 22 ppb. The cost of the project was held to $2 a month per household.

The Lake Washington success story closely parallels that of San Diego Bay (see Chapter 7). The requirements are few but difficult to generate: a small group or an individual with time, drive, and vision to convert public apathy into concern; and a public willingness to spend money in perpetuity to achieve a desired goal. Given these two requirements any community, large or small, can come to terms with its environmental problems.

THERMAL LOADING

Although water is the most abundant compound on earth, it has several strikingly different characteristics: one of the most unusual and certainly most useful is its high specific heat. By absorbing and releasing heat slowly, water becomes an excellent medium to effect heat exchange; hence, water has been used for heating and cooling purposes for thousands of years.

With the rise of industry in the eighteenth century, water came to play an increasingly important role as a coolant in many manufacturing processes. With continued growth and diversification of industry, the demand for cooling water has soared. Today over 80 percent of the water used by industry in the United States is used for cooling. Of all industrial water users, the electric power industry is the largest, requiring 111 billion gallons per day. When limited, this loading of waste heat into a natural aquatic system caused few problems, but when industrial plants began to use almost the entire flow of a river for cooling purposes, environmental damage resulted.

Heat as a Pollutant

Why is addition of heat energy to an aquatic system even considered pollution? Although heat loading of a stream causes it to depart from its normal thermal character, which was our definition of pollution at the beginning of this chapter, there is a more specific reason to view thermal loading as a form of pollution. As the temperature of water increases, its ability to hold oxygen decreases (Figure 6.11). Since dissolved oxygen is the key to assimilation of organic wastes by microorganisms, any activity that might impair this assimilation can be labeled as pollution.

Temperature and oxygenation

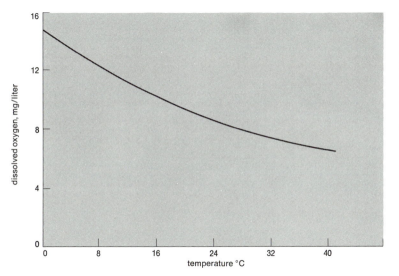

Fig. 6.11 • The quantity of oxygen dissolved in water is related to temperature, as shown in this graph. The higher the temperature, the lower the oxygen content of the water. Thermal pollution can lower the oxygen content below the point necessary to sustain many animals.

121

The presence of dissolved oxygen is probably the single most important factor in the biology of aquatic systems, and a great variety of physical and biological interactions stem from it. Aeration or oxygenation is derived from two sources: exchange with the atmosphere, and photosynthetic release by green plants. Depending upon the temperature difference between a heated effluent and the body of water into which it is being flushed, the heated discharge can either mix vertically and horizontally, or stratify in a layer above the cooler, denser receiving water (Figure 6.12). If the temperature difference is small and the receiving water shallow, complete mixing usually takes place. If the temperature difference is great and the receiving water deep, stratification usually results: the colder water beneath is covered with a blanket of hot water, which not only carries less oxygen because of its higher temperature but is less likely to exchange with the cooler water beneath because of its different density. Biological reduction of the oxygen content of the atmospherically unreplenished water below might then produce anaerobic or oxygenless conditions.

Fig. 6.12 • An infrared aerial photo of hot coolant water from a coal-burning power plant being dumped into a river. Since infrared film is sensitive to heat, the lighter areas are those having the warmest surface. Note the slow rate of mixing of hot and cold water. (*Environmental Analysis Department, HRB-Singer Inc.*)

With a given thermal load and assuming complete mixing, a stream or lake might have its temperature elevated by around 10°C (20°F), which could lead to a complete shift from cold- to warm-water forms of life. Whereas such a replacement might cause little disruption to the productivity of a body of water once the adjustment had taken place, in almost no instance would the production of heated effluent be constant for long enough periods to allow either adaptation or replacement to be effective. Shutdown of a power plant or factory for just one day in midwinter would have a disastrous effect on warm-water species introduced and maintained by the artificially high water temperature. Such an occurrence, perhaps extreme in its regular periodicity but illustrative of the problem, is seen

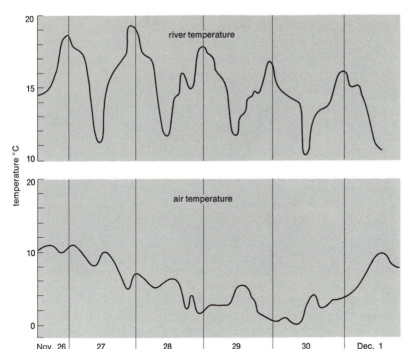

Fig. 6.13 • Periodic pulses of hot water flowing down the River Lea in England from a power plant make any adjustment of flora and fauna to either warm or cold water impossible. (*From Dysart and Kenkel,* Proc. 20th Ind. Waste Conf., *1965*)

in Figure 6.13. Heated effluent is dumped periodically into the River Lea from the Rye House Power Station in England, making adjustment by organisms adapted either to warm *or* cold conditions extremely difficult. Only those organisms without specific temperature preferences can survive such a change of temperature.

Much more unusual but occasionally observed is a cold effluent released into a warm river. The Fontana Dam, a unit of the Tennessee Valley Authority on the Little Tennessee River in North Carolina, releases cold bottom water from its reservoir downstream into the river, eliminating for many miles the warm-water fish usually characteristic of southern rivers.

Effect of Temperature Change on Organisms

As temperature increases, dissolved oxygen content decreases, as we have seen, but respiration and oxidation rates double for every 10°C temperature increase (Figure 6.14). In order to meet the respiratory needs of warm-water fish, the dissolved oxygen content of water should be at least 5 ppm during sixteen of every twenty-four hours and should not fall below 3 ppm at any time. Even carp, which have an exceptionally low oxygen requirement for fish, 0.5 ppm at 0.5°C (33°F), must have 1.5 ppm at 35°C (95°F).

But temperature interacts with other factors as well. At 1°C (34°F), carp are able to tolerate a carbon dioxide concentration of 120 ppm, but at 36°C (86°F) a concentration of 55 to 60 ppm is lethal. Similar interactions are seen with pH, salinity, and toxicity.

NATURAL TRAUMAS BECOME POLLUTANTS

Fig. 6.14 • Occasionally hot summer weather naturally heats shallow water to the point where it cannot hold enough oxygen to meet the needs of fish, causing them to suffocate. This is what killed these shad in the Anacostia River near Washington, D.C. (*U.S. Dept. of the Interior, Sport Fisheries and Wildlife*)

All organisms have temperature limits to their survival ability, but for most, acclimatization is possible to some extent if temperature change is gradual. It is the sudden change either up or down that is most often lethal.

The life cycle of many aquatic organisms is closely and delicately geared to water temperature. Fish often are distributed, migrate, and spawn in response to temperature cues. When water temperature is artificially changed, the disruption of normal activities and patterns can be catastrophic. Some animals spawn as the temperature drops in the fall; conversely, some spawn as temperature rises in the spring. Shellfish, such as oysters, are so delicately attuned to temperature change that they spawn within a few hours after their environment reaches a critical temperature. Trout eggs take 165 days to hatch at 3°C (37°F), only 32 days at 12°C (54°F), and will not hatch at all above 15°C (59°F). Lifespan is also affected by water temperature; water fleas (*Daphnia*) live for 108 days at 8°C (46°F), but only 29 days at 28°C (82°F).

Even more subtle effects involve competition among various species, predation rates, incidence of parasitism, and spread of disease. Any one or a combination of these more covert factors can destroy a population as effectively as the more obvious thermal shock.

MAN SIMPLIFIES THE ECOSYSTEM

Plainly, adding hot water to the environment is not the simple act of dilution it might appear to be. To put this subtle disturbance into perspective, we must view an organism together with its biotic and abiotic environment as an integrated system, an ecosystem.

Natural ecosystems, whatever their size, are complex; the larger they are, the more complex. Man's chief effect on ecosystems is to try, consciously or unconsciously, to simplify them. A potato field serves as a good example of this. To channel the productivity of a piece of land as directly as possible into a useful product—potatoes—man simplifies the ecosystem by removing all plants or animals that might compete with the potato plants and thereby decrease the yield.

Cultivated plants were selected to survive the stresses (drought, direct sunlight, flooding, and wind) that are characteristic of man-affected environments. Most native plants have requirements too specialized to tolerate such stress conditions. However, one group of plants, weeds, is not only able to tolerate stress but is able to outcompete man's plants and become superabundant. The appearance of weeds in a field is the first step in a natural process (succession) that reduces stress and ultimately leads to the return of a complex ecosystem capable of sustaining itself.

A typical nonpolluted stream may support as many as two dozen species of fish. After pollution, although there may be only three "weed" species, carp, killifish, and shiners, the *numbers* of fish may remain the same. Through the simplifying stress of pollution, the productivity, which in the normal stream was shared by small numbers of many species, has now been channeled into large numbers of a few species.

Even more basic are the effects of stress on food chains. The basis of all higher life is the green plant or, in aquatic systems, various forms of algae. Because of their ability to fix energy from sunlight, carbon dioxide, and available mineral nutrients (nitrogen and phosphorus especially) without dependence upon other organisms, we call them primary producers. The primary producers are in turn eaten by a series of consumers leading in a chain from tiny crustaceans and larvae to fish and, perhaps ultimately, to man. The productivity of a body of water depends upon the kinds and numbers of algae present.

There are three major groups of algae in fresh water from a food chain point of view: diatoms, greens, and blue-greens. While all types are eaten by some organisms, most consumers prefer diatoms to greens or blue-greens. Blue-green algae are eaten by few consumers, and they can be toxic to some aquatic and terrestrial consumers. These three algal groups differ in their temperature tolerances (Figure 6.15). With an increase in temperature the number of the more desirable diatom species decreases as the number of the greens and blue-greens increases. At a certain level the greens also decline, leaving the blue-greens most abundant at the higher temperatures. As these primary producers shift in their relative abundance, the consumers depending upon them for food are also affected. A water flea, for example, which might be able to tolerate the thermal extreme of 35°C (95°F), would probably starve to death if the diatoms on which it fed were unable to survive at that temperature. In turn, fish feeding on water fleas would be similarly hard pressed to survive,

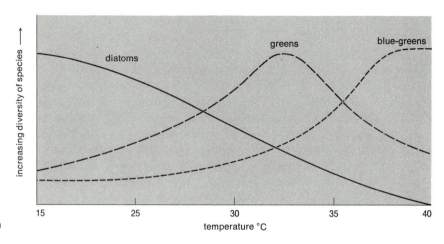

Fig. 6.15 • The three most important groups of common freshwater algae, (a) diatoms, (b) greens, and (c) blue-greens, have radically different temperature tolerances. (*Modified from J. Cairns, Jr., Scientist and Citizen 10, 1956.*)

regardless of their tolerance or adaptability to the high water temperature.

Another factor affecting food chains involves the large volume of water used in cooling systems. A plant circulating 500,000 gallons a minute may intercept and enclose a considerable portion of the flow of a small river. Because of the tendency of microorganisms to grow in the pipes and reduce the flow, periodic or continuous metering of chlorine is used to control such growths. This not only affects organisms where the coolant is returned to the river (the outfall) until the chlorine is diluted, but it may sterilize the water flowing through the cooling system. In some installations 95 percent of the organisms are killed during the cooling period. The thermal shock of a rapid 10°C (20°F) rise is destructive to many species. If the cooling system's intake uses more than a small part of a river's flow, a considerable portion of its microfauna and microflora could be eliminated, with grave effects on the food chain downstream, even to the point of eliminating certain populations of fish.

A heated effluent as an environmental stress, then, affects not only organisms directly but also the entire interaction of factors that make up the ecosystem. This makes it extremely difficult to predict the effects of thermal loading. With hydrologic models an engineer may be able to predict the degree of mixing, thickness, and the extent of stratification, but assessing the total and often subtle biological consequences is far more difficult than determining the thermal death point of one or two fish.

COOLING THE COOLANT

There are several current approaches to removing heat from cooling water before it is reused or discarded.

Wet Tower

The most common practice is to run water in a thin sheet over baffles in huge hyperbola-shaped towers relying on drafts of air entering at the base to remove heat by evaporation. A variant is to spray the hot water inside the tower in a fine mist. Cool air rising through the tower condenses the mist, releasing heat, which is carried by the air column out of the tower (Figure 6.16). In either case the cooled water is either discharged into the environment or recycled. Two problems associated with this wet tower technique are water loss (20,000 to 25,000 gallons per minute for a 1,000-megawatt plant), and fog formation on cold days. When the surface temperature is less than 0°C (32°F), the fog freezes on contact, forming hoar frost or rime ice, potentially damaging to vegetation and extremely dangerous on highways. The water loss not only vitiates somewhat the benefit of recycling but serves to concentrate whatever pollutants the effluent contains.

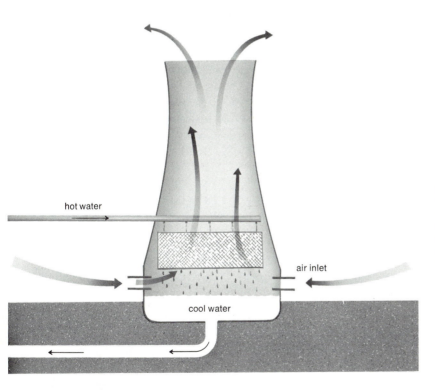

hot water

air inlet

cool water

Fig. 6.16 • In a wet cooling tower, the hot water is exposed to air circulating through the tower. As water evaporates, heat is lost. The cooled water is either recycled or released into the environment.

Dry Tower

In a dry tower the heated effluent is contained in a system of pipes much like the radiator of an automobile. Air is passed over the pipes by a large fan facilitating heat exchange by radiation and con-

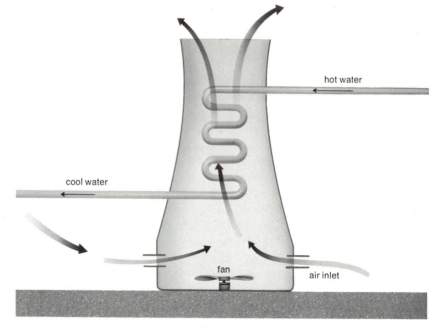

hot water

cool water

fan

air inlet

Fig. 6.17 • A dry tower resembles a huge car radiator. Contained in pipes, the water gives up its heat by radiation and convection to the air, avoiding evaporative water loss.

vection (Figure 6.17). Water loss and, to a lesser extent, fog are controlled by this method, but installation costs are much higher, perhaps two or three times the cost of a wet tower. Also, maintenance costs for one plant of 500-megawatt capacity were calculated at 3 percent of the power output or $500 thousand a year to keep the blower fans operating. If large enough the updraft of hot air could generate clouds.

Both types of towers are expensive to construct and since they depend on pressure and temperature differentials across the tower base for maximum efficiency, they work best in cool climates. The hot American summers make cooling towers less efficient than in the cool summer climates in Europe, where these towers have become conspicuous features of the landscape (Figure 6.18).

Fig. 6.18 • Large clusters of hyperbolic cooling towers, common features of the European landscape (these are in eastern England), suggest a modern-day Stonehenge.

Cooling Ponds

This approach takes advantage of the few positive aspects of thermal loading. Warm-water outfalls in winter attract fish (and fishermen) because they extend the feeding period and forestall ice formation, which often prevents adequate oxygenation of the water beneath. By combining waste water from a power plant and effluent from a sewage treatment plant, the heat and nutrients, considered a nuisance in most environments, can be put to good use. By supplying heat-tolerant algae as primary producers and introducing warm-water fish to harvest the algae, water would have time to cool, have its nutrient content sharply reduced, and its algae harvested in the form of useful protein, before being returned to the natural environment. One tropical species that might be used, milkfish (*Chanos*), is one of the few palatable fish that feeds on blue-green algae. This type of operation would probably have to be limited to northern latitudes, since heat tolerance of even tropical species would be exceeded in areas where heated effluents are released during long, hot summers.

There are, however, some pioneering attempts to make use of thermal discharges in the United States. Oysters are being cultivated year round in the thermal discharges of a power plant on Long Island (Figure 6.19). The intake water from the sea, which varies from 5 to 21°C (40° to 70°F) during the year is warmed by

Fig. 6.19 • (a) By placing small oysters in the heated effluent of a power plant, growth can be greatly accelerated. The frames between the piers hold the oysters. (b) Every week the frames are removed from the water so the oysters may be checked and screened for uniform size. After 8 to 10 weeks the oysters are transplanted to cooler, deeper water, where they mature. (*Long Island Oyster Farms*)

10 to 12°C (12° to 18°F) by being mixed with thermal discharge of 150,000 gallons per minute in a seven-acre lagoon. The young oysters are constantly produced in nearby hatcheries and are placed on racks in the discharge canal for four to six months and then transplanted into cold water for two years to mature. Although this seems a long time to wait for an oyster, the normal maturation period is five years.

A Texas power company raises catfish in cages exposed to an effluent of 22 to 24°C (70° to 75°F) in winter (Figure 6.20). With intensive feeding the experiment has yielded 100 tons per acre per year. However, all power plants have shutdown periods, sometimes at unpredictable times, which could be very destructive if alternate means of supplying heated water were not available; thermal effluents often contain toxic quantities of heavy metals, rust inhibitors, biocides, or low-level radioactive wastes; and the wastes of the food crop must be treated before discharge back into the stream or estuary to avoid eutrophication of natural environments.

Fig. 6.20 • (a) When catfish fry are suspended in cages in the warm-water effluent from the power plant in the background, the fish grow more rapidly than in unheated water. (b) The fish are fed through the hole in the top of the cage. (*Texas Electric*)

In terrestrial ecosystems, heated water could also be put to good use by applying it to crops. Preliminary evidence indicates that growth of some crops is accelerated by irrigation with warm water, suggesting the possibility of double cropping in areas of mild climate. Frost protection can also be achieved by spraying fruit trees with warm water. There are difficulties, however: frost is a problem in most orchards on only a few days each year, and irrigation is necessary for only a few months in the summer. A large power plant must cool 500,000 gallons per minute every day of the year. Also, the cooling effluent from nuclear powered plants contains slight amounts of radioactive minerals, which might be concentrated in crop plants irrigated with such an effluent.

These are problems that can be worked out; it is becoming increasingly evident that we cannot continue to inject our waste heat either into the water or into the air without causing serious environmental problems. There is a tremendous potential for ingenious uses of this heat energy, as these examples indicate—uses that must be exploited if the natural ecosystems are to survive until the day when more efficient power sources are found.

FURTHER READING

Beeton, A. M., 1970. "Statement on pollution and eutrophication of the Great Lakes." Special Rept. No. 11, Center for Great Lakes Studies, University of Wisconsin, Milwaukee. A brief review of pollution problems and their status in the five great lakes.

Bigger, J. W. and R. B. Corey, 1969. *Eutrophication: causes, consequences, correctives.* National Academy of Sciences, Washington, D.C.

Clark, E., 1967. "How Seattle is beating water pollution." *Harper's Magazine,* June, pp. 91–95. A model that perhaps can be applied elsewhere.

Hubschman, J. H., 1971. "Lake Erie: pollution abatement, then what?" *Science* **171,** pp. 536–540. Explores the possibilities of ultilization of long-term eutrophication.

Krenkel, D. A. and F. L. Parker (eds.), 1969. *Biological aspects of thermal pollution.* Vanderbilt University Press, Nashville. An interesting collection of papers on the general topic.

Kuentzel, L. E., 1969. "Bacteria, carbon dioxide and algal blooms." *Jour. Water Poll. Cont. Fed.* **41**(10), pp. 1737–1747. Elaboration of the thesis, hotly debated by limnologists, that carbon rather than phosphorus is the cause of eutrophication.

Likens, G. E. (ed.), 1972. *Nutrients and eutrophication: the limiting-nutrient controversy.* American Soc. of Limn. and Ocean.

Special Symposium, Lawrence, Kansas, Vol. I. Although the papers are often technical the discussions following each are extremely interesting. Gives much insight into the social and political background of what otherwise might be considered an academic argument.

Mathur, S. P. and R. Stewart (eds.), 1970. *Conference on beneficial uses of thermal discharges.* N.Y. State Dept. of Environmental Conservation, Albany. A collection of invited papers presenting current activities around the world.

McCaull, J., 1972. "Assault on a lake." *Env.* **14**(7), pp. 32–39. Although Lake Erie is heralded as *the* problem lake, Lake Michigan also has its problems.

Warren, C. E., 1971. *Biology and water pollution control.* Saunders, Philadelphia. A comprehensive and solid text on the topic.

Westing, A. H., 1969. "Plants and salt in the roadside environment." *Phytopath.* **59**(9), pp. 1173–79. A good overview of an increasingly serious problem.

Willrich, T. L. and G. E. Smith, 1970. *Agricultural practices and water quality.* Iowa State University Press, Ames.

Yee, W. C., 1972. "Thermal aquaculture: engineering and economics." *Env. Sci. & Tech.* **6**(3), pp. 232–237. Suggests we are on the edge of feasibility.

7

THE WORLD OCEAN:
ULTIMATE SUMP

NOT SO VERY LONG AGO, eutrophication was thought to be limited to small bodies of fresh water or streams, or, if found in the ocean, only in protected bays or harbors. While the experience of Lake Erie should have prepared us, it still comes as a shock to learn that the Baltic Sea is in serious environmental trouble. With a flushing cycle of thirty years, wastes pile up for years before they can escape through the North Sea into the open Atlantic. By the time they get there, the Atlantic may not provide much dilution—Thor Heyerdahl, that inveterate drifter, reported seeing oil slicks almost all the way from Morocco to the West Indies from his papyrus raft. Moreover, marine pollution problems are complex, involving coastal features as well as the open sea.

About 20,000 years ago, extensive continental glaciers had tied up tremendous quantities of water, exposing the continental shelves as broad coastal plains. As the ice slowly melted, the ocean level began to rise, first flooding the incised lower channels of large rivers, forming estuaries. These estuaries were covered as the sea advanced over the flat coastal plain, and a series of barrier islands formed, separated from the mainland by broad, shallow lagoons. Ultimately, when all the coastal plain is flooded and its uplands are encroached upon by the sea, another set of estuaries will be formed. But because there has been some crustal movement as well as sea rise, the effect in eastern North America is uneven. To the north all of the coastal plain is flooded; Maine and the Maritime Provinces of Canada have coastlines indented with fjords, the result of the sea flooding the coastal headlands. Further south, parts of whole river systems have been flooded, forming Delaware and Chesapeake Bays. From Norfolk, Virigina south to Yucatán, Mexico, the sea has advanced only part way across the coastal plain, hence the characteristic barrier-island–lagoon dominates. Should the sea rise another hundred feet or so, the character of the southern coastline would change, with the flooding of its flat coastal plain, from the barrier-island–lagoon to the estuarine type. Since the sea is continuing to rise at the rate of less than three feet per thousand years, we have about fifty thousand years to make plans.

If the sea level were rising very much faster than it is, all coastal cities would be in imminent danger (Figure 7.1). Why have so many great cities been built on or near estuaries? Since estuaries are usually drowned river mouths, they form natural harbors, protected from the oceans yet connecting the ocean and a river-drained hinterland. Because of this vital connection, estuaries serve as a focus of transportation. In addition, estuaries, combining characteristcs of both fresh and salt water, are extremely productive.

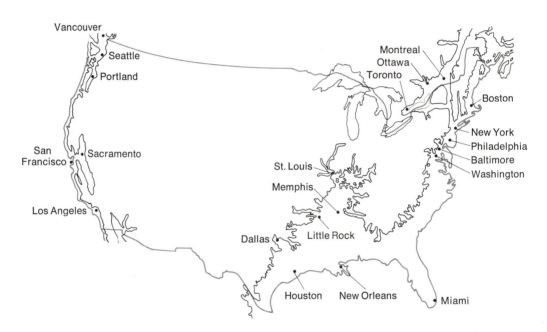

Fig. 7.1 • If all the ice now present at the poles should melt, the sea would rise by about 500 feet, flooding much of the eastern United States and most of the population centers of the Pacific coast.

WETLAND PRODUCTIVITY

This unusual productivity of estuaries and tidal marshes results in part from their ability to trap nutrients. The clay sediments kaolinite and montmorillonite, which are quite abundant in most estuaries, have a very high sorptive capacity—that is, because of their large surface area the clay particles pick up and hold loosely a variety of ions. As nutrients are removed from the water by plankton, new supplies are released from the clay sediments that act as a nutrient buffer.

A good measure of this productivity is a comparison of dry organic matter produced in a terrestrial system with an estuarine system. A typical wheat field produces 1.5 tons per acre per year (including straw and roots); a coastal marsh near Sapelo Island, Georgia produces ten tons per acre per year. The amount of this production that is actually used, however, is low and diffused through a large food web, with man harvesting only a small fraction in the form of shellfish, crabs, and fish. Broadly speaking, the great productivity of salt marshes and the estuaries, bays, mud flats, and tidal creeks with which they form an inseparable unit, is based on three communities of plants: cord grass (*Spartina*) in the marshes (Figure 7.2), mud algae on the creek banks, and microscopic plants called phytoplankton in the water.

Fig. 7.2 • A healthy stand of cord grass in one of the few remaining salt marshes in Connecticut.

Nutrient Cycling

Less than 5 percent of the cord grass is eaten in place, since few insects are able to survive both salinity and tidal flooding. Even the few hardy grasshoppers sometimes seen in the salt marshes excrete, undigested, two-thirds of the material they eat. As the grass is crushed and fragmented in the fall and winter by storms, it is broken down by microorganisms into fine organic particles containing much protein, carbohydrates, and vitamins. These materials are distributed by tides, but are retained and cycled in the ecosystem by various organisms. For example, phosphate is more abundant in estuarine ecosystems than in rivers or the ocean itself. The Altamaha River in Georgia contains 0.1 micrograms per liter, while coastal tidal creeks contain nearly 2–4 micrograms per liter. The reason for the abundance is found in the ecology of one of the most common animals in the southern estuaries, the ribbed mussel. As it filters the water for its food, the fine organic particles, which it does not eat, are stuck to-gether with mucus into little particles called pseudofeces (pseudo, because they do not pass through the digestive system of the mussel but are eliminated before entering the mouth). These particles, rich in phosphate, remain in the area and are fed upon first by fiddler crabs, then by other animals ultimately eaten by man, such as blue crabs and striped bass. Here is a classic demonstration of an economi-cally worthless organism performing an extremely valuable role in the cycling of phosphate, a most important mineral nutrient.

Food Chains

The mud algae include dinoflagellates—microscopic algae en-closed in sculptured jackets with a whiplike tail or flagellum—and diatoms, mentioned in Chapter 6. These organisms are usually so abundant that they give the muddy creek banks a characteristic brown or yellowish appearance. They are adaptable to the changing season and so are able to photosynthesize either at high tide in the summer when cooled by the tidal water, or at low tide in the winter when the creek banks are warmed by the sun. By reproducing at a rapid rate and washing off into the water, the diatoms and dino-flagellates provide a year-round food source not only for filter feeders such as oysters, clams, and mussels, but for tiny larvae and small crustaceans that are the basic food for most fish. The importance of the reproductive rate of these mudbank algae cannot be over-estimated, for we see only a very small fraction of the total algal pro-duction per year sticking to the mudbanks. Although the population on the mudbank remains constant, huge numbers are sloughed off. Much of the productivity of the ecosystem depends upon this process.

Because of the vertical movement of water masses of different salinities in estuarine areas, mineral nutrients tend to be trapped. Also, daily tides continuously provide food and oxygen while remov-ing waste products. This results in good production of phytoplankton and supplements the mudbank algae as a primary food source for the various consumer levels.

Another important plant-based food web is found in the man-grove marshes along parts of the Florida coast. As with cord grass it is the leaves of the red mangrove (Figure 7.3) that provide the base for this web. Three tons of leaves are shed per acre per year. Initially attacked by crabs and amphipods, a group of small crustaceans, the litter is quickly broken down. After nine months, the fallen leaves have been reduced to fragments about one millimeter wide. As this breakdown progresses, bacteria and fungi gradually cover the pieces, thereby raising both the protein and vitamin content of the fragments. While fresh leaves have a protein content of 6 percent, after twelve months of decomposition the protein content of the leaf fragments is increased to 22 percent. This enriched supply of leaf pieces or

Fig. 7.3 • By extending their aerial roots into the soft bottom sediments, red mangroves gradually colonize extensive areas of mudflats along tropical coasts; here, near Parguera on the south coast of Puerto Rico.

detritus provides an important source of nutrition for many species of worms, crabs, shrimp, insect larvae, small forage fish, and copepods, another group of small crustaceans. Over sixty species of young fish also feed on the mangrove detritus, fish that are highly prized gamefish as adults: tarpon, snook, and ladyfish. Other valuable species—menhaden, spotted sea trout, red drum, striped mullet, and blue crab—feed on the small forage fish and invertebrates that feed on the detritus. The economic value to Florida of these five species alone was nearly $20 million in 1968.

Until this food web was worked out by E. J. Heald at the University of Miami just a few years ago, mangrove swamps were considered wastelands to be bulldozed, dredged, and filled as quickly as population growth would allow. Along with most of the other East Coast states with large tracts of coastal wetlands, Florida has passed legislation to exert some state control over the development of wetlands which, we are finally discovering, play such an important economic role in coastal fisheries.

Utilization of Estuarine Productivity

The great productivity of estuarine areas has long been overlooked and underused because, unlike land areas, production and harvest in estuaries rarely take place in the same spot. These processes are constantly being moved here and there by the tide, and are separated not only by space but by time. There are two approaches toward better use of estuarine production: we could overcome our narrow prejudices and eat more of the elements of the food web than we consider edible today, or we could simplify the ecosystem somewhat by channeling more of the production into acceptable harvestable units or by introducing species more efficient at harvesting producers than those now naturally available. Ecosystem simplification, however, as we saw with the potato patch in Chapter 6, has its dangers. Although formidable, they are not insurmountable.

The point is that tidelands, regardless of the way in which their productivity is used, are extremely valuable as producers of food and as spawning grounds for many economically valuable species. This is seen in the productivity values of the continental shelf (1.3 tons per acre per year) and the open ocean (0.5 tons per acre per year), compared to the 10 tons per acre per year of the estuaries. Too often a fisherman catching a striped bass sees his fish as existing in a biological vacuum, not remembering that fish must spawn to reproduce, that eggs must have the proper environment to hatch, and that the fry must have the right food sources to develop into catchable fish. Very often, too, he ignores the fact that these early stages in the life cycle of many ocean organisms may require radically different temperatures, salinity, mineral nutrient levels, and food supplies than are found in the open ocean—conditions that are unique to estuarine areas.

Unfortunately, the productivity and usefulness of estuaries are not as immediately obvious as for good farmland or a stand of prime timber. So the estuarine complex remains wasteland to most people, to be tolerated until it can be made suitable for development.

Misuse of the Estuaries

Because of tidal flushing, estuarine areas are usually able to handle a level of mineral cycling that would turn a lake into a pea soup of eutrophication. But unfortunately the same factors that make an estuary productive make it vulnerable to pollution. The clay particles that concentrate nutrients gather hydrocarbons, biocides, isotopes, and heavy metals, and the filter feeders do the same. The ions of heavy metals, sinking rapidly into denser salt layers, may be carried back into an estuary rather than out to sea.

Because of the assumed regular removal of wastes by the tides, efforts to control pollutants are either not made or are initiated so late that recovery is extremely expensive. As a result, the future of estuaries and their adjoining wetlands as a continuing resource is mixed. Large tracts of marshes on the west coast of Florida have been dredged and urbanized, and the process, if slowed, still continues (Figure 7.4). If this type of development continues around the perimeter of the Gulf of Mexico without a sustained effort to preserve

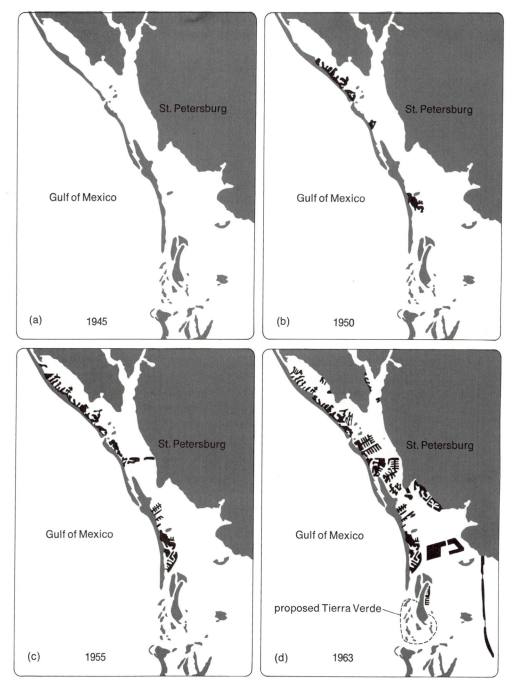

Fig. 7.4 • Boca Ciega Bay off St. Petersburg is rapidly being converted by dredging and filling from a productive estuary into a typical suburban sprawl. (*Modified from G. H. Lauf (ed.), Estuaries, 1967. Copyright by the American Association for the Advancement of Science.*)

some marshes for wildlife "nurseries," the harvest of various types of seafood in the Gulf, particularly shrimp, is bound to decline sharply.

ABUSE OF A BAY

Great South Bay, like Lake Erie, stands preeminent as a well-polluted body of water. In this instance, however, a sea edge is being affected, with the clear implication that in time the open ocean itself could be similarly polluted. Great South Bay, on the south shore of Long Island, New York, is a lagoon some twenty-four miles long and up to three miles wide, with an average depth of four feet, protected from the open Atlantic by the barrier of Fire Island (Figure 7.5). Just a few years ago most of Fire Island was incorporated into Fire Island National Seashore to assure the survival of a wild and relatively untrammeled stretch of sand dunes less than fifty miles from New York City. The bay, however, has been less fortunate.

Fig. 7.5 • Great South Bay on the south shore of Long Island, New York, over twenty-four miles long and three miles wide, is protected from the open Atlantic by Fire Island. Its only connections with the ocean are Fire Island Inlet to the left and Moriches Inlet, several miles off the right edge of the drawing.

Because of its small area and porous soils, Long Island has few streams of any size. The streams that there are flow from the Ronkonkoma moraine, a line of rubble marking the southernmost extent of glaciation, across an almost flat plain, and then into various bays, including Great South Bay, as a series of small estuaries. The combination of fresh water and tidal flushing makes these streams ideal for raising ducks.

The Long Island Duck

Although ducks have been raised on Long Island since its early settlement, wide-scale duck farming did not develop until the 1920s and 1930s. In 1965, thirty-two farms in the Great South Bay area produced over 3 million ducks, 60 percent of the national production in the United States, making the Long Island Duck a household word (Figure 7.6).

Fig. 7.6 • Duck farms lining the shores of Great South Bay have seriously polluted the water. Because Great South Bay is closely bounded by Fire Island and the mainland, there is little flushing of the bay by clean ocean water and wastes accumulate.

At first, ducks were penned on the rivers entering the bay and all duck wastes entered the water either directly or were flushed by rain from the stream banks. Later, small ponds or lagoons were constructed to settle out some of the solid wastes, but the soluble materials still drained into the streams and then into the bay.

The raw wastes of 1000 ducks per day produce a total of 5.7 pounds of nitrogen (in the form of uric acid, which is converted into ammonia by bacteria), 7.6 pounds of phosphate compounds, and 3.6 pounds of soluble phosphate. To get rid of these wastes, 14–120 gallons of water a day per duck are required. Larger farms use 2–3 million gallons of water a day. Altogether, each day the duck industry on the bay produces 3300 pounds of nitrogen, 5600 pounds of phosphorus, and 55,600 pounds of suspended solids in a total effluent of 133 million gallons.

The solid materials give a gray turbidity to the stream water and the heavier of these settle out in ten-foot-deep deposits of sludge in some places, suffocating almost all the bottom life and greatly reducing the dissolved oxygen in the water. With oxygen in short supply, anaerobic conditions encourage sulfide bacteria to produce hydrogen sulfide. This gas then bubbles to the surface, buoying solids with it, forming lumps or rafts that drift about. Finer material is precipitated by contact with saline water in the river mouths. Soluble nitrates and phosphates, of course, are distributed widely in the bay.

A Eutrophic Bay

If Great South Bay behaved in typical estuarine fashion, even this large quantity of mineral nutrients could be flushed into the sea by tidal action and diluted, sparing the bay the effects of overfertilization. But Great South Bay has only two inlets allowing entrance of sea water, Fire Island Inlet at one end and Moriches Inlet at the other. Flushing time is four to ten days and circulation is weak. Prevailing winds and currents from the southwest tend to confine the waste-enriched water to the east end of the bay farthest from the source of clean sea water. For these reasons, Great South Bay is unsuitable for the assimilation of large quantities of waste from any source.

Once in the bay, the wastes contribute nitrogen and phosphorus which stimulate the growth of a very small green alga, *Nannochloris* (2–4 *millionths* of a centimeter in diameter), previously so rare that plankton surveys in the early years of this century failed to notice it at all. Normally oysters feed on diatoms, particularly *Nitzschia,* which has quite different ecological requirements than *Nannochloris. Nitzschia,* for example, requires a salinity of at least 15 parts per thousand, grows best at 10 to 20°C (50–68°F), and is unaffected by changes in the nitrogen-phosphorus ratio. *Nannochloris,* on the other hand, grows well in a variety of salinities, prefers temperatures in the range of 10 to 26°C (50 to 78°F), and doubles its growth rate when the water is enriched with ammonia-nitrogen, which suggests that nitrogen rather than phosphorus is a limiting factor. Apparently phosphorus regenerates more rapidly from decomposing organic matter than ammonia.

Because of poor circulation, bay waters have lower salinity, higher temperature, and higher nitrogen and phosphorus content than ocean water. Each of these factors selects for *Nannochloris* against *Nitzschia.* At times *Nannochloris* becomes so abundant that there may be 3–10 million cells per milliliter of bay water, reducing visibility through the water to less than one foot.

Oysters versus Clams

Oysters feed by straining water through their gills and directing the food to their mouth, where it is ingested. But oysters feed efficiently only if the water is relatively clear. As the quantity of suspended particles increases, the feeding rate decreases. In plankton-rich water, the gills become covered with the tiny algal cells of *Nannochloris* that interfere not only with normal feeding but respiration as well, leading to starvation or suffocation. As a direct result of eutrophication, oyster production declined sharply between 1940 and 1960, from 600,000 bushels a year to, in some years, zero bushels. Although the duck farm wastes have had a very deleterious effect on the oyster industry, production of hard clams (*Mercenaria*) has soared, for this species is able to feed upon the pollution-induced *Nannochloris*, whereas the oyster is not. If the problem were simply the shifting of productivity from oyster to clam, we might simply dig for clams instead of diving for oysters. But, the problem is not that simple. Besides their nutrient content, duck wastes contain a variety of bacteria similar to those found in man. They are collectively called coliform bacteria, and are used as an index of pollution by the Bureau of Public Health. One of these bacteria, *Salmonella,* often found in clams growing in polluted water, causes an intestinal infection so violent that the victim may be cured of eating raw clams for life. As a result of such bacterial contamination, 40 percent of the clam beds in Great South Bay have been closed. When you consider that this closed area can produce 300,000 bushels of clams a year, worth over $2 million, the economic result of bay pollution comes into focus.

Pollution from People

Although we have emphasized duck farms, population growth in Suffolk County has also been a factor in bay pollution. Most of the housing disposes of its wastes through cesspools or septic tanks, which percolate the mineral nutrient and bacteria-rich wastes through the very porous Long Island soil into the water table draining into the bay. In all probability, if the duck farms were to disappear tomorrow, the unsewered wastes of central Long Island could easily keep Great South Bay oversupplied with both mineral nutrients and bacteria. Further organic pollution is contributed by the increasing number of pleasure craft being used in the bay (see Chapter 6).

Nannochloris is not the only alga to be stimulated by overfertilization. A filamentous green alga (*Cladophora*) forms great floating mats, at times almost thick enough to stand on. When these finally sink during the fall and winter they accumulate in the holes left in the bay bottom from sand dredged for beach stabilization. Anaerobic conditions lead to production of hydrogen sulfide, which then bubbles to the surface the following summer, often in such quantity as to blacken housepaint on shorefront homes. A third plant, eelgrass (*Zostera*) is also affected. Normally eelgrass is found in small clumps scattered over the bay floor. This not only stabilizes the sandy bottom but provides shelter for the highly prized bay scallop. But with over-fertilization, eelgrass spreads rapidly, covering the bay bottom for miles. Storms and the changes of season loosen the bladelike leaves of this plant, which then wash up on shore to decay anaerobically because of their sheer volume; again hydrogen sulfide is produced, and the beaches are covered with a slippery organic slime (Figure 7.7).

Fig. 7.7 • Windrows of eelgrass washed up from Great South Bay. When these masses of dead plants decompose, hydrogen sulfide is produced in such quantities that the beach is unusable. (*Photo by Dan Jacobs*)

The Solution of the Bay Problem

Great South Bay could be rid of the pollution caused by the ducks if they were grown like chickens or turkeys in pens or houses without access to stream water, and if the wastes were removed mechanically and treated or disposed of. But ducks do not feather well unless they are allowed to swim, and since the sale of duck feathers often represents the difference between profit and loss, the duck grower feels he cannot separate his ducks from the outside environment. With such a narrow profit margin, farm sewage treatment plants are probably not economically feasible. However, it is only a matter of time before the rolling lawns of suburbia supplant the duck farms of Suffolk County as they have the potato farms of Nassau County; so the issue of duck farms is perhaps only of passing concern.

Increased flushing of the bay has been suggested as a solution, but present inlets must be carefully studied from all ecological angles before being supplemented by man-made inlets. New inlets should not be constructed at the expense of organisms that cannot tolerate the salinity that full-strength ocean water pouring through them would supply. The duck wastes deposited as shoals of sludge in rivers and at their mouths must be dredged so that their mineral nutrient content is permanently taken out of circulation. Finally, onshore facilities must be developed at marinas, yacht clubs, and town docks to handle wastes generated and stored in holding tanks aboard the ever-growing fleet of leisure-time craft. Great South Bay and other estuaries with similar problems can be cleaned up, not easily or cheaply, but there is no inherent ecological reason why oysters, ducks, and people cannot coexist in an ecosystem satisfactory to all.

Encouraging evidence that once polluted bays can be cleaned up is provided by San Diego Bay (Figure 7.8). Today the bay is clean and sparkling; in 1963, it was in the typically brownish-red advanced stage of eutrophication. The change was accomplished by bonds costing $1.50 a month per household in San Diego, a price the people were obviously willing to pay. San Diego Bay, like Great South Bay,

San Diego Bay

Fig. 7.8 • San Diego Bay, virtually surrounded by the city of San Diego and a chain of federal installations, was rapidly becoming a sewer in the early 1960s when a strenuous effort was made to construct sewage treatment plants with outfalls well offshore. Today the bay is again the recreational focus of San Diego, as it should be. (*Photograph by NASA*)

is tailor-made for problems—fifteen miles long and up to two miles wide, it has a single outlet to the sea and is the home port for one-fourth of the Navy as well as 100 tuna boats. The eutrophication was due to more than 60 million gallons of sewage, mostly untreated, dumped into the bay per day. Effluent from tuna canneries and kelp processing plants and the high concentration of ships in the harbor also contributed to the waste load. Water clarity was less than six feet and the coliform bacteria count was ten times the permissible level, forcing the closing of many beaches. After a couple of false starts the $40 million bond issue was approved by the voters in 1960 and construction began on interceptor lines around the bay leading to a new treatment plant on the ocean with outfalls two miles out to sea. The completed system serves seven cities in the region and all the naval shore installations. The improvements after completion of the system in 1964 were surprisingly quick: fish began returning to the bay, clarity ranged up to thirty feet, the bacteria count fell so that the beaches were reopened, and dissolved oxygen rose again to a safe range. What happened in San Diego Bay can happen in any similarly polluted body of water if public concern can be channeled into willingness to be taxed or assessed to make the improvements.

OIL ON THE WATER

Pollution of the ocean surface by oil slicks is a relatively recent problem; it was not until after World War I that oil began to replace coal as a ship fuel. The discovery of huge reserves of oil in the Near East and in South America stimulated a tremendous increase in the quantities of crude oil shipped to North America and Northern Europe for refining. A typical tanker carrying crude oil during World War II held 16,000 tons of oil. By 1965, average capacity had almost doubled. New tankers today carry 326,000 tons, and some tankers are on order that will hold 720,000 tons. There is talk of a ship having a one-million-ton capacity; each of these tankers would hold almost the entire daily production of crude oil in the United States. This giant ship sweepstakes was forced, perhaps prematurely, by the blocking of the Suez Canal in the 1967 war between Egypt and Israel. Obliged to ship oil from the Near East around the Cape of Good Hope, the oil companies naturally wished to make each load as large as possible. Having done so, they found it was cheaper than paying the Suez Canal toll. Even if the canal were to reopen soon, the giant tankers have too deep a draft to use it. With this increase in long-distance shipping of oil, the danger of marine oil spills increases daily.

Oil Spills

Just such an accident occured on March 18, 1967. The Torrey Canyon, a 970-foot tanker carrying 117,000 tons of Kuwait crude oil from the Persian Gulf to Milford Haven, England, ran aground fifteen miles west of Land's End in Cornwall, seven miles northeast of the Isles of Scilly. The oil was contained in eighteen storage tanks each holding 6500 tons. As a result of the grounding, six of these tanks were ruptured, releasing by March 26 some 30,000 to 40,000 tons of crude oil into the western end of the English Channel. Initially there was hope of refloating the Torrey Canyon, and of salvaging both

Fig. 7.9 • The Torrey Canyon, split in two, hangs on a reef off Cornwall, England, while thousands of tons of crude oil pour into the sea from her ruptured tanks. (*Popperfoto*)

ship and cargo, but heavy seas soon broke the back of the vessel, dooming any hopes of recovering either (Figure 7.9). For the next six weeks, until drained of her oil and able to sink, the Torrey Canyon continued to spill the remaining 75,000 tons of oil she contained. For several days the oil slick was kept well offshore by storm winds, but with a change of the weather the oil began to pile up on the coasts of England and France, causing great consternation to governments and holiday-seekers alike.

But the Torrey Canyon episode was only a reminder that oily wastes are constantly being discharged into the sea from a great variety of sources, which together pose a sizable problem. Some pollution is produced by all ships large or small, whatever their cargo. Slop water from routine ship maintenance, which picks up oil from machinery lubrication, bilge water, oily wastes from engine rooms, all find their way into the ocean. But the tanker is the big problem. Let us look at a typical modern tanker carrying 300,000 barrels of oil—a barrel is forty-two gallons—in thirty-three tanks in three rows of eleven tanks each. After emptying her cargo the ship still contains 1700 barrels of oil, 500 in the pipelines and 1200 coating the insides of the tanks. Oil in the pipes is drained into one tank and pumped ashore. In turning around for another load of crude oil, ballast water is pumped into several central tanks from the sea to stabilize the otherwise empty ship for the return run to the oil field. The mixture of oil and water in the ballast tanks soon separates; the water below

147

is pumped into the sea and the oil above pumped into a collection tank. While this is going on, the other tanks are cleaned with steam. This mixture is also pumped into the collection tank. By the time the ship has traveled a few thousand miles her tanks are cleaned and filled with sea water, and the 1200 barrels of waste oil from the previous trip are in the collection tank to be pumped ashore at the next stop.

This is what modern tankers are supposed to do. Outmoded tankers, of which there are many, or modern tankers operating on a very slim profit margin, may not adhere scrupulously to this procedure and may simply dump the 1200 barrels of crude waste overboard at the first opportunity. The ocean is large, the nights are dark, and detection is difficult. Prosecution under existing antipollution laws is hampered by the need to prove willful negligence, a provision that is almost impossible to enforce.

Offshore Wells

A third source of waste is rapidly becoming a major problem. At one time most oil wells were located on land or in very shallow bays or estuaries. But with improvement of deep drilling techniques over water, off-shore wells are being drilled on continental shelves all over the world. The North Sea, the Gulf of Mexico, the Arctic Ocean, and the Pacific Coast of the United States have suddenly opened up to oil exploration and exploitation. One of the most recent areas to be developed is in the Santa Barbara Channel off the coast of Santa Barbara, California.

On January 28, 1969, workmen removing a drilling bit from a 3500-foot well were greeted with a gush of oil indicating a "blown" well. Normally, well holes are cased with pipe which allows the well to be capped should any accident threaten uncontrolled flow of oil Santa Barbara Channel from the well. In this instance the well casing extended only thirty-nine feet below the 200-foot channel bottom. Apparently under great pressure, the oil penetrated a nearby fault in the rock below the casing level, thus bypassing the pipe which was supposed to contain it, and flowing to the surface without restriction (Figure 7.10). While no one was sure how much oil was actually flowing from the well, flow rates of nearby wells suggested a figure of about 21 thousand gallons per day. Despite frantic efforts to cap the well by pouring 8 thousand barrels of drilling mud and 900 sacks of cement down the well, the flow continued unabated for eleven days. Then it stopped, started again at a lesser rate of flow, and although control efforts are still being made, the area is continuing to discharge some oil. The oil slick covered 800 square miles at its greatest extent (Figure 7.11) and probably produced an even greater uproar in southern California than the Torrey Canyon wreck produced in Europe.

Before we examine the effects of marine discharges of oil on

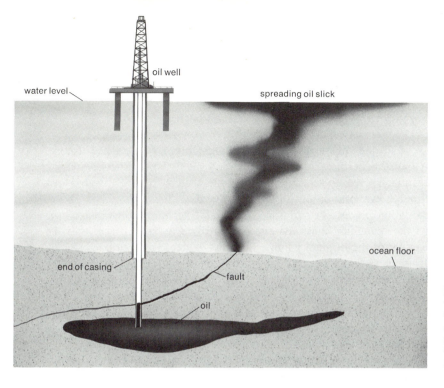

water level

oil well

spreading oil slick

ocean floor

end of casing

fault

oil

Fig. 7.10 • A diagram showing the probable explanation for the Santa Barbara oil spill in January, 1969. Incomplete casing allowed oil to escape to the surface along a fault line in the channel bottom.

Fig. 7.11 • Viewed from directly overhead the magnitude of the Santa Barbara oil spill can be better appreciated. The oil slick is the light gray area surrounding the well in the center of the photo. (*U.S Environmental Protection Agency*)

both oceanic plants and animals and the techniques used to cope with the problems, we must look more carefully at what happens to oil when it is released into the oceanic environment.

Oil and Water

Oil does not float around unchanged until it is washed ashore. All oils, regardless of type, contain some volatiles that evaporate readily. Hence up to 25 percent of the volume of spilled oil is lost through evaporation during the first few days. Then there is a photo-oxidation process in addition to bacterial decomposition that works on the remaining mass. By the end of three months at sea, only 15 percent of the original volume remains—a dense asphaltic substance. It is these black, tarry lumps that most frequently wash up on beaches and stick to one's feet and clothing. With massive discharges close to shore, however, there is not enough time for much decrease in volume to take place, and a thick layer of sticky oil is deposited on anything solid in its path. This was true of both the Torrey Canyon and Santa Barbara spills.

Several approaches were followed in trying to clean up the aftermath of the Torrey Canyon. At sea, the slicks were bombed in an attempt to ignite the oil and burn it off. Because the volatiles that would carry a fire had evaporated, aviation fuel was dumped on the slicks to try to ignite them, but without much success. The French used a different technique: they dumped quantities of ground chalk on the slicks in an effort, generally successful, to absorb, then sink the oil and so prevent its being cast up onshore. The technique used on the largest scale, mostly by the British, was to disperse the slick with detergents and emulsify it into small droplets that could be more readily attacked and decomposed by bacteria. Detergents have no effect whatever on oil other than dispersing it. Over 2.5 million gallons of detergents were used in a dispersion attempt both at sea and ashore. Despite the quantity used, the results were disappointing; not only were the slicks not effectively dispersed, but the detergent had a more negative effect on the flora and fauna than that which might have been caused by the oil.

Until quite recently it was thought that once the aromatics evaporated, oil was not particularly harmful to marine organisms. But a relatively small spill, 700 tons of #2 fuel oil (oil burner fuel) into Buzzards Bay off Cape Cod in 1969, has forced a reevaluation of that comforting notion. For the first time the modern techniques of gas-liquid chromatography were used by a research team headed by Max Blumer from the Woods Hole Oceanographic Institution, fortuitously located nearby. Their study of this pollution episode showed that oil emulsified by sea water and weighed down by mineral particles quickly contaminated the bottom sediments. Because the oil reduced the cohesion of bottom sediments, the spread of contaminated sediments over the sea floor was accelerated. By the use of gas-liquid chromatography, the individual hydrocarbons that comprise the fuel oil could be identified and followed in organisms over time, much the way a fingerprint can be traced on objects handled by any given person.

Right after the spill, 95 percent of the animals characteristic of the bottom community were killed; the same was true of the salt marshes and tidal creeks contaminated by the oil. But after a few days most of the oil had evaporated and some measure of recovery had begun. At this point most previous studies of oil spills proclaimed that the worst was over. But when the hydrocarbons involved were traced with chromatographic techniques, it was found that after eight months unaltered oil could still be recovered from sediment near the point of spillage. One year later bacterial degradation of the oil components had only begun, but it was the least toxic hydrocarbons that disappeared first.

Once on the bottom the oil entered the fat and flesh of marine organisms. When contaminated oysters were placed in clean water they lost their oily taste, but when checked with gas-liquid chromatograph techniques fuel oil was found in both the same concentration and composition as at the beginning of the period of flushing. The evidence suggested that the hydrocarbons were not only *not* flushed out but may remain intact throughout the oyster's lifetime. Obviously, taste cannot be trusted as a criterion of contamination.

Other research has indicated that low levels of hydrocarbons found in oil may affect the behavior as well as the physiology of marine organisms. If the oil either blocks or mimics natural stimuli, sex attraction, food finding, homing, and escape from predators may be affected. Finally, oil may concentrate various fat-soluble compounds such as DDT or PCBs (see Chapter 13) into the surface film of the water.

At any given moment 3 to 4 percent of the sea is covered with whitecaps, which generate 100,000 breaking bubbles per second per square meter. When the bubbles rise to the surface they break and jet a tiny volume of water into the air at 110 miles per hour. It has been calculated that this seemingly inconsequential mechanism injects 300 billion tons of sea water into the atmosphere per year. When the water evaporates, the salt particles, with whatever organic materials were found in the surface film that the bubble broke through, travel on the wind, form condensation nuclei for raindrops, and return to earth usually quite far from the source of origin. A specific example of this phenomenon was discovered south of Genoa, Italy, where pine trees were found dying because the needles were covered by a film of oil blown in from a slick offshore. Plainly, air and water pollution are not discrete problems!

Environmental Repercussions

The murre (*Uria*), a strange little bird weighing about two pounds and vaguely resembling a penguin, is the most abundant sea bird in the northern hemisphere (Figure 7.12). Altogether, there are probably 50 million murres in the world that eat 50 million pounds of fish per week. During the summer they have what may be a critical effect on mineral cycling, and hence productivity, in the Arctic. By feeding on bottom fish, murres, in their excretions, release to the surface water mineral nutrients such as nitrate and phosphate that ordinarily would sink to the bottom of the sea as these bottom fish or their predators died. This fertilizing effect is extremely important in maintaining the fertility of surface Arctic waters. The situation is especially critical off the southeast coast of Newfoundland, where many tankers find it convenient to flush their wastes into the sea. The resulting slicks kill at least 1000 birds a day in this region and perhaps as many as a million a year. Once again man is interfering with a biogeochemical cycle without knowledge of the ultimate effects of his actions.

The effect of detergents used in enormous quantities, however, is obvious to anyone. Although detergent used far away from the shore affects organisms only in the immediate surface layers, treatment of polluted shores has killed large numbers of organisms of varying kinds, as well as producing secondary effects on species not directly affected. Unfortunately, the detergents used in the Torrey Canyon cleanup, unlike those used in household products today, were of the hard type, that is, nonbiodegradable. As with the oil, it is the volatile fraction of the detergent that is most toxic. For this reason, on the open sea, detergent toxicity to organisms declines sharply after a few days.

In addition to barnacles, crabs, and snails, the chief victim of detergent toxicity on rocky coasts is quite ironically the limpet—one of the few organisms observed to remove in its feeding activities the very oil film that generated concern. Another unexpected effect of the Torrey Canyon cleanup was the great increase in the green alga, sea lettuce or *Enteromorpha,* which grew out of control following the death of the snails that normally keep it in check. This recalled the similar recovery of heavily grazed pastures when the rabbit population of Great Britain was sharply reduced by a disease a few years ago (see Chapter 14).

Nobody's Ocean

The Santa Barbara spill raised other questions of perhaps more basic concern. All accidents cannot be prevented; every oil well has a certain probability of polluting sometime in some way. The circle of

Fig. 7.12 • Northern murres are one of the most abundant sea birds in the northern hemisphere. (*Karl Kenyon from National Audubon Society*)

potential damage is greatly increased, however, when the oil well is perched on the continental shelf surrounded by very pollutable water. When a drilling site is not only crisscrossed with faults but is highly subject to earthquakes, one wonders how much the oil beneath the surface is really worth. It has been calculated that just two of the many pipelines connecting the offshore drilling platforms with the coast contain as much oil as was recently spilled in the Santa Barbara Channel. To engineer these pipelines adequately against the event of their breaking in an earthquake would be prohibitively expensive. Indeed, if these costs were carried on corporate ledgers, drilling for oil in the Santa Barbara Channel would be at best a marginal undertaking. What usually happens is that the costs of environmental protection are either disregarded or passed on to the public at large, and the oil companies continue their highly profitable business of resource exploitation.

The crux of the problem of ocean pollution is that the ocean, unlike Great South Bay or Lake Erie, belongs to no one. Beyond the continental shelf where most nations exert some proprietary interest and perhaps even concern, the ocean is up for grabs. There is no supranational body to pass in judgment on the ocean's future; that alone would seem to seal its fate. But if the ocean belongs to no one, it also belongs to everyone. E. B. White reflected this growing concern for the environment at large when he reacted to a report that the Atomic Energy Commission had authorized the dumping of radioactive wastes into the ocean: "I sometimes wonder about these cool assumptions of authority in areas of sea and sky. The sea doesn't belong to the Atomic Energy Commission, it belongs to me. I am not ready to authorize dumping radioactive waste into it, and I suspect that a lot of other people to whom the sea belongs are not ready to authorize it, either."[1]

FURTHER READING

Andrassy, J., 1970. *International law and the resources of the sea.* Columbia University Press, New York. An excellent brief review of the problem.

Blanchard, D. C., 1972. "The borderland of burning bubbles." *Sat. Rev.* 1 January, pp. 60–63. Describes some very interesting research with some startling implications.

Blumer, M., H. L. Sanders, J. F. Grassle, and G. R. Hampson, 1971. "A small oil spill." *Env.* 13(2), pp. 2–12. A pivotal paper on the long-term effects of oil on marine organisms.

Davenport, S., Jr., 1972. "Pave the wetlands or let them be?" *The New York Times Magazine,* 16 January, p. 16. A case history of

[1] Hardin, G., 1969. "Finding lemonade in Santa Barbara's oil." *Sat. Rev.* **52**(19), p. 18.

**NATURAL TRAUMAS
BECOME POLLUTANTS**

Great Salt Meadow in Stratford, Connecticut and some of the problems in administering a wetlands conservation law.

Friedman, W., 1971. *The future of the oceans.* Braziller, New York.

Heald, E. J., 1971. "The production of organic detritus in a South Florida estuary." Univ. of Miami Sea Grant Tech. Bull. No. 6. A beautifully executed study.

Lauf, G. H. (ed.), 1967. "Estuaries." AAAS Publ. 83, Washington, D.C. Many papers in this volume are specialized; but some are general and give valuable insight into the effects of man upon these important geographical features.

Marx, W., 1967. *The frail ocean.* Coward-McCann, New York. A broad survey of man-ocean interactions.

Odum, W. E., 1970. "Insidious alteration of the estuarine environment." *Trans. Amer. Fish Soc.* **4,** pp. 836–847. An excellent review paper.

Ryther, J. H. and W. M. Dunstan, 1971. "Nitrogen, phosphorus, and eutrophication in the coastal marine environment." *Science* **171,** pp. 1008–1013. Some interesting evidence is presented suggesting that phosphorus from detergents plays little role in coastal zone eutrophication.

Steinhart, J. and C. Steinhart, 1972. *Blowout, a case study of the Santa Barbara oil spill.* Duxbury Press, Belmont, California. A thorough and fascinating biography of an oil spill and its implications.

Teal, J. and M. Teal, 1969. *Life and death of the salt marsh.* Little, Brown and Co., Boston, Massachusetts. A general account of salt marsh ecology and the fate of coastal marshes.

8

THE AIR AROUND
AND IN US

ON THURSDAY, DECEMBER 4, 1952, a large high-pressure weather system spread slowly southeast across the British Isles.

The system brought with it light variable winds, dry air, and rather frigid temperatures. . . . Considerable fog began to form late Thursday evening and during the early hours of Friday morning. At first it remained comparatively clean and harmless—but not for long. As the city awoke, tons of smoke from millions of domestic chimneys were hurled upward into the cold motionless, foggy air. Huge power stations added still more tons of coal smoke and sulfur oxides to the atmosphere. Cars, trucks, buses, and a variety of factories and industrial plants all contributed their pollutants. In a short time the fog had become massively contaminated by a mixture of smoke, soot, carbon particles, and gaseous wastes. Now yellow, now amber, now black, the great killer smog held London in its grip, and by early evening, only twelve hours after its onset, the first of the city's inhabitants began to die.[1]

Lest we think air pollution a contemporary phenomenon, John Evelyn in January 1684 wrote: "London, by reason of the excessive coldness of the air hindering the ascent of the smoke, was so filled with the fuliginous steam of the sea coal, that one could hardly see across the streets, and this filling the lungs with its gross particles, exceedingly obstructed the breast so as one could hardly breathe."[2] Edward I (1307–1327) ordered a man put to torture for burning coal and fouling the air, an act sure to be recalled with some relish when one is trapped between two chain smokers on a long overseas flight.

Is it possible to go far enough back in time to encounter a halcyon period when the air was pure? If we think of pure air as we think of pure water, probably not. Air is always a mixture of many things. In addition to the three gases, nitrogen, oxygen, and carbon dioxide, air also includes varying amounts of water vapor, dust, and

[1] Wise, W., 1968. *Killer smog: the world's worst air pollution disaster.* Rand McNally, Chicago.
[2] Ibid.

every conceivable element and compound that wind or man distributes.

Long before man, however, dust storms, fires, volcanoes, and ocean storms polluted the air with vast quantities of particles and impurities of various sorts. The pollution continues: in 1883, a volcanic explosion pulverized the island of Krakatoa in the East Indies and threw very fine dust high into the atmosphere to circle the earth for several years before finally settling out. This suspension, probably without lasting effects, did produce a long series of lovely sunsets through the reflection and diffraction of the sun's rays by the many dust particles.

Man's appearance, of course, compounded the picture. While man controlled fire to his ends, one of its natural products, smoke, has been associated with man since that first fire back in the cave. The problem has been due not only to the increasing number of fires built by ever increasing numbers of people, but to the increasing complexity of the materials combusted and increasing numbers of uses to which the fires are put.

SULFUR OXIDE SMOG

For thousands of years wood, or perhaps dried dung or peat, was the only fuel used, but at some point it was discovered that the black, shiny, strangely lightweight stone that we now call coal could make a very hot flame that lasted longer than a wood fire. Apparently coal was used, if sparingly, for some time in the Far East prior to its "discovery" by Marco Polo. Supposedly, Polo introduced coal, or at least the concept of its combustibility, into Europe. Considering the decimation of forests around population centers and the growing scarcity and expense of wood and charcoal, one would think the idea of using coal would have been welcomed with open minds. But this was a Europe where the Church considered almost any new idea heretical.

When coal is burned, sulfur compounds, especially sulfur dioxide, are released in the smoke. While this was unpleasant enough in itself, perhaps the fumes suggested to the medieval mind the sulfur and brimstone associated with Lucifer. Whatever the reason, coal replaced wood with surprising slowness. But the gradual enlightenment of the medieval mind and perhaps the greater effect of the soaring price of wood finally brought coal into common usage throughout Europe; with that, of course, came the sulfur dioxide fumes. John Evelyn, in 1661, recognized the problem and proposed a reasonable solution for his time: banish the offending industry, although the individual coal hearth pushing its sulfur-laden smoke through thousands of chimneys all over London was equally to blame. Evelyn's suggestions were ignored and for almost 300 years nothing was done about the problem. Once accepted and even romanticized by nineteenth-century novelists, the beloved British hearth was not to be easily given up.

But all Englishmen did not ignore the problem. In 1905 a London physician, Dr. Harold Des Voeux, described the combination of smoke and fog as "smog." However, it was not until years later that the word was again used—and misused at that.

The Effects of Sulfur Oxide Smog

The Meuse Valley in Belgium is heavily industrialized; there are coke ovens, blast furnaces, power plants, glass factories, and steel mills. During the first week of December 1930 the valley was covered with a stagnant air mass that entrapped a smothering blanket of smog. Over a thousand people became ill and sixty died, ten times the normal death rate. In late October 1948, Donora, a small mill town on the Monongahela River in Pennsylvania, similarly surrounded by low hills and having a steel mill and a zinc reducing plant, reported seventeen deaths after three days of similar conditions. An additional 42 percent of the population became ill. The normal daily death rate was two, hence, the increase could only be attributed to the smog. The symptoms in all examples, London, Meuse Valley, and Donora, were throat irritation, hoarseness, cough, shortness of breath, nausea, and feeling of chest constriction.

The major cause of each of these disasters was the sulfur oxides. These compounds are found in greatest concentration in the air surrounding the major industrial cities of the world. This does not mean, however, that the amounts were lethal. Rather, as the concentration rose the sulfur oxides, in addition to many other pollutants, produced lethal results.

Sulfur dioxide, for example, is not an especially toxic gas, but in a humid atmosphere it is converted into sulfates or sulfuric acid, and adsorbed onto the fine particles or fly ash that also result from combustion. Most of the human body is protected from the environment by thick skin, and much of that in turn is protected by clothing. The major point of vulnerability thus lies in the delicate membrane lining the eyes, nose, and respiratory tract, which is far more sensitive to injury than the skin and much more absorbent. Aerosols or fine particles carrying damaging compounds are inhaled into the lungs.

Particulate matter is usually eliminated by the cells lining the walls of the respiratory system. Each cell has a hairlike cilium, which beats twelve times per second. Beating in waves, these cilia work dust particles and other foreign material into the mouth, where it can either be expelled or swallowed. A polluted atmpshere containing sulfur dioxide, ozone, or nitric oxide can inhibit cilia action, allowing particles to remain in the respiratory system and cause damage.

Four major types of respiratory damage from air pollutants are

157

bronchitis, bronchial asthma, emphysema, and lung cancer.

Chronic bronchitis is characterized by permanent damage to the bronchial tubes, resulting in reduction or failure of ciliary action, and overproduction of mucus by gland cells. Because cilial action cannot dislodge the extra mucus, a chronic cough develops. The mucus also constricts the opening of the bronchial system, causing shortness of breath.

Bronchial asthma is usually the result of the allergic reaction of the bronchial membranes to foreign protein or other materials. The membranes swell and make difficult the expulsion of air from the lungs. This explains the characteristic symptoms of wheezing and shortness of breath.

Emphysema follows the constriction of the finer branches of the bronchial tubes, the bronchioles (Figure 8.1). When air is exhaled, more air remains in the tiny air sacs of the lungs (alveoli) than should; when new air is inhaled, the overinflated sacs balloon larger and larger until they explode. This causes two adjacent sacs to unite. The gradual reduction in the number of air sacs destroys the capillaries through which oxygen is taken up by the red blood cells and slowly pushes out the chest, giving a characteristic "pigeon-chested" appearance. The loss of oxygen exchange capacity of the lungs leads to slow oxygen starvation of the entire body and chronic shortness of breath.

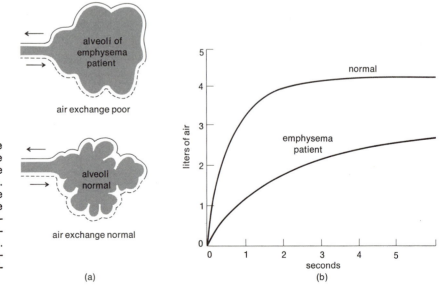

Fig. 8.1 • The lung alveoli are tiny air sacs providing surface area which facilitates uptake of oxygen by the bloodstream. The graph shows that the amount of air held in the lungs of an emphysema patient is far less than that inhaled by a normal person. (*Courtesy of the Oregon Tuberculosis and Health Association.*)

Lung cancer is stimulated by various substances called carcinogens: one, benzpyrene, is characteristic of coal smoke, others, like cigarette tars, are inhaled deeply by millions of smokers. It has been calculated that a reduction of benzpyrene from 6 μg to 2 μg per 1000 cubic meters could reduce the incidence of lung cancer in urban areas by 20 percent. Both tars and benzpyrene remain in the lungs in part because of the inhibition of the cilia, which might otherwise remove them. Another prime source of benzpyrene is charcoal-broiled steaks. The singed fat drips onto the coals and produces smoke rich in aromatic hydrocarbons, including benzpyrene, as much per steak as is contained in 600 cigarettes. This then coats the meat. But you need not give up barbecued meat; just trim off most of the fat before cooking.

When you consider that over a lifetime, the average person breathes 580 million times, sampling 8 million cubic feet of air, it is not surprising that when air contains pollutant particles various chronic lung conditions may develop. In Great Britain, where a large proportion of the population lives in very smoggy cities, fully 21 percent of men from forty to fifty-nine years old have chronic bronchitis. This condition causes 10 percent of all deaths in Great Britain.

The connection between asthma and air pollution was clearly demonstrated after World War II among servicemen stationed near Yokohama, Japan. Individuals with no previous history of asthma developed the classic symptoms of wheezing, coughing, shortness of breath, and sleeplessness. Even transferral to Tokyo did not improve the condition. Once out of the Kanto Plain where both Tokyo and Yokohama are located, most cases rapidly cleared up. There was even one case where a soldier, racked by asthma in Yokohama, spent six asthma-free years in the United States only to have a recurrence of asthma on a later visit to Yokohama. Even the common cold and other upper respiratory tract diseases are more frequent in cities, showing a distinct correlation with increase of pollution level in the late fall and winter, abetted quite probably by the extremely dry condition of most houses and apartments during the winter season (see Chapter 18). By suppressing antibody production, air pollutants may lower the body's resistance to infection and increase allergen sensitization.

Despite the strong circumstantial evidence linking air pollution to chronic disease, many are not convinced. A vice-president of a large oil company recently wrote:

This complex and troubled world we live in is so full of significant, pressing, and perplexing problems that we can ill afford the luxury of wasting effort on imaginary problems or trying to discover problems where none exist. Because human health is of such vital personal concern to each of us that we are naturally inclined to get emotional about it, we should be particularly cautious in ascribing or assuming a cause-and-effect relationship between air pollution and health until scientifically reliable supporting evidence is available.[3]

This somewhat cautious view was not shared by the Surgeon-General of the United States.

[3] *Scientist and Citizen* **7**(3), p. 3.

. . . much of the speculation and controversy about whether or not air pollution causes disease is irrelevant to the significance of air pollution as a public health hazard. We are accustomed to thinking that a disease state is brought about by a single cause—a carryover from a period in public health history when virtually total emphasis was placed on the bacterial or viral agent which had to be present before a communicable disease could be recognized and dealt with. . . .

New criteria must be employed in assessing the damage of air pollution—criteria which include statistical evidence that a disease condition exists in a population, epidemiological evidence of the association between the disease and the environmental factor of air pollution, reinforced by laboratory demonstration that the air pollutants can produce similar diseases in experimental subjects. . . . But the qualitative message at hand conveys a clear message. There is no longer any doubt that air pollution is a hazard to health.[4]

Sulfur dioxide and plants

Sulfur dioxide affects plants as well as people. In the form of sulfite, at concentrations of less than 0.4 ppm, it kills leaf cells, causing large red or brown blotches on leaves between the veins. Since sulfur often accompanies ores of silver, copper, and zinc, smelters which roast these ores to recover the metal have until recently poured huge quantities of sulfur dioxide into the atmosphere with disastrous effects on the surrounding vegetation. In the early 1900s, Ducktown, Tennessee, located near a copper deposit, released forty tons of sulfur dioxide a day into the air of this southern Appalachian valley. Not only were seven thousand acres of the vegetation killed and another seventeen thousand reduced to sparse grass, but most of the topsoil, exposed to rain without protection from vegetation, washed away, leaving a desert in the midst of one of the lushest forests in North America (Figure 8.2). When controls were finally instituted and the sulfur dioxide converted to the useful by-product sulfuric acid, the return on the once-wasted sulfur dioxide proved, ironically, greater than the copper. The desert remains.

[4] Ibid.

Fig. 8.2 • When the vegetation of this once lush hillside was destroyed by SO₂ fumes from a copper smelter, erosion removed the topsoil, making revegetation of the area extremely difficult and slow.

Materials do not escape the effect of sulfur oxides either. A combination of fly ash and sulfur dioxide in a humid atmosphere greatly accelerates the rusting and corroding of metals. The agent of this increased level of corrosion is rainfall which has become increasingly acid over the past 15 years. Rainfall is usually slightly acid because of the carbon dioxide that dissolves in the raindrops as they fall, forming a very dilute carbonic acid solution (See Chapter 3). But in Europe and parts of the northeastern United States the pH of rainfall has dropped from around 5.7 to 3, a result of rising levels of sulfur dioxide, which absorbs moisture to form a sulfuric acid aerosol that is washed out of the air by rainfall. The effects of such acid rainfall on building materials are obvious and ubiquitous (Figure 8.3). Less obvious but far more serious are the effects on natural ecosystems: changes in leaching rates of nutrients from foliage or soils, and small but important changes in the acidity of rivers and lakes. Since removal processes normally take several days, sulfur-contaminated air can be moved downwind during that time hundreds or even thousands of miles.

The problem will increase in importance as man's contribution to the sulfur cycle increases. At present, man releases about one-half the amount of sulfur naturally generated. By the year 2000 man-generated sulfur will increase, to equal, and in some places in the Northern Hemisphere, surpass, that produced by natural processes. To understand the magnitude of this increase we will have to clarify the rate at which sulfur dioxide is converted to sulfates in both polluted and unpolluted air, the efficiency of precipitation in scrubbing these compounds out of polluted air, and the amount of sulfur compounds generated by biological systems (the biogenic factor). Recent research in Salt Lake City, Utah indicates that at certain seasons biogenic sulfur equals that produced by industrial sources. Until these unknowns are clarified we cannot be sure of the importance of man's role in the sulfur biogeochemical cycle. This is all the more reason to control sulfur oxide emissions wherever possible.

The Control of Sulfur Oxide Smog

Efforts to control air pollution are usually concerned with the visible smoke, comprised of small particles of fly ash that can be easily controlled. Electrostatic precipitators installed in factory or power plant chimneys can remove up to 99 percent of the fly ash usually scattered into the environment (see Chapter 21). High voltage wires charge the ash particles, attracting them in groups to a set of vertical plates. When these are tapped or vibrated, the ash falls into hoppers below and is removed. Although the market has not been fully exploited, by using fly ash to make concrete, bricks, and paving material, a power plant producing large quantities of fly ash should at

Sulfur oxide smog and materials

Fig. 8.3 • In the more than 300 years since its erection this sandstone figure at Oxford University has been badly damaged by increasingly acid rainfall.

least be able to return the installation and operating cost of the precipitators, since it costs at least $2 a ton just to dispose of the fly ash. In the United States, which produces over thirty million tons of fly ash a year, only 18 percent is recovered.

The shift from coal to oil in many homes and industries has resulted in a particulate pollution decline in most cities in the United States. Dust fall in Chicago in 1928 was 395 tons per square mile per year. In 1962 this had dropped to 43 tons per square mile per year.

But few people realize that a smokeless chimney can be just as serious a polluter of the air as one belching clouds of black smoke. Many poisonous substances are colorless—sulfur dioxide, for example. At the moment there is no one economical method for removing sulfur dioxide, especially in small quantities, from effluent gases. If the sulfur dioxide content is large enough, wet scrubbers such as a sodium carbonate solution can be used to remove sulfur dioxide. Installation costs for scrubbing equipment at one power plant were $10 per kilowatt capacity; subsequent maintenance cost $1.17 per ton of coal used. The system removed up to 91 percent of the sulfur dioxide from the flue gases. If powdered limestone with small amounts of iron oxide as a catalyst is injected into the flame region of power plant boilers, a plant burning 5,000 tons of coal, containing 3 percent sulfur per day can reclaim 300 tons of sulfur dioxide a day at the cost of 465 tons of powdered limestone.

Another approach, adopted by New York and Chicago, is to limit the sulfur content of the fuels, rather than to try to trap the sulfur dioxide as it leaves the chimney. Pretreatment of both coal and oil, although expensive, is certainly practicable.

Utility companies are often reluctant to go into the chemical business, but if you produce 300 tons of sulfur dioxide a day, you *are*, like it or not, in the chemical business. Another problem is that many plants are too small to accommodate a complex sulfur recovery system that may be larger than the original plant itself. In addition, the variable nature of power plant operations, with units starting and stopping in response to power demands, makes it difficult to send a steady supply of by-product chemicals to a market.

In European cities, where the sulfur oxide problem is aggravated by home fires, a possible solution is conversion to natural gas or oil. From the point of view of sulfur dioxide reduction, electricity would be the ideal solution, but most people cannot afford this type of heating. Natural gas would be the next best choice, and low sulfur fuel oil after that. Until this is done air pollution control can only be partially achieved. The recent discovery of large natural gas deposits in the North Sea has certainly helped; because of this nearby source of gas, London heating systems have been almost entirely converted from coal to natural gas. The result has been an amazing decrease in the density and frequency of late fall and early winter smog episodes like that described at the beginning of the chapter, and a substantial increase in the frequency of sunny days. However, natural fuels are in short supply and the need for a cheaper and more abundant source of fuel is acute (see Chapter 10).

PHOTOCHEMICAL SMOG

In 1859 Colonel Drake drilled the first oil well near Titusville, Pennsylvania. With the increasing pace of technological development it was only a few decades more before the internal ("infernal" as some pessimists would have it) combustion engine was developed, and soon after that the automobile.

A New Kind of Smog

Because of the relatively late appearance of the automobile, most older cities had somehow to fit it into the scheme of pre-existing roads, urban centers, and suburbs. But Los Angeles grew up with the automobile delaying, perhaps forever, effective mass transit systems and making the city dependent upon the automobile. To accommodate this crush of traffic a freeway system, attracting still more cars which required still more freeways, developed, a transportational illustration of Parkinson's second law—expenses rise to meet income. At the height of this activity a haze, at first light and occasional, began to spread over the city. People started to complain of eye irritation and shortness of breath, and certain crops were blighted, particularly leafy vegetables and flowers whose value depended upon their unblemished leaves and petals. Somehow people picked up Des Voeux's descriptive word, smog, to describe the phenomenon. However, the phenomenon in Los Angeles involved neither smoke nor fog.

Believing otherwise, the city cracked down on incinerators and industrial smoke producers. The smog remained. Next the oil refineries were put under a strict emission control; the smog got worse. Finally Haagen-Smit, a chemist from California Institute of Technology, identified the problem—automobile exhausts in combination with the unique geography of the Los Angeles Basin.

Los Angeles is a beautifully situated city: facing the Pacific in a broad fertile basin, surrounded by mountains up to 10,000 feet high, and favored with a warm, dry climate. In the early years, people were attracted by the prospect of warm, sun-filled days, groves of orange trees, lovely vistas of soaring mountains. Today the sun is often obscured by a yellowish-brown haze, the orange groves are either subdivided or killed by smog, and clear vistas are limited to occasional days in winter. Apparently the very factors which at one time made the city so attractive have set it up for one of the worst pollution problems in the country.

The city is surrounded on three sides by mountains, the Santa

The geography of photochemical smog

Monicas to the northwest, the San Gabriels to the northeast, and the Santa Anas to the southeast. Although these mountains effectively shield Los Angeles from the hot, dry winds of the Mohave Desert to the east, they also prevent circulation of air. In addition, the weather of most of southern California during most of the late spring, summer, and early fall is dominated by a large high-pressure area capping the basin with dry warm air. Because of the stagnant high pressure and the encircling mountains, the only air movement is a gentle sea breeze from off the cold Pacific Ocean.

The Role of Temperature Inversion

If we were to attach a thermometer to a balloon and have it radio back the temperature as it ascended, we would see a rather unusual profile (Figure 8.4). Generally, the air temperature decreases at a constant rate from the earth's surface well up into the upper atmosphere. But this is not always the case. Sometimes the temperature decreases steadily as the balloon rises only for the first 1500–3000 feet, then it *increases* for a few thousand feet, and finally it resumes a

Fig. 8.4 • On a clear day in Los Angeles, the temperature *decreases* with increasing altitude. When the area is capped by a mass of warm stable air, the temperature *increases* with altitude for several thousand feet, preventing the dissipation of pollutants generated from below. This increase is shown by the curve at the left of the graph.

steady decrease. This reversal of the usual temperature profile is called a temperature inversion. The initial decrease in temperature is a result of the cool ocean breeze flowing in at a very low altitude over the basin. The next layer is the warm, dry air of the stationary high which is several thousand feet thick; above this layer, the temperature again declines with altitude. The effect of the warm, dry air of the high-pressure mass is to act as a lid covering the entire basin. Pollutants generated in the basin rise in the cool surface air until they meet the warmer air above at about 1000–1500 feet. Since they cannot penetrate this inversion layer, they spread out laterally and accumulate in the surface of the basin. The only escape is a very gradual drift southeast where, under influence of sea breezes, the polluted air flows over some of the mountain passes into the Mohave Desert and has been traced as far east as Arizona. Between June and October there is a temperature inversion 80 to 90 percent of the time over Los Angeles.

There is a third factor. Because the inversion layer is warm and dry, it remains cloudless most of the time. This allows the energy in sunlight to interact with the pollutants generated from below, producing a second generation of new pollutants. For this reason the Los Angeles smog is principally photochemical, not a mixture of smoke and fog as it is in London. There are other differences between the two types of smog as well. Smog in London occurs mostly in late fall or winter in the early morning with temperatures in the −1 to 4°C (30 to 40°F) range; the humidity is high and fog is usually present. The major effect on people is lung irritation. In Los Angeles, smog reaches its peak during the summer at midday with the temperature ranging from 24 to 32°C (75 to 90°F); humidity is low and the sky cloudless. The effect on people is eye and nose irritation.

Photochemical smog is found in any city that generates a quantity of internal combustion engine exhaust and has a high incidence of sunny days (Figure 8.5).

Fig. 8.5 • Photochemical smog is not limited to Los Angeles. Here, for example, is New York City. Every large city with frequent periods of sunshine and with many gasoline-engine vehicles generates this type of smog. (*Aero Service Corporation, Division of Litton Industries*)

THE AUTOMOBILE AS A POLLUTION SOURCE

The internal combustion engine was developed with the goals of reliable performance, smooth operation, and low cost, without any regard for the waste products of the combustion process. In the early years of the automobile it mattered little, because the concentration of cars in proportion to the areal extent of cities was insignificant. But as the number of cities grew their collective combustible wastes grew as well, until pollution became of critical importance. In 1963 daily emission from all sources in Los Angeles totaled 2500 tons of hydrocarbons, 8000 tons of carbon monoxide, 700 tons of nitrogen oxides, 100 tons of particulate matter, and 130 tons of sulfur dioxide. The 3.25 million cars of Los Angeles contributed 70 percent of the hydrocarbons and nitrogen oxides, and 100 percent of the carbon monoxide. So today we are burdened with a highly sophisticated, powerful engine spewing out considerable quantities of pollutants which, once released into the air, are beyond our control.

There are three major sources of pollution from cars: evaporation from gas tanks and carburetors, crankcase blow-by, and tailpipe emission. Although gas caps nominally keep gasoline in the tank, the seal does not prevent the escape of vapors. You can easily demonstrate this by parking your car with a full gas tank in the sun on a hot summer's day. At least a pint, sometimes as much as a quart of fuel will soon be dribbling onto the ground. Evaporation also takes place from the carburetor. When the car is decelerated or idles, more gas flows into the carburetor than is burned; much of this excess gas evaporates. Finally, when the engine is turned off, some of the gas remaining in the heated carburetor evaporates. About 20 percent of the hydrocarbons emitted by an automobile is the result of evaporation; altogether, 2.5 percent of all gasoline produced evaporates between the refinery and its combustion.

Because the seating of pistons in the cylinders is not air tight, some combustion products slip past the pistons after the air-gas mixture is ignited by the spark plugs. This exhaust gets into the crankcase where it is released into the air by a breather tube. This source of hydrocarbons is called crankcase blow-by and at one time amounted to 25–35 percent of the exhaust products of the engine.

Most of the combustion products, however, are vented to the atmosphere through the tailpipe. Gasoline burns most efficiently at a 15:1 air-fuel ratio. But a high-compression engine requires a rich mixture of fuel with air: 13 or 12:1. This means that the hydrocarbons in the fuel are not completely burned. This type of engine also requires high octane gasoline, that is, gas with higher proportions of aromatics or volatile components that are also incompletely burned in rich mixtures.

The Composition of Gasoline

If gasoline were a single compound which was completely burned, the end products would be carbon dioxide and water, and there would be no emission problem; but gasoline is not a single compound, nor is it completely burned. A typical gasoline contains three kinds of hydrocarbons: paraffins and olefins, which burn well at low speeds, and aromatics, which work well at high speeds. But paraffins often explode spontaneously before ignition, causing engine knock and wasted power. To combat this, tetraethyl lead was added. Lead oxide then formed on the plugs and valves so ethylene dichloride and dibromide were added to clean up the lead deposits. In addition antioxidants, metal deactivators, antirust and anti-icing compounds, detergents, and lubricants have been added. Small wonder that when this incredible mess is burned the combustion products are more than just carbon dioxide and water!

When fuel mixed with air (containing 78 percent nitrogen) is combusted, nitrogen dioxide is formed. As luck would have it, the ideal conditions for nitrogen dioxide generation are high temperature followed by rapid cooling, precisely the conditions provided by an internal combustion engine.

Evaporation contributes volatile hydrocarbons of great variety; combustion adds carbon dioxide, carbon monoxide, and nitrogen dioxide, and of course a large amount of incompletely burned hydrocarbons. These materials, together with sulfur dioxide produced by industry, enter the atmosphere and interact not only with each other, but, more importantly, with sunlight.

Sunlight as a Reactant

Although we are most sensitive to the visible or light portion of the electromagnetic spectrum, it is the shorter wavelengths, or ultraviolet, which have more energy than the longer infrared radiation. It is this ultraviolet radiation that reacts with pollutants to form the secondary products that may cause more problems than the primary ones. Nitrogen dioxide, for example, is split by ultraviolet radiation into nitric oxide and atomic oxygen. Some of this atomic oxygen reacts with the nitric oxide, forming nitrogen dioxide again. This is called an autocatalytic reaction because nitrogen dioxide, once formed as a combustion product, can continue to regenerate itself in the atmosphere. In another important reaction, atomic oxygen combines with oxygen to form ozone. Ozone has great oxidative potential and is quite reactive; it attacks rubber, causing it to crack and ultimately to decompose. In fact, one of the clues suggesting the photochemical nature of ozone production in the atmosphere was the observation that rubber cracked only during the day.

NATURAL TRAUMAS
BECOME POLLUTANTS

The physical form of smog is an aerosol, that is, a fine dispersion of either particles or droplets that are less than twenty-five thousandths of an inch in diameter. At least one source of these droplets is the ultraviolet energized reaction of sulfur dioxide and various hydrocarbons, especially the olefins. Other secondary pollutants are the irritants formaldehyde, arolein, and peroxyacetylnitrate (PAN) derived from further reactions of nitrogen dioxide and nitric oxide.

A typical photochemical smog, then, contains the primary pollutants carbon monoxide, sulfur dioxide, nitrogen oxides, hydrocarbon fragments from incompletely combusted fuel, volatile hydrocarbons evaporated from gasoline, and finally materials released from combusted additives, lead, boron, bromine, and so on. Many of these primary smog components react with each other and with sunlight producing secondary and even tertiary pollutants: nitric oxide, PAN, ozone, acrolein, and formaldehyde, to name just a few. How do these materials affect the environment?

EFFECTS OF PHOTOCHEMICAL SMOG

The gross effects of photochemical smog result from its dispersion in the atmosphere as an aerosol. Experiments using aerosols produced from various combinations of pollutants reduced visibility in the atmosphere from 26 to 6.5 miles. If the particles were allowed to grow to a point of equilibrium, further reduction of visibility would have ensued. This reduction of visibility is not only aesthetically unpleasant but reduces sunlight received at the earth's surface (Figure 8.6), makes driving and flying more hazardous, and may act together with other pollutants to produce various adverse physiological effects in man.

Fig. 8.6 • The upper curve represents the ideal distribution of solar energy on an October day in Pasadena, California. The lower curve is the observed energy. The difference is a measure of the time and density distribution of photochemical smog generated by vehicles having gasoline engines. *(Modified from P. A. Leighton,* Photochemistry of Air Pollution, *Academic Press, 1961.)*

But photochemical smog has its greatest impact in the differing effects of its major components on man and plants.

Carbon Monoxide

Carbon monoxide is one of the least reactive of the pollutants in photochemical smog, and is tolerated by the body in concentrations up to 10 ppm without noticeable effect. It is potentially quite dangerous because it is odorless, tasteless, and colorless, unlike many other pollutants. In the body it has an especial affinity for hemoglobin, the substance in the red blood cells that normally combines with oxygen, distributing it throughout the body. As the level of carbon monoxide increases in the body, less oxyhemoglobin and more carboxyhemoglobin is formed. Since carboxyhemoglobin does not provide oxygen for the respiratory needs of the body cells, increasing concentrations of this compound endanger survival. A moderate smoker exposes himself to around 30 ppm of carbon monoxide, resulting in a 5 percent level of carboxyhemoglobin. When placed in an urban environment where carbon monoxide concentration in the air may average 20 to 30 ppm, the additive effect for a heavy smoker may approach the danger level of 100 ppm. Probably 20 to 30 ppm total should be the maximum allowed, for the physical effects of a decline in sensitivity to environmental stimuli and lack of energy and endurance begin to be felt above this level. Although at a busy intersection in a large city one may breathe 20 ppm of carbon monoxide, it is possible to get much higher doses, for example, 370 ppm behind a car stopped for a red light.

For all the carbon monoxide produced by man—270 million tons per year, natural sources produce half that found in the atmosphere, more than 3.5 billion tons in the Northern Hemisphere alone. Since the residence time is only a month or so, there are obviously important sinks that effectively reduce the ambient level of carbon monoxide in the atmosphere; soil fungi and bacteria have been suggested as possibilities, and there must certainly be others. The fears of world-wide build-up of man-generated carbon monoxide now appear groundless. Similar fears were expressed a few years ago that oxygen would be used up if combustion continued increasing, more forests were cleared, and more fields paved. Some calculations by a geochemist, W. S. Broecker, have laid that fear to rest also. Broecker estimated that if all known fossil fuels were oxidized, less than 3 percent of the earth's oxygen supply would be used up; that if all photosynthesis stopped, less than 1 percent of the world's oxygen supply would be required to decompose this enormous biomass completely. Apparently man's pollutants will become limiting long before his oxygen supply.

The Organic Components

The hydrocarbons and nitrogen oxide products such as acrolein, PAN, and formaldehyde irritate the eyes and nose and even attack nylon stockings. Any student who has dissected a frog recalls the tearful fumes of formaldehyde. Typical concentrations of PAN in smog range from 0.05 to 0.6 ppm. PAN enters the leaves of plants through the stomates, and causes the spongy mesophyll cells to collapse (Figure 8.7). The resulting air space gives leaves a silvered or bronze appearance, ruining their sale as leaf vegetables.

(a)   (b)

(c)

Fig. 8.7 • (a) Cross section of a normal leaf; (b) cross section of a leaf whose spongy mesophyll cells have been damaged by smog entering the leaf through the stomates; (c) cross section of a leaf whose palisade layer has been damaged, causing small flecks to appear on the leaf surface.

Orange trees, although more resistant than leafy vegetables, are also affected by high concentrations of PAN. The twigs are killed, giving the tree a moth-eaten appearance, and the number of food-producing leaves is so reduced that the crop of oranges is severely diminished. Because of smog damage, commercial orange production

170

has shifted from the Los Angeles basin to the Sierran foothills on the edge of the San Joaquin Valley. While 50 percent of the reactive hydrocarbons such as PAN are generated by motor vehicles, 40 percent come from industrial sources, the chief being solvent evaporation.

Ozone

Ozone is also an irritant, but it can impair vision and depress body temperature as well. Its threshold for direct action is around 1 ppm. While the natural concentration of ozone is about 0.02 ppm, it may range as high as 0.5 ppm or more in severe photochemical smog. Ozone attacks the palisade layer in plant leaves (Figure 8.7), causing brown flecks to appear, especially in tobacco leaves, which are very sensitive to ozone. The flecking considerably reduces the value of the leaves, particularly those used as cigar wrappers. Natural vegetation is affected by ozone as well; there has been pronounced dieback of ponderosa pine in the San Bernardino Mountains southeast of Los Angeles. These mountains, which form one rim of the basin, are subject to heavy concentrations of smog drifting out of the Los Angeles area. White pine in the eastern United States develops a similar browning of its needles from ozone. This effect has been noticed in the Cumberland Mountains of Tennessee near Oak Ridge and along parkways in suburban areas such as the Merritt Parkway in Connecticut.

CONTROL OF PHOTOCHEMICAL SMOG

Control the automobile exhaust and you control photochemical smog (Figure 8.8). Like all simple solutions to involved problems, there are hidden complications. As we have seen, the automobile was

Oliphant in The Denver Post

" . . . Then, when the bag is full of hydrocarbons and noxious gases, you simply take if off and throw it away!"

Fig. 8.8 • Editorial cartoon by Pat Oliphant, copyright, The Denver Post; reprinted with permission of Los Angeles Times Syndicate.

developed without any particular interest in emission control. Suddenly after eighty years the hue and cry goes up. Control emissions!

In 1970 Congress passed the Clean Air Act, which calls for a 90 percent reduction of hydrocarbons and carbon monoxide levels, using a 1970 model car as a standard, and a 90 percent reduction in nitrogen oxide emission, using a 1971 model car as a benchmark. By 1975, cars would be limited to 0.41 g per mile of hydrocarbons, 3.4 g per mile of carbon monoxide, and 3 g per mile of nitrogen oxides. In 1976 permissible nitrogen oxide emissions would be reduced to 0.4 per mile. Anticipating that some manufacturers might have difficulty meeting this schedule, a one-year suspension of standards was allowed in the Act, and in 1973, to no one's surprise, the deadline was extended for one year.

The auto industry has easily handled two of the three main causes of emissions. Crankcase blow-by was eliminated by recycling emissions back into the carburetor. European cars have long done this; American cars made the change starting with 1968 models. Soon to follow was the requirement of a vapor-tight gas cap. The remaining source, tailpipe emission, is by far the most difficult to control. Of the several possibilities, catalytic converters have been adopted by most American manufacturers. Catalytic converting works well when new, but becomes less effective as fuel additives coat the catalyst. In addition the catalyst converter causes reductions in gas mileage although air conditioners, automatic transmissions, and greater car weight have a collectively greater impact on gas mileage. After twelve thousand miles the catalyst must be replaced. For this reason the major auto manufacturers in the United States are lowering the compression ratio of their engines to eliminate the need for leaded gasoline. But even without lead additives in the gasoline, catalyst lifetimes beyond twenty-five thousand miles are unlikely. Then too, a second catalyst must be present to take care of the nitrogen oxides. Catalysts are probably a blind alley.

A more likely development is the greatly increased use of stratified charge and Wankel engines, initially in foreign imports, and shortly thereafter in American cars. In one version of the stratified charge engine, a rich air-fuel mixture is pre-ignited; the flame front then invades the main part of the cylinder, igniting a much leaner mixture of fuel and air, which drives the pistons. The leaner the fuel-air mixture, the lower the emission level. The Wankel, a rotary engine, is "dirty," producing twice the hydrocarbons, about the same carbon monoxide, and 20 to 60 percent lower nitrogen oxides than a standard engine; but because of its small size and high-temperature exhaust gases, a thermal reactor can be easily used to complete combustion of the hydrocarbons and carbon monoxide. Unfortunately the Wankel's gasoline consumption is 30 percent higher than a normal piston engine's.

The most reasonable alternatives to internal combustion engines are the Stirling-cycle engine, the turbine, the steam or Rankine engine, and electricity.

The Stirling engine works on the principle of constant burning of fuel, which makes it a closed-cycle external combustion engine. With large amounts of excess air involved in the combustion process, hydrocarbons and carbon monoxide levels are well below the 1976 Clean Air Act standards. With an exhaust gas recirculation system, nitrogen oxides could be kept below the 0.4 g per mile limitation.

A gas turbine consists, basically, of a compressor and a turbine wheel on a common shaft. The combustion chamber lies in the air-flow path between them. Since the turbine operates at very high air-fuel ratios, the emission of unburned hydrocarbons and carbon monoxide is quite low, but the production of nitrogen oxides is relatively great because of the high operating temperature of the engine. Reduction in the output of nitrogen oxides would require a catalyst active at less than 1093°C (2000°F); such a catalyst has not yet been found. Whereas the turbine has the advantages of quick starting, no warmup, low vibration and noise level, fuel consumption at low speeds is excessive, acceleration poor, and the engine cannot be used in braking. But these problems are not insuperable. Once a major manufacturer tools up, turbine cars could be mass produced at approximately the same costs as present cars. Because of the pressure imposed by the Clean Air Act on the American automobile industry, we may well see turbine-driven cars by 1980.

Steam engines are undergoing a renewal of interest. With modern technology and new materials, compact and low-maintenance reciprocating steam engines are quite feasible, from the standpoint of both performance and emission control. The transmission system required is relatively simple and the long-standing problem of warmup time can be licked by modern boiler technology. Even if steam automobiles are not developed, fleet vehicles such as trucks, buses, and taxis could be converted to steam.

Electric automobiles use energy stored in batteries. Regardless of type, batteries work on the principle of a reversible chemical action between unlike electrodes placed in a conducting solution or electrolyte. The problem with battery-driven cars remains their limited range. The still unattained solution, then, is a battery of long life and high power. In present batteries, as the power increases, the energy level falls off, reducing the range. New battery systems such as sodium-sulfur and lithiumchlorine show promise of combining both power and range, but both require rather high operating temperatures and are in early stages of development.

The most logical solution to the automobile emission problem, short of abandoning cars for mass transport (Chapter 20), would be to replace the standard internal combustion engine. Since this will take at least a decade to accomplish, transitional types such as the stratified charge and Wankel engines would buy time to develop alternatives to internal combustion, namely the Stirling or Rankine engines. For the first time in many years it looks as though the next decade may see a marked reduction in the frequency and density of photochemical smog in our cities. If the oil industry's alarums about impending shortages of gasoline (see Chapter 10) are taken literally and the price of gasoline increases substantially, alternate fuels may become economically attractive. One such possibility is hydrogen. Not suitable as a primary source of power because of the high energy cost involved in its production, hydrogen may be quite useful as an energy carrier, since it is efficient and, unlike electricity, easy to store. When burned, hydrogen produces water, and levels of nitrogen oxides lower than other fuels. This makes it an attractive energy source for transportation. However, it is bulky to store, and presently very expensive to produce by electrolysis. But the problems are economic rather than technical and the economics of fuel supply are clearly entering an area of rapid change.

CARBON DIOXIDE AND CLIMATE

Because of its natural presence in the atmosphere and its basic role in photosynthesis and respiration, carbon dioxide is not commonly regarded as an air pollutant. But since the beginning of the industrial revolution the use of fossil fuels for combustion has released enormous quantities of carbon dioxide into the atmosphere. At present, the atmosphere contains around 2.3×10^{12} tons of carbon dioxide (0.03 percent) and the ocean another 1.3×10^{14} tons. Exchange between the atmosphere and the ocean amounts to about 200×10^9 tons per year, so the ocean acts as a reservoir or buffer of carbon dioxide in equilibrium with the carbon dioxide content of the air. If amounts of atmospheric carbon dioxide decrease, more is released from the ocean; conversely, if the carbon dioxide content of the air increases, the ocean tends to absorb more. The rate of change, however, is slow. It takes at least 1000 years for 50 percent of the change, up or down, to be accommodated by the ocean.

The Greenhouse Effect

In the atmosphere, carbon dioxide does not affect shortwave radiation. Upon striking the earth, however, shortwave radiation is transformed into, and reradiated as, longwave radiation or heat. This heat is absorbed by the carbon dioxide molecules and transferred to the atmosphere (Figure 8.9). It is this heat transfer that is ultimately responsible for long-term climate and short-term weather. Because of the balance of incoming light and outgoing heat the mean temperature of the earth remains at about 14°C (58°F). This phenomenon of carbon dioxide absorption of heat is called the greenhouse effect. Shortwave radiation or light also passes easily through glass, and is converted to longwave radiation or heat in a greenhouse. Because of the lack of convection and the inability of the heat to pass out through the glass, the greenhouse stays warm. The same thing happens in a closed car on a hot day.

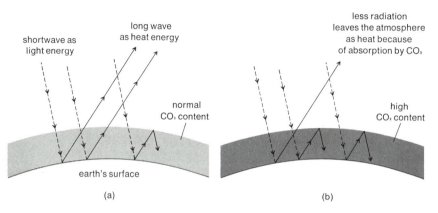

Fig. 8.9 • Shortwave radiation (light) strikes the earth and is transformed into longwave radiation (heat). Since carbon dioxide absorbs longwave radiation, the more carbon dioxide contained in the atmosphere, the more heat is retained and the warmer the atmosphere becomes.

A theory has been around for a number of years that as combustion pours more and more carbon dioxide into the atmosphere it will accumulate faster than it can be absorbed by the sea. More carbon dioxide in the atmosphere will absorb more outgoing longwave radiation and the earth's mean temperature will begin to rise. Since a rise of only a few degrees would melt the earth's icecaps and raise the sea level by 400 feet, the problem is not merely academic (see Chapter 7).

Before 1900 the average carbon dioxide concentration in the northern hemisphere was approximately 290 ppm. Today it is 330 ppm, an increase in proportion with the rate of fossil fuel combustion during this time. But at the same time we are dramatically increasing the quantity of dust particles in the atmosphere. This tends to reflect shortwave radiation back into space, thereby cooling the earth. No one is at all sure which of these processes is in the ascendancy or whether there is a balance.

Another experiment has been initiated by high flying jets. The condensation trails or contrails of jets are quite conspicuous in the sky (Figure 8.10). Usually they disperse quite rapidly; but sometimes they persist and occasionally they bleed together and cover an otherwise clear sky with a blurry cirrus overcast.

Fig. 8.10 • Jet contrails form patterns that may persist in the form of particles in the atmosphere. The effects of contrails are, as yet, unknown.

Because of concern over the unknown environmental effects of the supersonic transport, flying at 50–70,000 feet and because of economic considerations, development of an American SST was halted in 1971. One problem that has been brought up since then is the possibility that nitrogen oxides from SST exhaust could act as a catalyst, destroying the ozone layer that presently shields earth from dangerously high levels of ultraviolet radiation. A decrease of only 5 percent in the ozone concentration would elicit a 26 percent increase in the UV intensity.

NOISE

While we are unsure of the ultimate effect of contrails in the stratosphere, there is no doubt about the effect of one of man's activities in the first few hundred feet of the atmosphere—noise. Without question, noise can damage hearing; there is no threshold for ear damage (Figure 8.11). Claims for insurance compensation for loss of hearing have climbed to over a billion dollars a year in the United States alone, prompting industry to take noise reduction programs seriously. But more subtly, noise increases tensions already heightened by the other stresses of urban life. Noise can affect blood pressure and ultimately the heart. Unfortunately we have no earlids to shut out noise at night, hence we can be victimized at all hours.

The obvious solution is to identify the sources of noise and then work towards reduction in the level of sound from these sources. Unlike some other environmental problems, the technology for noise abatement has been available for some time. A good example is provided by construction equipment. In 1969 a major manufacturer

What can be done

50-hp siren (100') — 140
threshold of pain — 130
pneumatic chipper —
120
loud music (rock)
hammering on steel plate (2') — 110
thunder
boiler factory — subway passing
can-manufacturing plant — 100 riveting machine (35')
heavy truck — very loud music (classical)
90 — 10-hp outboard motor (50')
heavy street traffic —
inside bus — 80 — loud music (classical)
stenographic room — heavy traffic (25-50')
average factory —
average automobile — 70
conversation (3') — department store – noisy office
average office — 60 background music
quiet residential street —
50
average residence — minimum street noise
40 very quiet radio at home
very soft music — 30 — country house
quiet auditorium
20
quiet whisper — quiet sound studio
10 — leaves rustling
anechoic room —
0 — threshold of hearing

decibels

Fig. 8.11 • Noise is measured in dynes/cm², watts/cm², or, most commonly, decibels. Some common environmental sounds are listed here in ascending decibel values. (*Modified from C. Stark,* Stereo Review *23, 1969.*)

of air compressors brought out a new model that reduced the noise level from 110 decibels to 82 decibels. Because the quiet compressor cost 40 percent more than the older, noisier models, the construction industry was uninterested. Today, however, the cost differential has been reduced to 10 percent, the public is losing some of its apathy, and cities are beginning to consider ordinances that offer scientific standards for regulation and contain punitive "teeth" to bring compliance. In 1971 for example, New York proposed a noise control code that would set specific noise-level standards in decibels for garbage trucks, air conditioners, public-address systems, emergency sirens, and construction equipment. Unfortunately, subways, railroads, aircraft, and motor vehicles cannot be included in a city ordinance because of conflicting federal and state regulations. But the proposed code represents a great leap forward from the typical "nuisance" codes that were virtually unenforceable.

Subway noises have been sharply reduced by rubber-tired cars in some of the newer subway systems of the world. But even existing obsolete systems, such as those in Boston, New York, and Philadelphia, could be greatly improved with welded rails, aluminum-centered steel wheels, polyurethane foam between the ties, insulation pads under the tracks, and sound-deadening materials on car underbodies and station walls. The only thing standing in the way of such improvements is public apathy.

177

NATURAL TRAUMAS
BECOME POLLUTANTS

Much of the problem of urban noise lies in our inability to get away from it. New office buildings and apartment houses not only are notoriously transparent to outside sounds but also generate a complete range of noises of their own—elevator doors, air conditioning, light fixture hum, knocking radiator pipes, echoic halls and lobbies. Building code changes or stronger enforcement of existing codes could be very helpful in reducing the noise levels in new construction. Even more important is some form of land use planning that would isolate the noisier components of our urban society, such as airports, or control development near such facilities. One crude but simple method for estimating the potential noise that can be anticipated at a building site has been worked out by the Federal Department of Housing and Urban Development. One member of a two-man team reads aloud while the other moves away with a tape measure until he can hear only a few words in a 10-second period. If the distance measured is 7 feet or less, the site is unacceptable; 7 to 25 feet would preclude housing without extensive acoustic insulation; 26 to 70 feet is typical for residential sites; over 70 feet would be ideal.

Jets

One of the greatest sound generators is the jet airplane. Not that noise from piston aircraft was insignificant, but in the 1940s and 1950s before jets were introduced, airports were relatively small and flight volume moderate. Since the introduction of jets at O'Hare International Airport in Chicago in 1958, traffic there has grown to 1200 flights a day; at peak hours jets come and go every forty seconds, moving 23.5 million people through O'Hare every year. In 1960, sixteen airports were served by jets, in 1967, 150, in 1970, 350, and by 1975, 500 will be. The safest way for a jet aircraft to approach a runway is to follow a long, low trajectory and unfortunately this produces the most noise. Even more unfortunately, areas right up to the edges of large new jetports were allowed to be subdivided and are jammed today with residential housing.

Because of this lack of land use planning, the city of Los Angeles was required by landowner suits in 1971, to spend $200 million to buy up 400 acres of residential development on the edge of Los Angeles International Airport. The lesson was well learned, apparently, for strict land use controls are planned for the new airport under construction at Palmdale, California.

The sky and the sea have long borne the brunt of man's misuse, partly because their very immensity seemed to preclude any long-term effect by man, and partly because both belong to no one. But while it is possible to avoid a polluted sea by living inland, polluted air is in everyone's lungs. Slowly we are coming to realize the subtle cause-and-effect relationship between polluted environments and the increasing incidence of asthma, bronchitis, emphysema, hypertension, and heart deterioration—a relationship so long ignored because these debilities are not glamorous diseases, but the result of the body's inability to cope indefinitely with a broad spectrum of specific stresses.

The significance of all types of air pollution has begun to dawn upon us and the options are clearly focused, perhaps for the first time in man's history. Continued inaction in dealing with these problems is unconscionable for our generation, and suicidal for the next.

FURTHER READING

Alexander, G., 1971. "The little engine that could be an answer to pollution." *The New York Times Magazine,* 3 October, p. 18. A well-illustrated article covering the fundamentals.

Broecker, W. S., 1970. "Man's oxygen reserves." *Science* **168,** pp. 1537–1538. An authoritative put-down of reports that we are about to run out of oxygen.

Davitaya, F. F., 1969. "Atmospheric dust as a factor affecting glaciation and climatic change." *Assoc. Amer. Geog. Annals* **59,** pp. 552–560. A comprehensive examination of the problem of increasing dust levels in the atmosphere.

Goldsmith, J. R., and S. A. Landaw, 1968. "Carbon monoxide and human health." *Science* **162,** pp. 1352–1359.

Keeling, C. D., 1970. "Is carbon dioxide from fossil fuel changing man's environment?" *Proc. Amer. Phil. Soc.* **114,** pp. 10–17. Attempts a reasoned answer to the question by examining many of the complications.

Kellogg, W. W., R. D. Cadle, E. R. Allen, A. L. Lazrus, and E. A. Martell, 1972. "The sulfur sycle." *Science* **175,** pp. 587–596. A good review.

Landsberg, H. E., 1970. "Man-made climatic changes." *Science* **170,** pp. 1265–1274. Considers a number of factors and their interactions.

Malin, H. M. and C. Lewicke, 1972. "Pollution-free power for the automobile." *Env. Sci. & Tech.* **6**(6), pp. 512–517. Reviews the implications of the Clean Air Act and its likely effect on selection of new types of engines.

Norbye, J. P., and J. Dunne, 1973. "Stirling-cycle engine promises low emissions without add-ons." *Popular Science* **202**(2), pp. 72–75, 154. How it works and how it compares with the Wankel and standard engine.

Southworth, M., 1969. "Sonic environment of cities. *"Env. and Behavior* **1,** pp. 49–70. Some creative alternatives to the negative sound environment of most cities.

Stern, A. C. (ed.), 1968. *Air pollution.* 2nd ed., 3 vols. Academic Press, New York. The most comprehensive detailed work available on air pollution.

Still, H., 1970. *In quest of quiet.* Stackpole Books, Harrisburg, Pennsylvania. Probably as thorough a treatment of the whole subject of noise as is available for the lay reader.

9

RADIATION

RADIATION is the emission of radiant energy in the form of particles or waves from certain naturally occurring elements. It has always existed to some extent in the environment on earth. But man has, paradoxically, both concentrated and dispersed radioactive materials to an extent unknown in nature. Hundreds of thousands of people have died as a direct result of the concentration of this energy at Hiroshima and Nagasaki and millions in present and future generations are threatened by dispersed radioactive energy in the form of fallout or nuclear wastes and their effects. To understand the danger to life of both concentrated and dispersed forms of radiation it is necessary to know something about the nature of radiation and its manipulation by man.

THE NATURE OF ATOMS

Superficially, objects appear continuous; that is, they have a certain integrity: water is 100 percent water, wood is 100 percent wood, and stone is 100 percent stone. Democritus realized 2200 years ago that beyond the physical subdivision into fog, sawdust, or sand, each of these materials was discontinuous; that their superficial appearance bore little resemblance to the tiny units or atoms of which they were constructed. It wasn't until 1805, however, that John Dalton confirmed the concept of the atom as the basic unit of all matter.

The combining of atoms to form molecules was difficult to understand when atoms were conceptualized as exceedingly small spheres. Atoms actually combine by sharing pairs of electrons. An atom is composed of a nucleus of protons and neutrons surrounded by one or more electrons. The number of bonds that an atom is able to make with other atoms depends upon the number of electrons that it has to share. Hydrogen has but one electron to share. Two hydrogen atoms, each with a free electron to share, combine to form gaseous hydrogen, H_2.

$$H\bullet + \bullet H \longrightarrow H\colon H \quad \text{or} \quad H_2$$

Oxygen has two electrons to share, so it can form bonds with two hydrogens, H_2O.

$$H\bullet + H\bullet + O\colon \longrightarrow H\colon O\colon H \quad \text{or} \quad H_2O$$

Nitrogen has three electrons to share, forming NH_3, while carbon with four available electrons can form a great number of compounds, the simplest being methane (CH_4). Similarly, all elements have some electrons available for bonding.

The Nature of Radiation

But what is radiation? In 1895 Wilhelm Roentgen, working with an evacuated glass tube with metal electrodes at either end, found that an electric current applied to the electrode at one end of the tube emitted a mysterious ray through the electrode at the other end, which penetrated opaque substances, causing a zinc-sulfide screen to glow or emit light. Roentgen called his discovery the X-ray. Apparently the rays were given off by the metal of the tube plate when stimulated by an electric current. Not long after, Henri Becquerel placed some minerals containing uranium ore on a carefully covered photographic plate and found that they exposed the plate even in the dark by giving off rays similar to those noticed by Roentgen. Becquerel observed also that only minerals containing uranium showed this trait. Pierre and Marie Curie became interested in the phenomenon and found that in addition to uranium, thorium and radium were also radioactive or capable of giving off radiation.

Radium was of particular interest to physicists because it was found to give off three kinds of radiation, called alpha, beta, and gamma. Alpha radiation was later identified as positively charged particles or protons, beta radiation was found to be negatively charged electrons, and gamma radiation was similar to the X-rays discovered by Roentgen.

When radium gives off alpha particles, a gas called radon is formed. The atomic weight (protons plus neutrons) of radium is 226. If an atom of radium loses an alpha particle with an atomic weight of 4, the remaining atom, radon, should have an atomic weight of 222. It does. Later it was found that when an atom of lithium is struck by a proton, it splits into two alpha particles accompanied by an energy release. The splitting of the lithium atom or any other atom into fragments is called fission. Although lithium yields little energy, other atoms like uranium yield tremendous amounts of energy when split.

181

Many elements were found to have stable forms called isotopes, slightly different in atomic weight but otherwise identical to the regular element. Heavy hydrogen or deuterium, which will someday be a source of fusion power (see Chapter 10), is one example. Irene Curie and her husband, Frédéric Joliot, found that when they bombarded aluminum with alpha particles, neutrons were given off and a strange species of phosphorus appeared. Normal phosphorus has no isotope, but the new species disappeared in fifteen minutes leaving silicon! The Joliot-Curies had produced a new kind of isotope, an unstable one, called a radioisotope, which rapidly decayed into another element. Thus began a whole new era of artificially produced radioisotopes, an era that will continue as long as man manipulates atoms.

Isotopes

Today, radioisotopes have been produced for every known element, both intentionally for research and inadvertently in the fallout of atomic bombs. The variety of these radioisotopes is truly fantastic; not only do they vary in the kinds of radiation that are emitted, but in their longevity as well. Longevity of radioisotopes is described in terms of the length of time it takes for one half of their radioactivity to dissipate, the half-life. Carbon-14 has a half-life of 5600 years: that is, after 5600 years one half of the radiation will be left, in 11,200 years one half of one half or one fourth is left, and so on. Other radioisotopes have much shorter half-lives than carbon-14: for example, strontium-90, 28 years; cesium-137, 27 years; iodine-131, 8 days. Because of their variety and inherent radioactivity, radioisotopes can be introduced into biological or geological systems and their movement followed by sensitive radiation counters, allowing a better understanding of many processes that involve the movement or cycling of atoms. This is possible because such radioisotope tracers, despite their radioactivity, behave chemically exactly like their normal analogs.

ATOMIC FISSION

The 1930s were heady times for nuclear physicists; discovery piled on discovery, providing new insights into the nucleus and its organization, and also suggesting further manipulations of atomic and subatomic particles. In 1939, Hahn and Strassman, German chemists, who had been looking at the chemistry of uranium when it was bombarded with slow neutrons, were surprised to find barium as one of the by-products. Uranium has an atomic number of 92, barium 56. Usually when uranium or any other element is bombarded with various particles, the atomic numbers of the by-products are rather close to the original. But barium is remote in every way from uranium.

Then Lise Meitner, a German physicist, had the brilliant idea that maybe when uranium absorbs a neutron it splits into two fragments. After all, lithium was known to split into two alpha particles. If so, what was the other fragment? The atomic number of krypton is 36 (92 − 56) and krypton turned out to be the other fragment. But the most exciting part of this disintegration of uranium lay in tallying the neutrons involved. Uranium has 146 neutrons in its nucleus. It absorbs one, making 147; but the two fragments, barium and krypton, have 82 and 47 respectively, totaling only 129. This leaves 18 neutrons, neutrons that are released when uranium undergoes fission. Suppose these neutrons were absorbed by other uranium atoms and more neutrons and energy were released as a result; we would have a chain reaction generating much energy and still more neutrons. The means of unlocking vast stores of atomic energy seemed at hand.

But there are three isotopes of uranium, U-238, U-235, and U-234. Did all three undergo fission, and if not, which one did? Pure uranium contains 99.3 percent U-238, 0.7 percent U-235, and 0.006 percent U-234; of these it was found that only U-235 underwent fission by capturing a slow neutron. Although U-238 captured fast neutrons, fission did not take place. This posed a problem if a chain reaction was to be sustained.

The Chain Reaction

Slow neutrons cause U-235 to undergo fission but in doing so it releases 18 fast neutrons per atom. These are captured by U-238 rather than U-235, stopping the reaction. What of the U-235 which absorbs neutrons? It is unstable and forms a new element, neptunium. This too is unstable and forms still another element, plutonium. Fortunately, plutonium fissions as well as U-235 and can be derived from the abundant U-238 isotope, which is more easily obtained in quantity than U-235. A chain reaction once again seemed possible.

Why the excitement over this fission reaction? If one pound of U-235 or plutonium could be induced to undergo fission, it could release a sudden explosive force equivalent to 10,000 *tons* of TNT, or, if released gradually, power equal to 12 million kilowatt hours!

But *pounds* of U-235 or even plutonium are not easily obtained. Could neutrons somehow be slowed down, favoring their absorption by U-235 instead of U-238? By introducing such light elements as beryllium or carbon, which do not easily absorb neutrons, the velocity of a fast neutron can be slowed as it bounces off these light elements, thereby increasing the opportunity for its absorption by U-235.

Strangely enough, the capturing of neutrons by a fissionable material is related to its volume, while the loss of neutrons is related to its surface area. But as a quantity of matter grows in size its volume increases faster than its surface area. Therefore, the larger the piece of fissionable material, the greater the number of neutrons retained for fissioning the U-235. This brings up the important concept of critical mass. The fissionable material must be large enough to retain sufficient neutrons to sustain a chain reaction. Since U-235 is so scarce, and we never use more than is necessary, what is the critical size for U-235? It depends in part on the purity, for impurities usually absorb neutrons as well as U-235. Early calculations indicated a critical mass of 2 to 200 pounds of U-235. This was like asking for a ton of hummingbird tongues! Several possibilities were open: first, processing tons of uranium ore to get pure uranium, then separating U-235 from U-238; second, bombarding U-238 with fast neutrons to get the fissionable plutonium; or, third, introducing purified moderators (beryllium or carbon) into a purified mixture of U-235 and U-238 to produce the desired quantity of slow neutrons.

While all these avenues were explored, the third produced the first sustained chain reaction. In a squash court at the University of Chicago an atomic pile (quite literally) was constructed of bricks of a purified form of carbon (graphite) between which pieces of uranium were placed. To prevent the possibility of too vigorous a reaction, control rods of cadmium, which absorbs neutrons, were inserted into the pile. The graphite blocks slowed the neutrons so that they could be absorbed by the U-235 and enough slow neutrons were available to sustain the reaction. On 2 December 1942 controlled atomic fission was achieved.

Once a sustained chain reaction was obtained, fabrication of an atomic bomb was relatively simple. All that was necessary was to prearrange several noncritical masses of fissionable material, each supplied with a neutron source, fire them together into a critical configuration, and in a fraction of a second enough fission reactions could take place to cause an explosion of incredible dimensions (Figure 9.1).

The hydrogen bomb

Although a large atomic bomb is more than adequate to destroy any military target, a super or hydrogen bomb was developed, an extension perhaps of the "bigger and better" fixation with which modern society seems preoccupied. The hydrogen bomb involved a fusion principle, just the opposite of the fission process in the atomic bomb. Instead of supplying neutrons to split atoms, heat was supplied to fuse atoms of hydrogen into helium, releasing huge amounts of energy. To do this required a seemingly impossible amount of heat— 100 million ° C! Since this level of heat was possible on earth only in an atomic explosion, it was necessary to trigger the fusion reaction by a fission reaction. So the H-bomb was created, concluding, we hope, the evolution of such grim devices.

A few years ago our attention was riveted to the destructive aspects of atomic energy by Hiroshima, Nagasaki, and a long series of tests in Nevada and the Marshall Islands. More recently, with the

Fig. 9.1 • An atomic bomb test near Bikini Atoll, 1946. (*AEC*)

adoption by most nations of a test ban treaty, the peaceful uses of atomic energy have come to the fore, first with atomic fission power plants and then with the longer term potential of power from hydrogen fusion (see Chapter 10).

RADIOACTIVE FALLOUT

We still have a legacy from that weapons-oriented past, however. Exploding nuclear devices created large quantities of radioisotopes, 15 percent of the total bomb energy. Because of the variety of material in the bomb, and in the soil and water of the test sites, a broad spectrum of radioactive isotopes was produced with each explosion. Some of these are short-lived, lasting only fractions of seconds; others last minutes, hours, days. Still others last centuries, even millennia. But all emit radiation that is potentially harmful in a variety of ways.

Injected high into the stratosphere by the bomb explosions, the radioisotopes gradually sifted down, mostly over the Northern Hemisphere; a classic example of short-sighted experimentation without the least notion of the long-term effect on man or the environment.

Food Chain Contamination

Fallout from bomb testing programs in the 1960s was a particular problem in the arctic tundra, where the food chain is dramatically short. Caribou in the New World and reindeer in the Old World feed extensively on lichens in the winter. Because of the high amount of fallout in the Arctic (one-fourth to one-half of that in the temperate zone), and a slow rate of growth, lichens accumulate cesium-137 to a level well above that of other plants. When these radioactive lichens are eaten by caribou or reindeer the cesium-137 is concentrated still more in the meat of these animals. Should the caribou be eaten by a wolf or a man, the level increases again. The cesium-137 concentration doubles at each step in the food chain from lichen to caribou to man. This radiation concentration is especially clear in the arctic ecosystem because of the simplicity of the system and the limited socioeconomic development of the people—in the winter, caribou have only lichens to eat and the Eskimos have little to eat but caribou. Cesium-137 has a half-life of about thirty years; that is one half of its radioactivity will have dissipated every thirty years.

There are other examples of radioisotope concentration in ecosystems (Figure 9.2). Zinc-65 is readily absorbed by oysters, clams, and scallops in that order, not only by contact in their gill area but by ingestion of food as well, for diatoms like *Nitzschia,* which we

Fig. 9.2 • The radioactivity this deer has picked up from fallout is being measured by a whole-body counter. (*Colorado State University*)

discussed in Chapter 7, absorb zinc-65 quite readily. In late summer of 1956 after the spring H-bomb tests in the Pacific, some of the tuna caught by Japanese fishermen all over the Pacific were disturbingly radioactive. Radioactivity is measured by a device which counts the bursts of radiation from a radioisotope. One fish registered 4500 counts per minute (cpm) in its kidney, 1200 cpm in its stomach, 1800 cpm in its intestine, 2500 cpm in its liver, and 1200 cpm in its heart. Although there was little radioactivity in the edible flesh there was concern about the sources of this radioactivity. An expedition sent into the Pacific by the Japanese government discovered that sea water over 1300 miles northwest of the bomb site at Bikini registered over 100 cpm. But that did not explain the excessive radioactivity in the tuna. The answer lay in food chain concentration, leading to potentially dangerous contamination of the top consumer in the chain, tuna. Over 457 tons of tuna with counts above 100 per minute were destroyed, causing panic in the fish-eating Japanese public, and financial ruin to many in the tuna trade. Picture the consternation in the United States if beef were suddenly found to be radioactive!

Not all radioisotopes are produced in so straightforward a manner. In follow-up studies after bomb tests in the Marshall Islands, investigators were puzzled to find high levels of cobalt-60 in shellfish. But cobalt-60 was not one of the radioisotopes produced by the blast. Apparently some other blast-generated material, a neutron emitter, bombarded the natural isotope, cobalt-59, producing indirectly cobalt-60. The possibilities for harmful contamination are enormous.

Fortunately a test ban treaty has been signed by all nations testing atomic devices except China and France. Although 200 million tons of radioactive debris were thrown into the atmosphere to fall out over a period of several years, most of this material has finally come to rest since the virtual cessation of atmospheric testing. In 1970 the fallout of strontium-90 was only a twentieth of what fell in 1963. While fallout from testing is apparently in the past, the proliferation of nuclear and fossil fuel power plants threatens a continued source of airborne radioisotopes, although at a far reduced level from that of the 1950s and 1960s.

RADIATION SICKNESS

All of us have come into contact with radiation through fallout from bomb tests. The levels, however, are very low and the effect is probably quite long-term, if it is measurable at all. What about the more spectacular exposure to radiation that leads quickly to sickness and often death?

Ionization

As we saw earlier, there are three major types of radiation, alpha, beta, and gamma. The first two are particles which have relatively little energy; alpha particles are barely able to penetrate the skin and beta particles can penetrate only a millimeter or so. So under normal conditions these two types of radiation cause only skin burns, although when matter emitting either of these radiation types is inhaled or incorporated into the body significant damage may result. Gamma radiation, the third type, is the most energetic and potentially the most dangerous since it can penetrate most substances, certainly the human body, with ease.

Regardless of the form, alpha, beta, or gamma, radiation usually affects living tissue by causing ionizations that, in turn, cause cellular damage. A cell is composed largely of water. When a water molecule is irradiated, an electron is knocked out of orbit. The ejected electron may then become attached to a normal water molecule, making it unstable. These unstable molecules split into hydrogen ions (H^+), hydroxide ions (OH^-), and the free radicals $OH\cdot$ and $H\cdot$. Free radicals react with various molecules in the cell, which can then no longer function normally, and the cell dies.

Measurement of Radiation and Dose

Although there are many units used to describe radiation in its various aspects we need be concerned only with two types: disintegrations in a radioactive substance producing radiation and ionizations in either air or tissue caused by these radiations. To measure the first type, a curie is defined as 3.7×10^{10} disintegrations per second and is roughly equal to the radioactivity of one gram of radium together with its decay products. The radioactivity of materials is usually described in curies, millicuries (10^{-3}Ci),[1] microcuries (10^{-6}Ci), or picocuries (10^{-12}Ci) sometimes called micromicrocuries. But this gives us no information about dose. The roentgen (r), named for the discoverer of X-rays, is a measure of the number of ionizations caused by radiation in air and is defined as the amount of radiation causing 1.6×10^{12} ionizations in one cubic centimeter of air. Since we are more concerned with dose rate in tissues than air, two other units have been devised, the rad, which is 100 ergs (a unit of energy) absorbed by one gram of tissue, and the rem, which is a unit of absorbed dose taking into account the relative biological effect of various types of radiation. But for our purposes a rem or a rad is about one roentgen.

---

[1] This is a shorthand way of writing 1/1000 of a curie. A picocurie would be 1/1,000,000,000,000 of a curie.

A dose of 1000 r is fatal to all humans. A liver cell receiving 1000 r would be exposed to 2 million ionizations. Since a liver cell has about 2 billion protein molecules only one of every thousand protein molecules would be likely to be ionized. Yet this is enough to lead to the death of the cell.

About 50 percent of those people irradiated at 500 r die. This statistical point is called the LD-50 (lethal dose for 50 percent of the population).

At levels of 100 to 300 r, radiation sickness develops. The first symptoms are nausea, vomiting, diarrhea, and nervous disorders followed by a period of relative well-being. Then the secondary symptoms begin: falloff in red and white blood cell numbers, hemorrhages just below the skin, loss of hair, and ulcerations in the mouth and gut. These symptoms may also disappear, leading to slow recovery. Perhaps the most dangerous phase is during the period of low white blood cell level. Since the white cells defend the body against disease organisms, a person may survive the radiation only to die of a disease like pneumonia or influenza.

The tissues most sensitive to radiation are those which reproduce rapidly under normal conditions: the blood-forming tissues, bone marrow, lymph nodes, spleen, thymus in children, liver in fetuses, gonads which produce either sperm or eggs, epithelium which lines the intestinal tract, and all embryonic tissues which are in a state of flux. The symptoms of nausea, vomiting, and diarrhea are due to disturbances or death of the cells lining the digestive tract; the hair loss is due to injury to the hair follicles. The lens of the eye, which is covered with epithelial cells, is also quite sensitive to radiation and may develop a cataract or opacity.

Survivors of acute radiation in the 100 to 300 r range often suffer some permanent hair loss, cataracts, low fertility, and an increased tendency toward leukemia. Beyond this dose range, 600 r leads to irreversible destruction of bone marrow, 1000 r to irreversible intestinal damage, and greater than 3000 r, death of the central nervous system. Of three men involved in nuclear accidents, one received 800 r and died in 26 days, a second exposed to 1000 r died in 9 days, and a third received 12,000 r and died in 36 hours. Between 25 and 100 r there are no visible symptoms, just a decrease in the number of white blood cells. Below 25 r there are no measurable changes.

There is a greater potential and demonstrated effect of radiation on reproductive cells than on asexual or somatic cells. By damaging or even slightly altering the chromosomes in the egg or sperm, mutations can be caused that lead to abortions, deformed births, or genetic defects of a less obvious sort. Though only 10 percent of mutations have been calculated to be radiation produced, the mutational rate is proportional to the total amount of radiation absorbed by the sex cells. It makes little difference whether they receive 10 r in one year or 1 r a year for ten years. Most measurements show a linear relationship between dose and damage. So in a sense there is no such thing as a harmless or "safe" dose of radiation. This is why there is such concern in many scientific circles about what seem to be insignificant amounts of radiation, in fallout or in more mundane exposures such as X-ray examinations. James Crow, an eminent human geneticist, stated that even if the maximum fallout of the late 1950s were continued over a thirty-year period and amounted to a total of only 0.1 r in that time, 8000 children in the world's population of the next generation could be born with gross physical or mental defects as a result of that radiation.

Of all the radiation beyond the natural background, X-ray examinations far surpass that received from fallout. The United States population is probably the most X-rayed in the world, yet there is no overt evidence of any increase in mutations, higher incidence of birth effects, or shorter lifespans that can be directly attributed to radiation. Indeed, another human geneticist, Hermann Muller, regards the increased frequency of reproduction by people with genetic defects as a far greater danger to the genetic wellbeing of man than any level of radioactivity currently encountered.

Although we have recently heard much about fallout and artificially produced radioisotopes, radiation has always been with us. At sea level we all pick up at least 0.1 r a year and up to 50 percent more radiation at altitudes over one mile (a result of greater cosmic ray intensity). A quarter of this background level comes from cosmic rays zooming in from outer space, another quarter from various radioactive elements in the body, the rest from soil and rocks. Fallout amounts to 0.001 to 0.005 r per year for most people, less than 5 percent of the natural background. In some parts of India and Brazil, radioactive sands emit as much as 17.5 r per year and houses built on these deposits receive 2 r per year, compared with 0.69 r per year in New York City. Long-term studies have been initiated to determine if this higher than normal background radiation has had any measurable effect on the local population.

The role of radiation as an environmental variable is somewhat more obvious in plants. A reasonable picture of radiation effects in a natural ecosystem can be seen from an experiment done at Brookhaven National Laboratory by G. M. Woodwell. A piece of relatively homogeneous scrub oak-pitch pine forest was irradiated with a powerful source of gamma rays. The intense radiation killed plants, *but differentially*. It was discovered that pine trees were more sensitive than oak, oaks more sensitive than blueberry, and blueberry more sensitive than a particular sedge (a grass-like plant growing on the forest floor). The radiation was emitted from a point source about fifteen feet from the ground, hence the dieback was circular around the source (Figure 9.3). The trees and shrubs were generally more sensitive to radiation than the sedge, because the sedge buds were at ground level and somewhat protected; the buds of trees and shrubs

Radiation and plants

Fig. 9.3 • A radioactive source is located in the mast at the center of the circle of dead and dying trees in an experimental forest. This experiment determines the effect of radiation on a forest area over a defined period of time. (*Courtesy of Brookhaven National Laboratory, New York*)

were fully exposed to the radiation. Variations in radiosensitivity among the trees and shrubs were due, in part, to differences in the number and size of the respective plants' chromosomes. One theory, pioneered by A. H. Sparrow, suggests that cells with a small number of large chromosomes are more likely to be damaged by radiation than cells with a large number of small chromosomes, simply because the former offer a larger target. Moreover, the many small chromosomes of the latter possess more duplicate genetic information so that when one or two are damaged, a number of others can take over their function.

NUCLEAR POWER AND RADIOACTIVE WASTES

Despite the long recital of environmental and health problems elicited by a greatly augmented flow of various kinds of radiation into the environment, there has been a silver lining in the radioactive cloud. Some radioisotopes have been extremely useful in cancer therapy, cobalt-60 and gold-198 for example. Phosphorus-32, carbon-14, and oxygen-18 are used for tracer work in physiology, ecology, and geology. But the most spectacular beating of sword into plowshare has been the use of atomic energy in the production of power (see Chapter 10).

A pessimist might with some justification claim that technology, while seeking to solve old problems, always manages to find new ones. The new problems, incurred when radiation is used in power plants, are not insoluble, but as more nuclear power plants are built, they become increasingly important.

In 1957 the Atomic Energy Commission (AEC) released a document dealing with the theoretical possibility and consequences of a major accident in a large nuclear plant. "Large" in 1957 meant 200,000 kilowatts (kw). This theoretical plant was near water, thirty miles from a city of 1 million, and at the end of a 180-day fuel cycle—and thus invested with 400 million curies of radiation. Assuming general release of this radiation through massive failure of containment structures and procedures, 3400 people would die, 43,000 would be injured, $7 billion of property would be damaged, and over 150,000 square miles of agricultural land would be affected. Today's reactors are larger by a factor of ten, but though the 1957 report was updated a couple of years ago the new figures were never released to the public. The core density of a 1000-megawatt (mw) reactor is 40 percent higher than the 1957 theoretical model and the fuel cycle is longer, allowing a far greater buildup of fission products.

Are present reactors safe? Well, there is no chance of a thermonuclear explosion. But there is a chance of high-level fission products escaping into the environment. The precautions are great: a typical reactor is constructed on a nine-foot-thick slab of concrete and is protected by a dome (Figure 9.4) of vapor-proof steel covered with three feet of concrete. Considering all the high-pressure plumbing in a pressurized water reactor (PWR) and the critical importance of the coolant water, there is a chance, however small, that the coolant system could rupture and fail and the fuel rods heat up to the melting

Fig. 9.4 • The dome that shields this reactor is a characteristic feature of nuclear-fueled power plants, this one at Haddam Neck, Connecticut. (*Connecticut Yankee Atomic Power Company*)

point. Just such an accident happened against all odds in the Enrico Fermi fast breeder reactor near Detroit in 1966. While no radioactivity escaped from the plant, experts considered the incident a near miss.

The Fermi reactor accident was caused by sloppy workmanship in constructing the reactor. Unfortunately, there have been too many examples of poor design in reactors and components, loose supervision and quality control during construction, faulty construction materials, and inept handling of fission products. Most of these criticisms could be leveled at the construction of any building or factory in the United States today, but in the design and construction of nuclear reactors the awesome power of radioactivity is being risked.

Present-day reactors (see Chapter 10) are not necessarily unsafe; but they can be made far safer, starting with greater care in choosing a site. A few years ago a nuclear plant was almost constructed on the San Andreas Fault near Bodega Head in northern California. Recent estimates of underground siting suggest an added cost factor of only 10 percent, which raises the possibility of reactors being constructed underground, perhaps far closer to cities that now seems prudent. A general tightening up of standards for design, materials, and construction would certainly help, but a major improvement in reactor safety cannot be expected until today's water-cooled reactors are replaced with more advanced designs.

If that is the best we can expect from protection against accidental release of radioactivity, what about routine discharge of radioactive substances?

There are three kinds of radioactive pollutants from nuclear power plants: solid, liquid, and gaseous. Solid wastes may consist of such items as tools, reactor parts, and clothing, which may be quite radioactive depending upon their use and are usually buried in cement drums either in trenches on land or at sea.

Liquid wastes come in part from leaky seals and valves in the extensive plumbing of PWRs. These low-level wastes are stored and their radioactivity measured. Depending on their accumulated activity, the wastes are released on a batch basis. Other liquid wastes result from isotopes formed when impurities in the coolant water and corrosion products from the coolant pipes are bombarded with neutrons escaping from the core area. This can be controlled somewhat by demineralizing the coolant water before it enters the heat exchange area but there will always be some radioisotopes generated from this source. A reactor in the 150 to 300 megawatt range would generate 1 to 10 curies per year in the form of cobalt-58, chromium-51, manganese-54, iron-59, and molybdenum-99. Because of the low radiation level of these wastes, they are usually discharged into the environment.

The fuel elements themselves are an even more dangerous source of both liquid and gaseous radioactive wastes. Although clad in either stainless steel or zirconium alloy, carefully fabricated to minimize leakage, complete sealing of the fuel elements is apparently impossible to attain or to sustain. Minute cracks allow radioactive fission products to escape into the primary coolant. In some instances, the high radiation level causes the cladding to flake or weaken, allowing still more leakage. When the steam is condensed for return to the pile of a PWR, the gaseous fission products are separated and held in storage for thirty days or more. This allows time for many of the short half-life isotopes to decay. The remaining wastes are then vented through the reactor stack. A boiling water reactor (BWR) generates a greater volume of radioactive gases that are held for only thirty minutes before being vented through the stack. Consequently

Radioactive pollutants

BWRs release much larger quantities of gaseous radioisotopes than PWRs. Gaseous releases from both types of reactors could be substantially reduced with new systems developed by GE and Westinghouse that cost less than 1 percent of the capital cost of the plant.

One of these gases, you will remember, is krypton-83 (when U-235 undergoes fission barium-138 and krypton-83 are the two fission products). Others are xenon, argon-41, iodine-131, and nitrogen-13. Not all of these are fission products, of course; some are due to impurities of air and water within the reactor.

Disposal of Radioactive Wastes

The biggest problem lies in disposing of the fission products. Sooner or later, usually in one to three years, fission products accumulate to the point where the chain reaction stops or is poisoned, that is, the neutrons are being absorbed at greater rates by the fission products, leaving fewer to sustain the chain reaction. At this point the fuel elements, now extremely radioactive, are removed and shipped in specially cooled and shielded containers to a fuel reprocessing plant. Here they are chopped up and placed in concentrated nitric acid. The fission products and unreacted fuel are dissolved and separated. The unreacted fuel can be used again but the fission products, concentrated somewhat by evaporation, are stored as a liquid in underground stainless steel tanks. With a radioactivity that is usually between 100 and 1000 curies per gallon, enough heat accumulates to keep the tanks boiling like teakettles.

At one time the federal government contracted for all fuel reprocessing but now private enterprise has been allowed to enter the field. A plant planned for South Carolina would release from its stack 12 million curies per year of krypton-85 and 500,000 curies per year of tritium since reprocessing plants are not required to follow the strict standards required of nuclear power plants.

Assuming the present level of efficiency (25 percent), almost one and a half tons of fission products are produced per 1000 megawatts of power. Already hundreds of millions of gallons of high-level wastes are in storage. Since these wastes will remain highly radioactive for hundreds of years while the life of the tank is measured in decades, underground storage tanks represent at best a temporary solution to the waste disposal problem (Figure 9.5).

Though high-level wastes are considered too hazardous to be dumped into the ocean, intermediate- or low-level wastes are disposed of there, often without a clear idea of the hazards involved. One obvious hazard would be containers breaking and releasing their radioactive contents, which might be washed up on crowded beaches. Contamination of food chains, as we have already seen in Pacific tuna, could easily occur also. Concentration of wastes on continental shelves might make the future exploitation of oil, gas, or

Fig. 9.5 • Million-gallon tanks have been constructed at the AEC Hanford Works in Washington to contain high-level radioactive wastes. When completed the tanks will be buried in 7 to 14 feet of earth. (*U.S. AEC, Richlands Operations Office*)

metal nodules awkward. All too often, wastes are not dumped at either the required distance from shore or the proper depth. At too great a depth the containers may be crushed by sea water pressure, releasing their contents immediately. Also, it is not known how long the containers last on the sea bottom, and how far and in what direction they are moved by underwater currents.

In the future, before the quantity of wastes from increasing numbers of reactors overwhelms the possibilities of storage in liquid form, reduction to a solid state and incorporation into ceramics or glass, both environmentally resistant, may be possible. These stabilized solids could be permanently stored in dry, stable salt mines of which there are large numbers in New York, Kansas, and along the Gulf Coast. By the year 2000, twelve acres a year of such salt mine storage space will be needed to dispose of high-level reactor wastes.

Present disposal of radioactive wastes can best be summed up by citing the three basic approaches: dilute and disperse, used for high volume, low-activity wastes; delay and decay, used for medium activity wastes where slow movement through soil allows time for radioactive decay; concentrate and contain, reserved for high-activity wastes, now stored underground in tanks.

Radioactive wastes involve more than the reactor and its by-products: uranium must be mined, purified, and processed into fuel elements, and as we have seen, when the elements are exhausted they must be reprocessed. Each of these steps produces radioactive wastes. One example will suffice. Scattered around the Colorado River Basin are huge piles of uranium tailings (Figure 9.6). Ground to a sand-like consistency to remove the uranium, these tailings contain radium-226 with a half-life of eighty thousand years. Radium, like strontium-90, is absorbed by the bones. As radioactive dust from the piles blows into rivers and ultimately into Lake Powell and Lake Mead, the

Mine tailings

195

Fig. 9.6 • Abandoned uranium mine tailings are found throughout the western states, and have been used for construction in the form of fill, concrete, and mortar. Once thus fixed in a building, the tailings release radioactive gases that may accumulate to dangerous levels. (*The New York Times*)

radium-226 level has increased in places to twice the maximum permissible level suggested for human consumption. Near Durango, Colorado, a pile of over a million tons of radium-226 sits near the Animas River. As it erodes, this material will certainly increase the radium-226 content of the river. Crops grown on irrigation water from the Animas River already contain twice as much radium-226 as crops growing above the tailings. Since radium is concentrated in hay and alfalfa that is eaten by cattle and in turn by man, the possibility exists of yet another food chain being contaminated by radioactivity. Of course, the tailings should be stabilized by grading and planting, which would greatly reduce erosion into surface drainage. This approach has been followed to some extent by state and federal agencies, but apparently not at a rate that has kept up with continuing production.

Further problems have arisen from the tailing sand being used for children's sand boxes and by contractors as a base for concrete slabs or in the backfilling of basements. As the radium-226 decays, radon, a radioactive gas, is given off. Although radon has a short half-life, it has been established as the prime cause of lung cancer in uranium mine workers. As a result of the use of uranium tailings in construction, many people have been exposed in their homes to levels of radiation many times higher than the maximum dose allowed miners in a uranium mine, the equivalent of nearly 553 chest X-rays in one year. In some instances houses have had to be abandoned.

Once again, as with strip mine spoil banks, no one seems interested in the tailings once plants cease operation. The million tons of tailings in the Colorado River Basin may contain as much as 8000 grams of radium-226. When it has been carried by the rivers of the Basin and deposited into the bottom sediments of Lake Powell and Lake Mead, it may well be put out of circulation. But radium-226 may become a much larger problem than it is now as increasing use is made of river water in this arid region.

Since radioactivity is so easy to measure and its effects are well-known compared with those of many biocides, standards were set up, somewhat arbitrarily, well in advance of intensified radioactive waste production. Three categories, each with its own permissible level, were established by the International Commission on Radiological Protection. This group recommended a limit to workers in radiation-oriented industries of four rems (roughly 5 r) a year for blood and sex organs, thirty rems a year for the eye, skin, thyroid, or whole body radiation, and fifteen rems a year for other organs. The individual should receive only one-tenth of that permitted the occupational group and the general public one-thirtieth of the occupational level. This assigning of permissible dose evidently assumes a threshold for various types of radiation and places the permissible dose somewhere below this level despite evidence that *any* amount of radiation can cause damage. To be sure, avoidance of all radiation is impossible since background alone provides 0.1 to 0.5 rems a year.

Radiation levels

In the late 1960s two researchers at the Livermore Radiation Laboratory, John Gofman and Arthur Tamplin, caused a great stir when they announced that the maximum permissible dose of *man-made* radiation, 0.17 rems per year, was far too high and could result in 32,000 extra deaths from cancer per year if all Americans were exposed to that level of radiation. While that was unlikely to happen, the standard implied that the AEC was prepared to let it happen. In response to this claim the National Academy of Science appointed a panel, the Committee on the Biological Effects of Ionizing Radiation, to review the whole issue. While feeling that Gofman and Tamplin's claim was somewhat exaggerated the committee did state that perhaps 6000 extra deaths per year could result from the 0.17 rem maximum dose. The Environmental Protection Agency (EPA), which now seems to have effective control over setting radiation standards, is reviewing the report and may well abandon blanket exposure limits for the general population. It might make more sense to impose strict limits for specific sources which would discourage the natural tendency to irradiate a population up to an authorized level, presuming no other source of radiation in a given area.

It is senseless for anyone to expose himself needlessly to radiation. It is even more senseless for whole populations to be unwittingly subjected to any increase in radiation levels that can be prevented; and it is also unnecessary, because the technology exists to remove radioisotopes from effluent wastes, whatever their source. It has simply been cheaper and more expedient to disperse them into the environment. As earth's population continues to grow we can no longer allow the environment to slip from the desirable to the merely tolerable if life is to have any value beyond bare existence.

FURTHER READING

Calder, R., 1962. *Living with the atom.* University of Chicago Press, Chicago, Illinois. Easily read discussion of radioactivity in the environment.

Curtis, R. and E. Hogan, 1969. *Perils of the peaceful atom.* Doubleday, Garden City, New York. Interesting discussion of reactor problems.

Hecht, S., 1954. *Explaining the atom.* Rev. ed., Viking Press, New York. Beautifully written account of atoms, atomic energy, atom bombs; easily accessible to the layman.

Lapp, R. E., 1971. "The four big fears about nuclear power." *New York Times Magazine,* 7 February. A concise analysis of some of the major points of popular anxiety about the use of nuclear reactors.

Metzger, H. P., 1971. " 'Dear Sir: your house is built on radioactive uranium waste.' " *New York Times Magazine,* 31 October. A good case history of the uranium tailings problem.

Novick, S., 1969. *The careless atom.* Houghton Mifflin, Boston, Massachusetts. Exposé of the dangers inherent in nuclear reactors.

Tamplin, A. R. and J. W. Gofman, 1970. *Population control through nuclear pollution.* Nelson-Hall, Chicago. Rather sensational presentation of their basic thesis.

Tsivoglou, E. C., 1971. "Nuclear power: the social conflict." *Env. Sci. and Tech.* **5**(5), pp. 404–410. In contrast to some of the above, a calm, rational yet frank statement of some of the nuclear power problems.

Weinberg, A. M., 1972. "Social institutions and nuclear energy." *Science* **177**, pp. 27–34. Another rational note in a sea of emotion.

3

Man
Makes
New
Traumas

The World Health Organization spraying Kinshasa, Zaire with DDT in 1961 to control malaria-carrying mosquitos. There is no record of what was also controlled, for at the time the effects of DDT were only beginning to come into focus. (*UN*)

10

IN QUEST OF ENERGY

Energy is Eternal Delight.

WILLIAM BLAKE, 1793

ON NOVEMBER 9, 1965, for the first time in memory, the lights went out in New York City. Not just a block or two but all of Manhattan, Queens, and the Bronx were plunged into darkness. For thousands stranded in elevators and subways, the blackout meant suspended motion as well. Because of the unusually mild weather, the crazy novelty of the situations in which most people found themselves, and the restoration of power before the next nightfall, the city survived pretty much intact. The principal result seems to have been a minor population explosion nine months later, marking the night when hundreds of thousands of people, denied one form of entertainment, remembered another.

The cause of the blackout, which took much detective work to unscramble, was an extraordinary combination of human and mechanical failure; but in retrospect the incident seems to mark a turning point in the pattern of energy production and use. Until that November night, electricity was thought to be a commodity like water—inexhaustible and cheap. Since that night a combination of economic, political, and ecological events have indicated that Blake's statement is only a poetic metaphor.

Unlike materials, which can cycle or be recycled in the environment, energy flows in a one-way trip from source to sump (Figure 10.1). The source for most of our current uses is the sun, which through a complicated series of reactions throws immense quantities of energy out into space; the sump is the heat into which all energy is ultimately transformed. Although only a small fraction of the sun's energy is intercepted by the earth, it is enough to keep all the organic and inorganic cycles (Chapter 3) working. Through the photosynthetic process, plants can store some of the solar energy in the form of food. Although the efficiency of photosynthesis is quite

MAN MAKES NEW TRAUMAS

Fig. 10.1 • The various sources of the power produced in the United States, and their proportions, are indicated along the left side of this diagram. The fate of this power can be followed by moving to the right. (*Source: Earl Cook, Texas A&M University*)

low, trapping only a small percent of the sunlight striking the earth, the prodigious quantity of solar energy and the sheer numbers of plants have made possible the food webs of which we are a part today. So the internal energy we use to function as individual organisms flows from the sun through green plants to us. The external energy we use to function as a society follows a similar pathway, except for an enormous time lag. The fossilized remains of plants and animals alive hundreds of millions of years ago (Figure 10.2) have come down to us as coal, oil, and natural gas in quantities once thought to be inexhaustible.

If only a modest fraction of the earth's share of solar energy could be transformed into useful power, then the eternal delight of

Fig. 10.2 • When coal seams are as thick as this, it becomes profitable to strip off the overburden so that the coal can be removed. (*H. K. Svenson, courtesy of the American Museum of Natural History*)

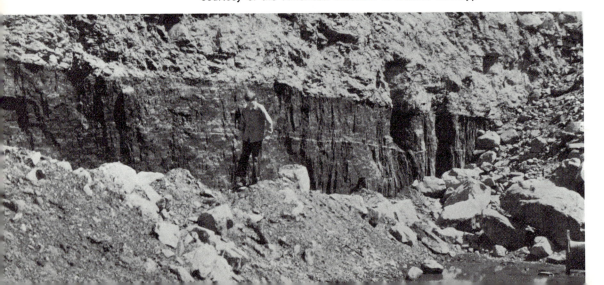

energy could be shared by all the earth's peoples. For the present, the average American needs twenty times the energy available to an Asian to support himself in the style to which he has grown accustomed, and his appetite has been increasing by 10 percent per year. Without direct access to the virtually unlimited supply of solar energy, it takes 35 percent of the world's available energy, mostly the fossil fuels, to supply America's energy needs.

As the energy demands of other industrially advanced countries continue to increase, the most convenient fossil fuels, oil and natural gas, are bound to become limited, then limiting. The threat of vanishing supplies of energy sources has given rise to much concern about an "energy crisis." Are we facing such a crisis, and if so, what are the prospects?

Energy crisis?

Rogers Morton, President Nixon's Secretary of the Interior, summed up the official view of the problem rather concisely when he described the current crisis as reflecting "the seasonal difficulties being experienced by certain areas in meeting peak demands for electricity in the summer and fuels for heating in the winter. These problems are still local or ephemeral in scope and duration. They are produced by a progressively failing supply of domestic natural gas; a steady decline in proved reserves and productive capacity in crude oil in the conterminous United States; by restrictions upon the production and use of energy in all forms imposed by environmental considerations, and finally by a growing dependence for energy upon sources over which our nation has no control."[1] But what of the longer term? The United States has within its territory all the resources for a continued supply of energy over an indefinite period. The difficulty seems to be marshaling the right technology at the right time; this requires foresight, planning, and perhaps a bit of luck.

Since research and deveopment of new technology depends on the rate of investment, which in a nondirected economy such as ours is always uncertain, let's examine each of the contemporary energy modes and evaluate its present and future potential as an energy source, then look at some more advanced concepts in energy supply.

COAL

Of all the fossil fuels, coal has been in use for the longest time. It made possible the industrial revolution and the urbanization on which that revolution depended. While the environmental price was high, it was paid, and the use of coal increased until the close of World War II. With more fuel available for nonmilitary uses after

[1] Quoted in Smith, W. P., 1971. "Now, Sythentic Fuels." *The New York Times,* 31 October, p. 1, "News of the Week in Review."

the war, and with the discovery of huge new deposits of oil and natural gas in the Middle East and Venezuela at about that time; coal began to lose favor and millions of home owners switched from the mess and labor of coal-fired furnaces to modern oil or gas burners. At the same time industries discovered the greater ease of handling oil and gas and also began abandoning coal. So as the emphasis switched to the glamour fuels, oil and gas, coal production languished. The postwar era also saw the beginning of the nuclear age, filled with hopes for the peaceful use of atomic energy. It was claimed that the dirty fossil-fueled plants would soon be replaced by clean nuclear power plants; this cheerful estimate ignored that it took almost fifty years to switch from wood to coal or coal to oil.

But by the late 1960s nuclear-fueled plants had run into snags— engineering, environmental, social, and political—and weren't being built nearly fast enough to keep pace with the surging demand for power, much less to replace the older fossil-fueled plants. Many power companies, desperate for power, discarded plans for the "new age" reactors and built huge fossil-fueled plants, two or three times the size of the old ones and requiring enormous quantities of coal. With these developments the market for coal picked up and a new phase of strip mining began, exploiting not only the older soft coal fields in Appalachia, but the large fields in Arizona. Plans were also made to develop the deposits in the Great Plains states of Montana and Wyoming.

The coal reserves of the United States are enormous—1.5 trillion tons, which at current or even accelerated rates of use should last for several hundred years, easily bridging the gap between present and still more extensive and efficient energy sources.

But neglect of coal technology for the past two decades has now come to haunt us. For while there is plenty of coal, the technology for recovery, combustion, and conversion of coal into other, more conveniently used forms is archaic. In addition, much of the available coal is of limited usefulness because it is either too high in sulfur or too low in energy value to be worth the cost of transporting to distant users. Construction of power plants close to the source of the coal is of course one solution to this problem, but since the demand for natural gas at present exceeds the supply, a great deal of attention is being given to converting some of the low-grade coal into more desirable products. Although work has just begun, three processes are beginning to look promising: gasification, fluidized bed firing, and magnetohydrodynamics.

Coal gasification

By grinding coal into a granular powder, treating it with steam and oxygen or air, then purifying the resultant gas, high-energy methane, free of sulfur, carbon monoxide, or free hydrogen can be produced. While the process is relatively simple in ouline, the details are more complex and the cost of the good-quality methane produced is high. To produce enough synthetic gas from coal to alleviate the natural gas shortage, the quantity of strip-mined coal would have to be doubled, with the environmental costs which that implies. Additional demands would be made on the environment for cooling water, about 20 percent of which would be consumed in the gasification process. However, since gasification is carried out in closed vessels, air pollution is prevented. But gasification research and development is just beginning, and it will be 1980 before gas from coal reaches the market in any quantity. Had the necessary research been undertaken during the 1950s and 1960s, we could easily be using coal now to help fill the need for natural gas. Hindsight, however, is always 20/20.

While coal gasification appears to be a practical method for reclaiming high-sulfur coal by conversion into natural gas, it is possible to combust coal directly in such a way that sulfur oxides are not released to the atmosphere, thus making available large reserves of high-sulfur coal for power production and relieving some of the drain on oil and gas supplies for producing power in areas where air pollution is critical.

Fluidized process

In most power plants coal is pulverized, then injected with air into a very large box whose walls contain tubes filled with boiling water. This allows for maximum transfer of heat from the burning coal dust to the boiling water. Traditionally the soot, fly ash, and sulfur oxides went up the chimney to be shared with the world. But modern air pollution control standards require that particles be trapped by installation of electrostatic precipitators and that the release of sulfur oxides be lowered by the use of various types of scrubbers. Both processes, however, are expensive to install and maintain, like the catalytic converter now required on new automobiles. A much more satisfactory solution would be to devise new ways to burn coal.

One very promising process places granulated coal on a moving grate, then feeds air upward into the bed so that the particles of burning coal float freely on the grate. When heated to 1100°C (2012°F) the coal ash becomes self-adhering and forms aggregates that are carried by the moving grate to an ash pit instead of being released up the chimney as fly ash. The excess level of carbon necessary to maintain the temperature at which ash adheres causes incomplete combustion, which produces carbon monoxide and hydrogen sulfide; but they can be converted by secondary combustion above the bed into carbon dioxide and sulfur dioxide. If limestone is added to the coal on the grate, most of the sulfur dioxide can be absorbed and removed from the combustion chamber. If the lime is roasted in the auxiliary chamber, the sulfur dioxide can be released in a high enough concentration to make commercial-grade sulfuric acid. The limestone, when divested of the sulfur, can then be recycled back into the combustion chamber. Unfortunately, conventional pulverized-fuel firing methods are so widespread that it will probably be some

Magnetohydrodynamics

time before the newer technique of fluidized-bed firing can be brought into broader use.

Another even more advanced technique that utilizes coal more efficiently for power generation is magnetohydrodynamics. In an ordinary steam power plant the steam drives a turbine that revolves metal armatures through a magnetic field, causing a flow of electrons, which constitutes captured current. An MHD generator is much simpler: a stream of very hot, partially ionized gases flows through an intense magnetic field, forming a current that is collected by electrodes in the sides of the chamber and bled off. Because of the direct generation process and the lack of moving parts, an MHD plant would be 10 to 20 percent more efficient than fossil-fueled steam plants or nuclear plants. Another advantage is that the seed particles of potassium sulfate or dioxide that are used to ionize the gases react with the sulfur in the gas, precipitating it out. Since the seed particles must be recovered and recycled for economic reasons, most of the sulfur is removed. This means that high-sulfur coal, of which there is a great supply in the eastern United States, can be very conveniently used. Since the exhaust gases are quite hot (the MHD generator must run at around 2400°C, 4352°F) a steam turbine might be used in tandem to generate more power. The high operating temperature of an MHD generator will of course generate nitrogen oxides. But research suggests that these can be reduced to acceptable levels by controlling the composition and temperature of the combustion gases.

At this point in development (the USSR has a 25-megawatt generator in operation, but the United States is just getting a research program together) the problem seems to be technological rather than environmental: the high corrosiveness of the 2400° gases, which affects electrodes and the generator walls; the expense of the large superconducting magnets required; and instability of the ionized gas. But none of these problems is insurmountable. The high efficiency and low emissions of MHD generators make them extremely attractive producers of power from existing supplies of fossil fuel.

OIL

Unlike coal, which occurs in well-defined strata whose geology has been relatively well worked out for years, oil deposits are confined by vagaries of bedding and pressure to portions of highly porous sedimentary rock layers, making the location of oil less reliable and predictable than the location of coal. Present reserves of oil in the United States are estimated to be about 100 billion barrels. Since our consumption of oil in 1970 was 5.4 billion barrels and is increasing at about 5 percent a year, we have about 20 years' supply of oil left if we depend exclusively on domestic supplies. Oil imports, which run about 20 percent of the total, have come mostly from Venezuela, with smaller amounts from Canada and the Middle East. In the next ten years, oil imports are likely to rise to about 60 percent of our needs, with the proportion of Middle Eastern oil rising from 4 to 46 percent. Imports are likely to increase as the United States seeks to delay total depletion of its domestic reserves; the shift from South American to Middle Eastern sources reflects a growing preference for the low-sulfur oil of Libya over the high-sulfur Venezuelan oil.

However, the immediate problem seems to be not a shortage of crude oil but a lag in refining capability, particularly for the low-leaded gasolines required by automobiles with pollution control devices. The reduction in gas mileage because of the increased weight of cars, the greater frequency of air conditioning and automatic transmissions, and pollution controls also put greater strains on refining capacity. These factors will increase the demand for gasoline by at least 20 percent in the next few years. If the current gasoline shortage should worsen and prices of gasoline continue to rise as a result, the restriction on use of cars that might be necessary could have a beneficial impact on both the energy crisis and the air pollution problem. Since a significant proportion of car usage involves trips of less than ten miles, a doubling of gasoline prices coupled with a decrease in gas mileage might encourage a number of drivers to walk or explore other modes of transport.

Extremely large deposits of oil sands in Alberta, Canada, and oil shales in Wyoming, Colorado, and Utah hold some promise for supplementing the supply of crude oil (Figure 10.3). The oil sands are saturated with a petroleum that is so viscous that it does not flow and so cannot be pumped out of a well like normal crude oil. The Athabasca deposit in Alberta covers an area of 9000 square miles, and together with a couple of smaller deposits nearby is estimated to contain around 300 billion barrels of oil. While exploitation of the oil sands has begun, there has been little economic incentive to persevere because of the competition from regular crudes. However, the magnitude of the deposit together with the similarity of the oil to crude oil means that refining can be done without major modification of existing oil refineries.

Oil shales are something else: rather than being in a liquid or viscous form the hydrocarbon content is in a solid form called kerogen. Although kerogen resembles crude it contains nitrogen as well as a great deal of sulfur, both of which must be removed in the refining process. The largest and richest deposits of oil shales in the United States are found in the Green River Formation that occupies the Piceance Basin of Colorado, the Uinta Basin of Utah, and the Green River Basin of Wyoming. The kerogen content of the shales is quite uneven, varying from zero to 1.5 barrels per ton. So although the total oil content of the Green River shales is estimated

Synthetic fuels

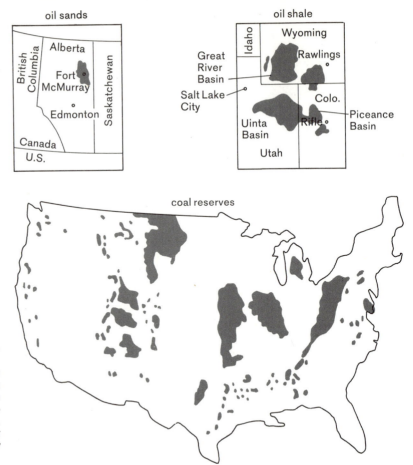

Fig. 10.3 • Although the deposits of oil sands and oil shale shown in the inset maps are small in comparison with the coal reserves shown in the large map, their energy content, if recoverable, is vast. (© 1972 by The New York Times Company. Reprinted by permission)

at 1.43 trillion barrels, perhaps only 80 billion barrels are likely to be recovered using contemporary techniques.

Suggestions that nuclear explosives be used to free the oil below ground are rather unrealistic. Project Gas-Buggy, designed by the AEC to free natural gas in rock strata 4000 feet below the surface, successfully liberated 300 million cubic feet of gas per year into a crushed-rock cavity formed by a 29-kiloton atomic bomb explosion. Assuming a constant flow of gas, 3 billion cubic feet of gas could be produced over a ten-year period. To equal the annual gas consumption of 23 trillion cubic feet, the yield of this one explosion would have to be multiplied 7000 times. Seven thousand atomic bomb explosions a year would very likely have profound environmental implications and be very expensive as well. Since the same technology used in the Gas-Buggy experiment would probably be applied to below-ground extraction of oil from either oil sands or shales, it seems that the peaceful atom may be of limited use as a cheap way to exploit the Green River shales.

The environmental impact of processing oil sands and oil shales is similar to the strip mining of coal, except that the waste from oil shale, because of the peculiarity of the processing, occupies more volume than the original shale. This presents a waste disposal problem of some magnitude if reclamation of the sites is to take place.

While many oil companies have formed subsidiaries or consortia to explore the possibilities of obtaining oil from shale, this interest is quite probably a hedge against future desperate need, or more likely, lower profits from or higher cost of Middle East crude. Hence the efforts to recover usable oil from sand or shale are likely to be somewhat desultory until the increase in demand and price occurs.

What can the individual do to reduce gasoline consumption? The United States Department of Transportation suggests keeping the engine tuned (a hopeless chore for most drivers, who take care of cars only when they stop running), keeping tires properly inflated, avoiding stop-and-go traffic if possible (the point of the interstate highway system, by the way), forbearance in the use of air conditioners, and slower driving. A car cruising at 50 mph uses 11 percent less gas than at 60. Since gas consumption at 80 mph is twice that at 50, the seemingly inconsequential factor of speed can be monumentally important. If everyone used 1 less gallon of gas a week, whatever gasoline shortage exists would be solved.

NATURAL GAS

Gas, together with salt brine, is an integral component of oil deposits; the gas-to-oil ratio is around 6000 cubic feet per barrel. In the early years of oil drilling the natural gas was blown off or burned, for there was no market or means of transport. But during World War II several pipelines were constructed to transport vitally needed oil from the east Texan oil fields to the Northeast and Midwest without having to run the gauntlet of German submarines, which made coastal transport risky. After the war some of the lines were converted and greatly expanded to bring natural gas to the urban centers of the eastern states. In a relatively short time a substantial market was made for gas; it has been increasing ever since. The gas is especially valuable as a source of heat because it has 1.6 times the energy value of oil. Present reserves of natural gas are estimated at 1000 trillion cubic feet. But only a portion has been exploited, apparently because of the lack of economic incentives.

During the halcyon years of plentiful supplies of gas and oil, when there was ample time to develop alternate sources of energy, there was virtually no commitment on the part of industry or government to explore and develop interim sources of energy that would buy time while the more remote sources were being explored. Research in alternate combustion techniques and gasification of coal, utilization of oil sands and shales, and liquefying gas for more economical transport should have been started after World War II.

211

Indeed in 1947 Congress authorized the Bureau of Mines to begin research in some of these areas, but appropriations were not approved by the Bureau of the Budget and it wasn't until almost fifteen years later that the Office of Coal Research was founded and provided with a rather modest budget—a typical example perhaps of the penny-wise, pound-foolish policies that governments, regardless of political persuasion, seem to follow.

Because of the lead time necessary to develop new technology it will be at least 1980 or 1985 before any alternate source of gas or oil can be developed beyond the pilot stage and so make even a minor contribution to the energy deficit that is beginning to appear, at least in some sectors of the energy market.

THE RISE OF NUCLEAR POWER

Through efforts of the AEC exercising its mandate to explore the peaceful uses of atomic energy, and of industry eager to use new technology to reduce costs and increase profits, a new energy source for power production was developed—nuclear fuel. There is a popular misconception that nuclear-fueled plants are radically different in their mode of power generation. They are not. Certainly, heat is generated from a radioactive source rather than a fossil fuel, but the rest of the power train is the same. The heat produces steam, which turns turbine blades, which run a generator, which produces electricity. The major improvement is the elimination of the combustion products of fossil fuels, which have contributed greatly to air pollution (see Chapter 8). The price paid for this boon was an increase of almost 50 percent in the already heavy demand for cooling water in the condensing system, and in the various radioactive wastes—liquid, solid, and gas—that were released. Use of nuclear fuel requires lower steam pressure and the less efficient use of steam results in more waste heat.

Also, 10 percent of the waste heat from fossil fuel plants goes up the smoke stack and is dissipated in the air, whereas in a nuclear plant all the waste heat must be transferred to the water coolant. Hence, nuclear plants produce 10,000 Btu per kilowatt-hour rather than the 6000 produced by fossil-fueled plants.

A typical reactor uses as its fuel source a mixture of U-235 and U-238 packed as uranium oxide pellets into stainless steel or zirconium alloy rods about one-half inch in diameter and several feet long (Figure 10.4). These are grouped together in subassemblies close enough to sustain a controlled chain reaction. Reactor fuel is not pure enough to form a critical mass, so there is no danger of a nuclear explosion. Altogether, a reactor core may contain hundreds of thousands of pounds of uranium oxide. The chain reaction is moderated by control rods of cadmium and the core is kept at a

The light water nuclear reactor

Fig. 10.4 • This bundle of uranium oxide fuel elements is being readied for installation in a nuclear reactor. (*General Electric*)

temperature of 551°C (1000°F) by a cooling system containing water. The light water reactor is so called because plain water, rather than heavy water (which is enriched with heavy isotopes of hydrogen), is used as the coolant. Up to this point all reactors are the same. The variations involve the way the primary coolant is used. In the boiling water reactor (BWR) the primary coolant water is heated to boiling in the reactor vessel and the steam is tapped to drive a turbine (Figure 10.5). With a pressurized water reactor (PWR), the

Fig. 10.5 • The fuel elements in a nuclear reactor are usually cooled by a primary coolant in a closed system. To cool the primary coolant a secondary coolant is necessary. It is the dumping of this secondary coolant into the environment that causes thermal loading problems. (*Modified from* Environment *11, 1969*)

213

primary coolant is prevented from boiling by sufficiently high temperature and pressure. The pressurized, heated water passes into a heat exchanger, which converts the secondary coolant water into steam that drives the turbine. Light water reactors (LWRs) are very inefficient sources of power: less than 1 percent of the potential energy in the fuel is utilized, and because of the lower temperature of the steam produced, the efficiency of the thermodynamic steam cycle is low also—32 percent compared with 38 percent for the large fossil-fueled plants. But LWRs were never intended to be more than an interim source of energy. The LWRs, which took advantage of the technology developed for the nuclear submarine, were scheduled to be replaced by breeder reactors. However, because of technical problems with both the LWRs and the experimental breeders, it looks as though we will be stuck with the LWRs for several more decades. If dry cooling towers are used (see Chapter 6), their 12 percent greater cost could be returned by the far greater flexibility allowed in siting. The present generation of water-cooled plants is much too closely tied to large sources of cooling water, necessitating construction on lakes, rivers, and the seashore, all potential prime recreation areas. Efficiency could be significantly increased by using a high-temperature gas-cooled type of reactor. These reactors, by using a gas such as helium to drive the turbine directly, could avoid both the inefficient steam cycle and the thermal loading problem.

Although fission power plants have been highly touted as pollution-free, at least in the lack of combustion products, the gas diffusion plants in Oak Ridge, Tennessee that concentrate and enrich the uranium necessary to fuel the nuclear plants produce effluents and consume huge quantities of electric power produced by the Tennessee Valley Authority's huge coal-burning plants. These not only pollute the air but indirectly wreak havoc on the landscape of Kentucky, the source of their strip-mined coal. Despite these problems, twenty-five LWR plants are in operation and 117 more are under construction or planned.

Breeder reactors differ from LWRs in several particulars. Most importantly, they produce a greater neutron flux, which means that there are neutrons left over from the chain reaction business to convert isotopes conveniently placed in the reactor into fissionable fuels. In this way cheap and more abundant fuels can be produced; for example, the isotope thorium-232 or uranium-238 is converted into the fissionable U-233 or Pu-239. In the fast breeder reactor the excess neutrons are not moderated or slowed but are allowed to react directly with the breeder materials. Consequently, fissionable fuel is produced at a faster rate and the ultimate cost of the power produced is lowered. The fast breeder will allow a doubling of the fissionable fuel in ten years.

The breeder design under most intensive development in the United States is the liquid-metal-cooled fast breeder reactor

Breeder reactors

(LMFBR), Figure 10.6. The liquid is sodium, which is highly reactive in air and prone to form bubbles that can block the tubing and affect the cooling of the core elements. If the core overheats the containment vessel may rupture and leak radiactivity. While loss of the sodium coolant by accident is less likely than loss of the water coolant in a LWR, coolant loss in a LMFBR is far more dangerous. Since the LMFBR produces a much greater flux of neutrons it operates much closer to the limits of an uncontrolled chain reaction than the LWR.

Fig. 10.6 • One of the first breeder reactors, the Enrico Fermi Power Plant in Michigan. Plagued with problems, the reactor was permanently shut down in 1973. (AEC)

By rushing the LMFBR the AEC has put all its eggs in one basket, for there are other promising breeder types, including the gas-cooled and thermal breeders. The LMFBR is being counted on for significant power generation by about 1990, but since it is as yet unproven in scaled-up commercial power plants, there could be a serious power shortage if the LMFBR design gives us as much trouble as did the LWR.

The advantages of the breeder reactor are the 50 to 80 percent utilization of fuel, the elimination of the power-hungry gas diffusion plants, the greater thermodynamic steam efficiency (40 percent), and the ability to produce fissionable fuels from abundant and relatively inexpensive isotopes of thorium. Unfortunately plutonium-239, the fissionable fuel that is bred, is exceedingly toxic and has a half-life of 24,400 years. Coolant loss, which might lead to escape of radioactivity, is another omnipresent danger. Fears have also been expressed that as breeder plants begin to produce large amounts of plutonium, worth $10,000 per kilogram, a black market may develop and fissionable materials may get into the hands of saboteurs or terrorists. Simple designs for atomic bombs have been published in Europe, and with a few kilos of plutonium such devices can be put together in a basement or garage.

NUCLEAR FUSION

With the development of the hydrogen bomb came the realization that if the fusion reaction that characterized the bomb could be controlled, mankind's need for energy could be very well satisfied for an indefinite period. The fusion process takes place when a suitable material, usually a heavy isotope of hydrogen, either deuterium or tritium, is heated to the point where the nuclei and electrons become dissociated from each other and form plasma. Dissociation of deuterium into plasma begins to take place at 5000°C; but if the plasma is then heated to the incredible level of 100 million degrees centigrade, the nuclei, which normally repel each other, become so energized that they slam into one another and fuse, releasing enormous quantities of energy. The trick is to keep a relatively low-density plasma at a temperature high enough so that more energy is harvested than is invested into the process.

Viewing the rapid evolution of a fission chain reaction in a graphite pile just three years after the theory was developed, it was at first thought that control of fusion might be equally rapid. But plasma technology was virtually unknown and has had to be painstakingly worked out before much progress could be made with containment problems. For if plasma must be heated even for a very short interval of time, to 100 million degrees centigrade to initiate fusion, then the container of that plasma must be radically different from those technology has produced in the past. Two approaches seem to hold promise. The first is a magnetic "bottle" that by means of its magnetic field suspends the fuel, obviating destructive contact with any container wall. Of the three types of magnetic field designs now being intensively worked on—the doughnut, pinch, and magnetic mirror—the doughnut or toroidal shape, particularly the Russian tokamak version, seems the most promising.

The second approach to a controlled-fusion reactor involves lasers. A laser is simply a machine that produces a pure stream of light concentrated in a single wavelength. Because of the lack of interference from other wavelengths, the beam is very intense and sharply focused. When a laser beam is focused on a small pellet of deuterium and tritium, the pellet heats up to the fusion point before it has a chance to explode. In other words, the inertia of the expanding pellet confines the fusion reaction. In another laser approach, the fuel pellet is hit from several sides, heating up the outer part of the pellet. This layer ionizes and blows off at such a high velocity that the rest of the pellet reaches the fusion point from the resultant high pressure that is directed inward. This means that the laser power necessary for fusion can be reduced by 1000 times, to only one million joules.[1] But the most powerful laser available so far can only generate 600 joules. More powerful lasers are under development. A major advantage of the laser approach to fusion control is that the scale can be small enough to power communities, ships, or factories, and there is a possibility of direct conversion of nuclear power into electricity. The magnetic field approaches, in contrast, are likely to be gargantuan and would produce power by pumping neutrons into a lithium shield that in turn would heat water to steam to drive that familiar turbine.

Whatever approach finally proves effective, laser or magnetic field, it seems likely that a sustained reaction might be possible before 1980, with commercial power available, after extensive development time, between 1990 and 2000. So while fusion power is certainly a longer shot than the other devices and techniques already described, it does have two major advantages over fission reactors: the fuel resources are virtually unlimited, and the danger of a non-nuclear explosive accident is remote, at least with the laser approach. Deuterium, one component of the fuel, is extremely abundant in seawater and could be extracted very cheaply. Tritium would be produced in a fast breeder fission reactor. As with fission reactors there is no release of combustion products into the atmosphere. The efficiency of a fusion reactor has been estimated to be about 60 percent.

There are of course environmental problems with fusion reactors—you can't get anything for nothing. Because of the nature of the fuel, the by-products of the reaction are very high-energy neutrons, which are quite damaging to the structural materials in the reactor, and tritium, the third isotope of hydrogen. Tritium is a volatile and highly reactive isotope with a half-life of twelve years. Although its radioactivity is confined to low-energy beta particles, because tritium is a light gas and can easily replace hydrogen, it must be regarded as a biological hazard. Tritium release might amount to sixty curies per day but could possibly he held to as little as one curie per day for a large reactor. This is a range comparable with the tritium release during the reprocessing of fuel rods of a fission reactor. Despite these problems, the fusion reactor, if and when it becomes available, offers far more promise than the light water fission reactors now is use and the fast breeder reactors under development, from the point of view both of the availability of fuel and of environmental pollution.

[1] One million joules is equivalent to about one-quarter of a kilowatt hour.

WATER POWER

Although water is the oldest means of generating energy, its proportional contribution has declined steadily over the years despite some rather spectacular installations that generate up to 1000 megawatts. The maximum possible water power capacity for the United States is 161,000 megawatts, of which only 28 percent has been realized. This figure is unlikely to increase substantially in the future, for while water power is nonpolluting, the reservoirs that provide the head of water to turn the turbines must be regarded as temporary, since many may well be filled in 100 years or less by sediments. Furthermore, there is growing resistance to the damming of virtually every free-flowing river for power dams (Figure 10.7), since rivers that are most attractive to the engineer looking for an exploitable fall of water are also the wild rivers flowing through canyons, with falls or extensive areas of white water, that are most attractive to the recreationist. As with other environmental manipulations there is a trade-off between environmental goods and economic goods; the problem is to be sure that we have an accurate balance sheet before a project is begun. Environmental impact statements required by the federal government for public projects and by some states for private projects certainly provide some more accurate means of evaluating environmental impact probabilities. But great care must be taken that these evaluations do not degenerate into a perfunctory check-off of phantom problems, or become regarded as just another bureaucratic hurdle whose purpose is to inhibit "progress."

Tidal Power

Hydroelectric power makes use of a unidirectional flow of water; tidal power, on the other hand, takes advantage of the two-way flow of the tide by channeling it into basins so constructed that the tidal amplitude is great. Although tidal power projects are locally attractive and may be capable of competing in output with other types of power installations, the world potential for tidal power is a piddling 13,000 megawatts, compared with even the 2,900,000-megawatt possibilities of hydroelectric power.

For many years the governments of the United States and Canada have toyed with the idea of a tidal project in Passamaquoddy Bay, off the Bay of Fundy, which is well known for its extreme tidal range. With an area of 262 square kilometers and an average tidal range of 5.52 meters, the bay is capable of producing 1800 megawatts of power. But the best design submitted could capture only 11 percent of this potential power, so the project was never funded. The dam would work by means of gates to be closed at dead low tide, then

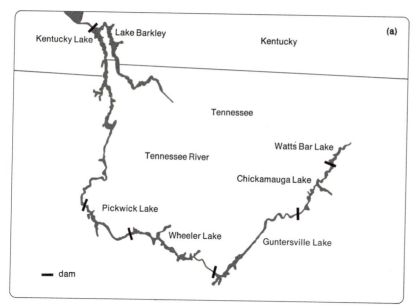

(a)

Kentucky Lake
Lake Barkley
Kentucky

Tennessee

Watts Bar Lake

Tennessee River

Chickamauga Lake

Pickwick Lake

Wheeler Lake

Guntersville Lake

— dam

(b)

Lake Sakakawea

North Dakota

Missouri River

Oahe Reservoir

South Dakota

Lake Sharp

Lake Francis Case

Lewis and Clark Lake

— dam

Fig. 10.7 • (a) The Tennessee River has been almost totally converted from a river into a series of pools. (b) These huge reservoirs have caused the once-free Missouri River to become a series of lakes in the Dakotas.

opened at full high tide. When the tidal basin had filled the gates would again be closed. On the next low tide the gates would be opened again. Since the basin would continue to flood and drain, the impact of the dam on the estuary would be less than with a uni-directional dam, which would create a permanent reservoir. But the transfer of nutrients important to both the productivity of the bay and the offshore water might be curtailed, and the delicate adjustments that intertidal organisms have evolved to tidal levels could be disrupted by sudden raising and lowering of the bay level.

Geothermal Power

In general, there is an increase of 0.5°C (1°F) for every 100-foot descent into the earth's crust, so that very deep mines are usually uncomfortably warm not only because of poor ventilation but from the earth's inherent heat. In most areas of the crust this temperature gradient is much too gradual to permit efficient utilization. However, as there are deposits of ores in the earth's crust so there are deposits of heat, which cause ground water to boil to the surface as hot springs and geysers (Figure 10.8), particularly in areas of geologically recent volcanic activity and at the boundaries of the major crustal plates. The possibility of utilizing this natural heat has long been recognized and electric power from geothermal energy is being produced in several countries. At present in the United States geothermal power is being produced only in California, although geothermal deposits exist in several other western states. The steam from drilled wells is filtered to remove abrasive particles and is then passed through special turbines designed to operate efficiently at the pressure and temperature of the steam, which is much lower than that produced by fossil-fueled plants. The exhaust steam is condensed into water, which is used in the cooling towers; but because of trace chemicals and salts about 20 percent of the condensed water is injected back into the ground (Figure 10.9). Although the lower temperature and pressure of geothermal steam makes the power plants one-third less efficient than fossil-fueled plants, the geothermal plants are cheaper to build and operate.

Despite their convenience and ability to compete economically with fossil fuel or nuclear power plants, geothermal steam plants are not without environmental problems. The salinity of the waste water is often six times that of sea water, and the waste water may contain toxic trace elements in addition. Not only is there a potential environmental problem, but the salts can precipitate out in pipes and valves, causing continuing maintenance problems. Large volumes of water would be removed to propel the turbines, but by injecting the saline or toxic wastes into the ground, subsidence might be avoided. This must be done with some care, however, to avoid the generation of earthquakes that have plagued some deep well injection projects

Fig. 10.8 • Geothermal energy is widely distributed over the western states, not all of it in national parks such as Yellowstone.

high tension
wire towers

steam well

heat
exchanger

turbine

generator

condenser

cooling tower

gas
separator

steam

other
gases

water

reinjection well

to chemical plants

Fig. 10.9 • This schematic diagram shows how geothermal steam could be utilized as an energy source. When chemical plants are not available to make use of the impurities in the steam, expensive pollution control measures must be taken to avoid air and water pollution. (*This diagram is adapted from an illustration which originally appeared in "Power from the earth," David Fenner and Joseph Klarmann, Environment 13(10), p. 19. Copyright © 1971, Scientists' Institute for Public Information*)

(see Chapter 6). Since the hydrogen sulfide associated with most geothermal sources is quite soluble in water, air pollution is still another problem. Sulfur oxide production in some proposed geothermal plants has been estimated to be equal or greater than that produced by a fossil-fueled plant of equivalent size.

SOLAR ENERGY

Efforts to tap solar energy have been made, fitfully, from time to time, but the attempts have been hopelessly inefficient or impractical. The real problem was probably lack of economic incentive to pursue yet another energy source when others were more available. But the galloping demands for power are about to bring us to a rational exploitation of even this last frontier.

While the total radiation that the earth receives is mind boggling, sunlight is notoriously intermittent in all but a few fortunate spots, and as air pollution continues to increase, it is becoming more and more diffuse everywhere. Nonetheless, solar energy will surely come to play a much greater role in the total energy picture over the long run. This is likely to happen in two phases: first, the development of solar-powered systems for home heating and cooling, and later, the use of solar heat to produce electricity directly by photovoltaic cells or other means.

Solar heating and cooling systems

Most solar heating systems devolve to an application of the greenhouse effect: a black metal surface, shielded with glass to reduce heat loss, absorbs sunlight and transfers the heat to water, part of which is circulated during the day and part stored for release during cloudy weather or at night. Conversely, an air conditioner that uses an ammonia-water system has been devised at the University of Florida. Solar energy drives the ammonia from the water and the ammonia is collected and condensed; when cooling is desired the liquid ammonia, by evaporating and expanding, cools the system that contains it and, through convection, the air. The ammonia vapor dissolves back into the water to repeat the cycle. If such systems could be perfected, they might very effectively reduce the peak loading factor that consumes 20 percent of the electric energy produced during the summer season. Because of rigid building codes and the initially higher cost of such systems, public acceptance of modular solar heating and cooling devices is likely to be slow.

Solar power plants

Sunlight, if focused with a Fresnel lens onto a pipe, can be concentrated up to ten times its normal intensity. By covering the pipe with special coatings that reradiate only 5 to 10 percent of the energy they absorb, and by placing the pipe in a vacuum chamber, heat can be conducted, either directly by conduction or indirectly by gas pumped through the pipe, to a central heat-storage medium of molten salts. Connecting a steam boiler and turbines to the heat-storage unit would allow continuous generation of power. A 1000-megawatt power output would require 13.5 square miles of collecting surface.

A third possibility would catch the sunlight falling on a square-mile surface and reflect it to a solar furnace on a 1500-foot tower placed in the center of the collecting array. The solar energy could heat the water to 1700°C and use this hot water to produce power by an MHD generator. Nine square miles of collecting surface would generate about 1000 megawatts. Power coud be stored by using some of the electricity generated to split water into oxygen and hydrogen. By recombining these elements in a fuel cell, or by burning hydrogen directly as a fuel, electricity could be generated at night or during cloudy weather.

While economic as well as technological problems are large, the rising costs of traditional fuels and decreased costs of mass-produced solar power components could make such systems competitive. Most of the traditional environmental problems are avoided by solar power techniques, but of course subtle new ones are introduced. Earthbound solar power plants only utilize energy that has already entered the earth's atmosphere. Even though efficient energy absorbers collect more energy than would a comparable area of the earth's highly reflective surface, this loading factor is probably not critical. But a large array of solar energy receivers deployed in space could conceivably increase the earth's heat load, because heat that might otherwise be reflected from the earth's atmosphere or surface is being transported to earth in the form of electricity, which when consumed ends up as heat. As it is, by the year 2000 man-generated heat may begin to approach sun-generated heat in the United States.

Perhaps the ideal answer to conversion of sunlight into electricity lies in photovoltaic cells (PVC). Developed to provide power for space vehicles and satellites, photovoltaic cells enable the efficiency-lowering thermodynamic cycle of steam and turbine to be bypassed. But the cost of PVCs is way too high, efficiency is low, their lifetimes are too short to be amortized over a period of years as is usually done, and they cannot store energy.

In principle, PVCs absorb solar photons and generate positive and negative charges that diffuse across the cell and recombine to develop about a half a volt per cell. To obtain high voltages, large numbers of cells must be connected in series. The direct current output would then be converted to alternating current for consumer use. The three materials most used in PVCs are silicon, cadmium sulfide, and gallium arsenate. Silicon cells must be made out of single silicon crystals, which are expensive to make: the cost of very high-grade silicon necessary for crystals free of impurities is $60,000 per ton. Cadmium sulfide cells can be made from very thin films instead of crystals, but the cell efficiency is only 6 percent and they break down rapidly in sunlight. Gallium arsenite cells are highly efficient (18 percent), but gallium is too rare a metal to allow the production of huge panels with millions of cells.

THE CONSERVATION OF ENERGY

With a rising demand for energy (Figure 10.10) and a stable or decreasing supply of at least some conventional fuels, it has finally become necessary to begin thinking about stopping the lavish wasting of power that, as much as anything else, characterizes our culture. Without involving the threat of a change in life style that seems to traumatize our society so, it should be possible to reduce the demand for power. This will require some amount of foresight and planning, to be sure, but if unduly postponed, obligatory changes in our way of life might become necessary, a prospect most of us would find difficult and unpleasant. What, then, are the possibilities of conserving energy?

Collectively, industry uses about 42 percent of the nation's energy supply; the principal users are the manufacturers of primary metals, petrochemicals, food, paper, glass, and concrete. About 17 percent of the energy is used to produce steam for various processes, 11 percent produces direct heat, and 8 percent is used to drive

Industry

machinery; nearly 10 percent of this energy is supplied by gas, 26 percent by coal, 17 percent by oil, and only 11 percent by electricity. There has already been a significant rise in the efficiency of most industrial processes that is reflected in reduced consumption of energy. Between 1960 and 1966 the steel industry managed a 13 percent decrease in the energy necessary to produce a ton of steel. This wasn't done to decrease energy consumption per se, of course, but to cut manufacturing costs to meet the competition of cheap steel from Germany and Japan. But as energy costs begin to climb more sharply, further cuts, some likely to be rather sizable, are expected.

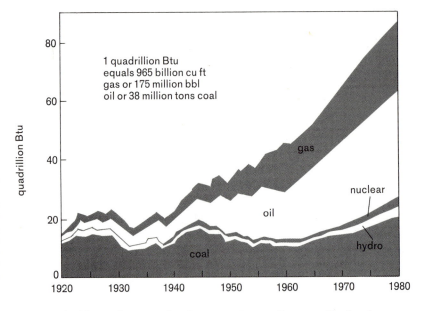

Fig. 10.10 • Not only is consumption growing, but also the spectrum of energy sources is slowly changing over the years. (*Reprinted from* Env. Sci. & Tech. *5(1), p. 30. Copyright 1971 by the American Chemical Society. Reprinted by permission of the copyright owner.*)

A significant increase in the rate of recycling would also have an important effect on industrial consumption of energy as well as the solid waste problem (see Chapter 21). Since finished products have already been processed their recycling often takes less than one-fifth the energy required to manufacture them from virgin raw materials to begin with. Of course there are exceptions, such as cans.

If the various transportation modes are examined as energy consumers, the most efficient—trains, barges, mass transport systems—are being replaced by the less efficient trucks, cars, and planes (Table 10.1). As this shift has taken place the energy consumption of transportation has increased significantly. Of the 3 billion barrels of oil required for transportation in 1970, 55 percent was used by cars, 21 percent by trucks, and 8 percent by planes. The remaining 16 percent was split among trains, buses, barges, pipelines, and so forth. Since the average American car gets only 12 miles to the gallon, half what the average European car gets, the demand the automobile makes on petroleum supplies seems likely to

Transportation

Table 10.1

Energy and Price Data for Intercity Freight Transport, 1970*

MODE	ENERGY (BTU/TON-MILE)	PRICE (CENTS/TON-MILE)
Pipeline	450	0.27
Railroad	670	1.4
Waterway	680	0.30
Truck	2,800	7.5
Airplane	42,000	21.9

* Data taken from Hirst, Eric, 1973. *Energy Intensiveness of Passenger and Freight Transport Modes, 1950–1970.* Oak Ridge National Laboratory, Oak Ridge, Tennessee.

increase, at least over the short run. Removal of lead from gasoline would also increase the use of crude oil by a million barrels a day by 1975.

While the car-buying public is perhaps ultimately responsible for the excessive consumption of fuel by cars, since it seems to prefer dinosaurs to lizards, the federal and state governments have consistently encouraged the use of cars, trucks, and planes by direct and indirect subsidies of expanded road and airport construction, while at the same time allowing outmoded rules and controls to stifle railroads, waterways, and mass transport. As a result, railroad stations have become the dismal symbols of urban decay and neglect. Reversal of governmental policy is always unlikely. But if billions of dollars were poured into mass transit systems, enabling them to compete with the automobile, and if railroads and waterways were rejuvenated so they could compete with the ubiquitous truck, the consumption of crude oil could be reduced significantly over a ten- to fifteen-year period (see Chapter 20). The creation and subsidy by the federal government of Amtrak, the agency which now operates rail passenger service, is a tentative step in the right direction. Reducing the weight and horsepower of cars and paying greater attention to fuel efficiency in the development of engines to replace the notoriously wasteful internal combustion engine will help also. But the greatest impetus toward fuel economy by automobiles will come from the increased prices for gasoline. These high prices will be with us perhaps only for a short period, perhaps for good.

Just as automobiles are rarely designed to run efficiently on low-grade fuel or to burn that fuel cleanly, so are buildings seldom designed to conserve energy. The major design requirement, at least for commercial buildings, is the maximum square footage of space for the minimum price. While the construction of these buildings and the manufacture of the materials requires only about 8 percent of the electricity produced in the United States, maintenance of the

Buildings

buildings already constructed takes a whopping 50 percent of the available power. This drain could be reduced by at least 25 percent if greater attention were given to energy conservation during the production of materials, assembly, maintenance and operation, and finally demolition. All these aspects are related to the design of the building, which in turn is the responsibility of the architect. But the architect must satisfy his client, who supposedly pleases his tenant, and so the buck passes back to the consumer.

Selection of materials is largely influenced by the practical experience of what best provides the greatest margin of safety in the completed structure. Rather than making a scientific analysis of how much should be used of what materials, builders have calculated safety margins rather generously. The generosity of the calculations has been climbing as the overall quality of construction declines. Masses of various materials, especially concrete, have commonly replaced careful design and construction. Prefabricated buildings, which require less material and take less time and energy to be put up or taken down, are one solution—but their full acceptance would necessitate revising building codes that favor buildings constructed in the field. Since concrete requires much energy to produce, large reductions in the use of concrete could result in saving almost 20 billion kilowatt hours per year, enough electricity for 3 million families.

Synthetics, plastics, and aluminum as building materials require far more energy to produce than steel and wood. To produce the aluminum skin for a skyscraper, for example, requires almost three times the energy needed to produce and use stainless steel. Once such a building is constructed there are continuing and heavy demands for power to light, heat, and cool the structure. In the last several years the minimum lighting level recommended by the Illuminating Engineering Society has been exceeded in most new construction, often for a purely cosmetic effect. Hallways and corridors are often grossly overlit. Although the fluorescent lamps that are used almost exclusively consume only one-fourth as much power as the incandescent lamps still used in most homes, the excessive lighting of most new commercial buildings more than makes up for the economy. Again, cheap electricity appears to be the cause; suburban banks and shopping centers blaze all night long, wasting large amounts of power for no purpose except security or perhaps advertisement. Rising power rates will probably be effective in reducing this abuse.

Heating is another vast source of excessive power consumption. Underinsulation and a trend toward electric heating have both contributed to this problem. Doubling the insulation presently considered adequate would in many instances halve the energy requirements of heating and cooling, and pay for itself over the lifetime of the building. While electric heating is a rather efficient way to heat a given unit of space, it is very inefficient to produce and transmit. The increasing popularity of electric heat is due in part to the cost-conscious builder who finds it cheaper to install than a furnace, and in part to building managers who find that there is much more profit in the rental business if the tenant foots the heating cost as part of his electric bill.

Insulation is an important aspect of energy conservation in any structure, but it is especially so in suburban houses that, since they are free-standing, have a greater surface area over which heat may be lost or absorbed. This heat load can be effectively countered by air conditioning, but the energy cost is high, higher than it need be. The energy cost can be reduced if adequate insulation is emplaced during construction or remodeling, if the roofing material is highly reflective, and if attic fans are installed to prevent heat buildup beneath the roof. When air conditioners are used some attention must be paid to their efficiency, which can vary from 4.7 to 12.2 Btu per watt hour (the more Btus per watt hour the higher the efficiency). Customarily ignored by both manufacturers and consumers, inefficient air conditioners have inflated electric bills needlessly during the cooling season. Today, however, the more enlightened manufacturers list, and wise consumers request, the cooling capacity of their air conditioner in terms of Btus per watt hour.

Other small but potentially significant energy drains involve pilot lights on appliances, which can be replaced by electronic ignition systems; frost-free refrigerators, which use twice the energy of manual-defrost units; fireplace flues left open when the fireplace isn't being used; and incandescent lamps, which consume four times the energy of fluorescent lamps.

Life style of course can play a surprisingly important part in energy conservation. It has been estimated that if everyone set his thermostat 1°C(2°F) higher in the summer and 1°C lower in the winter, 219 million barrels of oil could be saved per year. But this would be possible only in well-insulated houses. Those that are poorly insulated allow less flexibility in controlling energy loss. Mobile homes, which are increasing in popularity as permanent housing (one of every four new houses is a mobile home), are prefabricated, so there is an energy saving in their construction. And because of their small size, heating and cooling should be more efficient. But poor insulation makes many of them greater consumers of heat-control energy than they need be.

Perhaps the most logical way around the inefficiency of heating, cooling, lighting, and steam generation is to integrate these services into a total energy system. A centrally located, efficiently run power plant could supply heat, air conditioning, steam, hot water, and electricity to a compact group of buildings: a shopping center-apartment house complex, for example, or a subdivision with homes, schools, churches, shopping center, and recreation center. A pilot project exploring the possibility of such a community system is being evaluated by the National Bureau of Standards in Maryland.

ECONOMICS

Although economics is closely related to all environmental problems, the clearest link is with the energy crisis. This can be seen in the current shortage of natural gas and the impending shortage of electric power and gasoline. Many people have doubtless wondered why a country with huge reserves of coal, oil, and gas, together with the technological base to develop more exotic alternatives at will, could possibly be faced with an energy shortage. Yes, fossil fuels are a nonrenewable energy source compared with fusion or solar power, but surely a government concerned with our welfare has an energy policy that would plan the rate of development and exploration so that, as one energy source is depleted, others would be available to supply whatever energy is required. Unfortunately such a national policy does not exist, and oil companies are allowed to increase profits through the depletion allowance while the natural gas industry has its profits reduced by being forced to maintain unrealistically low prices. The impact of such policies has been to overproduce domestic oil and underproduce gas. The Federal Power Commission, which is charged with regulating the power industry, has no independent survey capability, depending instead on figures obligingly supplied by the industry it is supposed to regulate. Small wonder that many feel that the natural gas shortage has been "arranged" by an industry sitting on an ample, if not abundant supply, whose value is not reflected in the present rate schedule. According to one scenario, when the pressure is sufficiently high the FPC will be forced into increasing natural gas prices and the industry will "discover" they had more gas reserves than they supposed and the short-term shortage, at least, will be over.

A similar economic explanation has been suggested for the current gasoline shortage. The large consortia that collectively exploit the oil resources of the Middle East are seeing their profit margin steadily whittled away by the increasingly aggressive producer countries, eager for control of their resources as well as increased profits. Indeed, the Organization of Petroleum Exporting Countries plans to demand not only increased prices for Middle Eastern oil when contracts are renewed in 1976, but also the options of buying into the oil companies with the ultimate goal of 51 percent ownership. Faced with the specter of lower profits and potential loss of control, the oil companies would understandably like to minimize their dependence on Middle Eastern oil by rapid development of the continental shelf deposits of Alaska and the Atlantic coast. But environmentalists concerned over the Santa Barbara oil spill have forced a government slowdown in the leasing of these offshore areas and the general uproar over construction of the Alaska pipeline has also delayed this plan. A gasoline shortage could serve as a very effective lever: if it inconveniences the public sufficiently, it could pressure the government into overruling the environmentally concerned "obstructionists," allowing a more rapid development of oil deposits with less stringent environmental safeguards.

The upshot of such a situation would be to isolate people concerned with ecological as well as economic values and impose an artificial either/or solution to the energy problem: either the power industry be allowed to exploit oil or gas deposits wherever and whenever it pleases, and construct refineries and deepwater ports in the most convenient locations, or the country will suffer from shortages of electric power, natural gas, and gasoline. The tactic is widely used: for years the Corps of Engineers said, either dams or floods; the biocide industry, biocides or starvation. But there are *always* alternatives, and they can be economically sound, if not universally popular. Perhaps the quickest way to solve the energy crisis would be to increase substantially the price of fuel of all kinds. Waste of energy is so great in the United States because the cost is so low compared with other goods and services.

Less than 5 percent of the typical American family budget is spent on energy; it represents only 1.5 percent of the total value of manufacturing goods. Indeed, until 1971 the cost of electricity declined in relation to the cost of virtually all other materials. Rate structures that favor the large user (and waster) of power with lower rates have supplied no incentive for more economical use of energy. The very structure of the electric utility is based on a rather strange kind of economics. Capital costs of power plant construction are exceptionally high, so the larger the plant, the greater the profit from ultimate sale of power. This results from a curious tax ruling that bases permissible profit on the percentage of capital investment. So it is often more profitable for a utility to build a new power plant than to purchase power from another company. The expansion is justified by promotion of the use of electricity. One way around this would be to create a national power grid, which could facilitate the distribution of power from power-rich to power-poor regions, and might reduce the need for new generating capacity by 20 percent. Naturally such a plan is strongly opposed by the electric power industry, because it would threaten its death hold on local power supply. Electric power has always been a sellers' market; this explains the enormous rate differential around the country. Allowing utilities to deduct the cost of advertising promotion of their product from taxes rather than profits has further inflated the growth of power demand, to the point where many urban power companies are very close to being unable to meet the demand they have stimulated. While no one would suggest that a company discourage the public from using its product, it is unconscionable to stimulate the use of electrical gadgets merely to fill the valleys between the daily or seasonal peaks of energy consumption.

As the cost of power and fuel increases, people are likely to take greater interest in their rate of energy use. We could achieve perhaps as much as a 25 percent reduction in our total energy consumption by simply turning off unused lights and air conditioners, loading dishwashers and washing machines fully, or taking public transportation instead of using cars. The eased demand for power would allow time for more careful exploitation of fossil fuel deposits; for siting refineries, pipelines, and deepwater terminals as they become necessary; and for the development of new technologies that may have a lesser environmental impact than those in present use. The government must play a prime role as well as the consumer, for a major factor in the "energy crisis" has been lack of a coherent long-term plan for energy resource development. Only the government is in the position to chart such a course by its control over power and fuel rates, review of environmental impact statements, leasing of offshore oil and gas deposits, its role in supporting transportation modes, and support of research and development of energy sources that are more efficient, less environmentally damaging, and indefinitely renewable.

There is no reason whatever for the present "shortages" to be continued into the indefinite future when so much could be done by a concerned government with the active support of the people to increase supply as well as reduce demand.

FURTHER READING

Fenner, D. and J. Klarman, 1971. "Power from the earth." *Env.* **13**(10), pp. 19–26, 31–34. Discussion of the possibilities of geothermal power.

Garvey, G., 1972. *Energy, ecology, economy.* W. W. Norton, New York. An interesting perspective on the economics of the energy problem.

Hammond, A., W. Metz, and T. Maugh, II, 1973. *Energy and the future.* AAAS, Washington, D.C. An excellent overview of the current and future sources of energy.

Hubbert, M. K., 1969. "Energy resources," from *Resources and man.* Committee on Resources and Man, NAS/NRS. W. H. Freeman, San Francisco, pp. 157–242. While the data are a bit old, the view is broad and authoritative.

Hirst, E. and J. C. Moyers, 1973. "Efficiency of energy use in the United States." *Science* **179**, pp. 1299–1304. Examines the sources of energy waste.

Lincoln, G. A., 1973. "Energy conservation." *Science* **180**, pp. 155–162.

Mills, G. A., 1971. "Gas from coal, fuel of the future." *Env. Sci. & Tech.* **51**(12), pp. 1178–1183. The details of coal gasification.

NSF/NASA Solar Energy Panel, 1972. *Solar energy as a national energy resource.* Dept. Mech. Eng., University of Maryland, College Park. A thorough overview.

Rocks, L. and R. P. Runyon, 1972. *The energy crisis.* Crown, New York. A hard-hitting popular account of the "energy crisis" and its background.

Soucie, G., 1972. "Oil shale—Pandora's new box." *Audubon* **74**(1), pp. 106–112. Good discussion of the costs and benefits of oil shale exploitation.

Squires, A. M., 1970. "Clean power from coal." *Science* **169,** pp. 821–828. Clear discussion of the fluidized-bed technique of coal combustion.

Stein, R. G., 1972. "A matter of design," *Env.* **14**(8), pp. 16–20, 25–29. The prospects of designing buildings to use energy more efficiently.

11

INORGANIC POLLUTANTS

BECAUSE OUR BODIES are made of organic matter, we are especially conscious of the effects of organic chemicals on the complex biochemistry of life. But inorganic chemicals often escape the attention they deserve; it is usually long after their use has been initiated that their danger to man's health is recognized.

Of the many inorganic substances that interact negatively with physiological systems, mercury, lead, cadmium, beryllium, and asbestos are potentially the most dangerous because they are stored in the body and have a cumulative effect.

MERCURY

The Hatter was the first to break the silence. "What day of the month is it?" he said, turning to Alice: he had taken his watch out of his pocket, and was looking at it uneasily, shaking it every now and then, and holding it to his ear.

Alice considered a little, and then said "The fourth."

"Two days wrong!" sighed the Hatter. "I told you butter wouldn't suit the works!" he added, looking angrily at the March Hare.

"It was the *best* butter," the March Hare meekly replied.

"Yes, but some crumbs must have got in as well," the Hatter grumbled "you shouldn't have put it in with the breadknife."[1]

In Lewis Carroll's day, hatters *were* often mad, a condition that was noticeable enough to be enshrined in a popular expression, but it was not until recently that the madness was recognized as an occupational hazard of hatters or people who worked with animal skins. At one time fur pelts were treated with mercuric chloride in processing and subsequently anyone who worked with the skins ran the risk of mercury poisoning, especially hatters who, in fashioning men's hats from beaver skins, were constantly exposed to mercury fumes. To-

[1] Carroll, L., 1939. *The complete works of Lewis Carroll.* The Nonesuch Press, London.

232

day, mercury compounds are no longer used in processing furs, nor are beaver hats wildly popular, so the phrase "mad as a hatter" has passed from common speech and even understanding. Mercury poisoning, however, remains a hazard.

Because of its initially vague symptoms—fatigue, headache, and irritability—mercury poisoning often goes unrecognized at first. At a later stage, arm and leg numbness develops, followed by disruption of balance, blurring of vision, deterioration of muscular coordination, emotional disturbances, and finally wasting away of muscles.

Mercury in naturally occurring stable compounds causes no problems, but its liberation in the form of soluble salts from a group of mercury compounds with industrial or agricultural uses is beginning to cause some concern.

Inorganic Mercury Compounds

Inorganic mercury compounds are used by over eighty industries including plastics, industrial chlorine, and electronics. Until recently most cases of mercury poisoning were derived from occupational exposure: accumulation of inorganic mercury in the kidneys affected readsorption and secretion of sugar, protein, and salts; and accumulation in the brain caused a loss of coordination. But in 1953 a number of people living in the vicinity of Minamata Bay, Japan, became ill with a strange series of disorders. Ultimately over a hundred cases were reported, many ending either in death or permanent disability. Investigations finally linked the disorders with seafood taken from Minamata Bay, giving the disease the name, Minamata disease. One case history gives some idea of the nature of Minamata disease:

A 14-year-old boy who had been agile and bright before his illness is said to have eaten a large number of crabs and small fish from a posted area of Minamata Bay during a ten-day period in July, 1958. A few weeks later he noted numbness around his mouth and in his hands and feet. He did not have fever, headache, or a stiff neck. He . . . became clumsy in buttoning his clothes and handling his chopsticks; his family observed that he staggered slightly when walking. His auditory acuity and attention span diminished, and he developed the mannerisms and behavior of a younger child. . . . His memory for most recent events was adequate, but he was unable to perform calculations beyond the eight- or nine-year old level. When observed in 1960, he was still hospitalized in the Minamata City Hospital. . . . His physical disability was considered mild, but his inability to calculate and remember complex Japanese written characters made it impossible for him to continue in school.[2]

Of fifty-two cases originally studied, seventeen died, and twenty-three were permanently disabled.

Other circumstantial evidence linking the disease to the eating of

[2] Kurland, L. T. *et al.*, 1960. "Minamata disease." *World Neurology, 1,* pp. 370–395.

seafood from Minamata Bay was the death of large numbers of cats belonging to afflicted families. Judging from the nature of the symptoms, the illness was a form of mercury poisoning. In Minamata City, at the head of the bay, a chemical plant had greatly increased its production of vinyl chloride; simultaneously the number of cases of Minamata disease increased.

Mercuric chloride is used as a catalyst in the production of vinyl chloride. Considering that sixty grams of mercuric chloride are "lost" per ton of vinyl chloride manufactured, it has been calculated that in one year nearly 2000 kilograms of waste mercury were generated: 1000 kilograms were washed from the vinyl chloride produced and another 1000 kilograms of spent catalyst were discarded. Although the source of the mercury was inorganic and insoluble, when it settled in the bottom mud of aquatic systems, it was converted into water-soluble methylmercury, which accumulates in organisms. Although the Minamata plant had settling basins and improved waste treatment facilities, enough inorganic mercury was transformed into organic forms to enter the bay's food chain and poison the bottom mud, shellfish, fish, and finally people. The organic mercury content of mud from the bay had a range of 12 to 2010 ppm, shellfish had 38 to 102 ppm, and people who died from Minamata disease, 13 to 144 ppm in kidney tissues.

These instances of the conversion of inorganic mercury to organic mercury, which is poisonous, are dramatic but they are geographically limited to the vicinity of plants that use inorganic mercury compounds in their processing. Of still greater concern to the general public is the entry of mercury into the environment directly through organic mercury compounds.

Organic Mercury Compounds

There are two types of organic mercury: aryl salts of mercury (those with carbon rings attached), which break down into inorganic mercury in the body, and the alkyl (straight chain hydrocarbon) salts of mercury, particularly methylmercury, which are able to diffuse easily through membranes and spread throughout the body. Methylmercury also generates a somewhat different symptom picture than inorganic compounds of mercury.

Mercury in the Environment

In 1970 a number of reports of mercury accumulation above 0.5 ppm in marine fish, especially swordfish and tuna, caused much concern, followed by much confusion when 90-year-old and 2000-year-old oceanic fish tissue was found with mercury levels as high or even higher. The results can be interpreted in several ways: that since there has always been mercuric contamination of fish we needn't worry about the small additional amount man adds through his activities (less than 1 percent of the present mercury concentration in the sea); or that because oceanic fish already have naturally high levels of mercury in their tissues and the difference between tolerably high natural levels and harmful levels is very small, we can ill afford to increase that body burden by carelessly adding more mercury to aquatic systems (Figure 11.1).

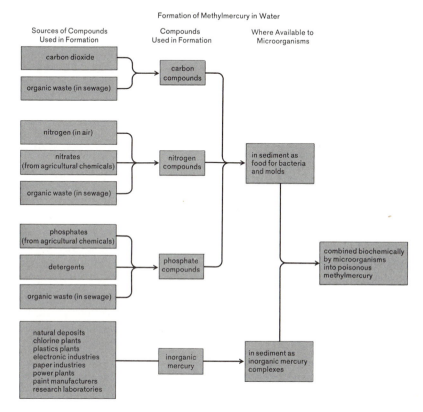

Formation of Methylmercury in Water

Fig. 11.1 • The occurrence of the biologically active methylmercury in water depends on many factors besides the manufacture and release of methylmercury itself. (*Adapted from an illustration that appeared in "A progress report on mercury," John M. Wood, Environment 14(1), pp. 36–37. Copyright © 1972 Scientists' Institute for Public Information*)

The most common pathway for entrance of mercury into ecological systems is in the widespread use of fungicides, for mercury compounds are potent killers of fungi. Organic compounds of mercury, especially methylmercury, have been widely used by pulp mills to keep fungi, bacteria, and algae, which thrive on wood pulp, from clogging up the machinery. By 1970 methylmercury released into the environment from this source had accumulated in freshwater fish to such an extent (more than 0.5 ppm) that people were advised not to eat fish caught from the St. Lawrence, Oswego, and Niagara Rivers as well as Lakes Erie, Ontario, Champlain, and Onondaga. In response to this problem most pulp mills have either eliminated or drastically curtailed use of fungicides containing mercury. Another use of mercury-containing fungicides has been the treatment of seeds to prevent or inhibit the growth of molds during storage and after the planting of seed in the spring. In 1970 the tragic poisoning of an entire family who ate pork fed on methylmercury-treated grains served as an illustration of the danger of such a practice.

Swedish farmers, until recently, regularly dusted up to 80 percent of all grains sown with a fungicide containing methylmercury. The poisoning of birds eating such treated seed led to investigations that showed that eggs produced in Sweden contained four times the mercury of eggs from other European countries, 0.026 ppm contrasted with 0.007 ppm. Although these levels are not considered dangerous, the World Health Organization (WHO) warns against more than 0.05 ppm. When hens were fed treated seeds, the methylmercury content of their eggs rose to 4 ppm. The question then arose whether grain ultimately developing from treated seeds would retain any of the methylmercury and pass it along in a food chain.

Experiments in Sweden indicated that seeds of plants grown from methylmercury-treated seeds contained 0.030 ppm mercury, while the plants from untreated seeds produced seeds with only 0.014 ppm mercury. The eggs from hens fed seed from treated seed contained 0.29 ppm mercury; the eggs from hens fed seed from untreated seed contained 0.010 ppm mercury. As a result of such data, changes in the formulation of the fungicide and in the rate of its application were recommended to Swedish farmers. A less toxic form of organic mercury, methoxyethylmercury, replaced the highly toxic methylmercury, and only seeds containing mold were treated before planting, reducing the percentage of treated seeds from 80 percent to 12 percent. It was found that the yield of grain was not reduced by these practices, and the mercury level in eggs decreased from 0.029 ppm in March 1964 to 0.010 ppm in September 1967.

In the United States, methylmercury fungicides are widely used in agriculture, especially in the wheat-growing areas. Although preliminary testing by the United States Department of Agriculture has not corroborated the Swedish finding that there was an appreciable amount of mercury found in seeds of plants grown from treated seeds, further testing might well demonstrate this.

Treated seed is routinely eaten by game; as a result in 1969, pheasants in Montana contained between 0.05 ppm and 0.45 ppm of mercury. In California that same year, pheasants with a mercury level of 1.4 to 4.7 ppm were found, and a general level of at least 1 ppm in the pheasant population of Alberta, Canada caused the pheasant-hunting season to be canceled in that province.

Although the Food and Drug Administration allowed *no* mercury residue in or on foods until 1970 (in compliance with the much-debated zero tolerance rule or Delaney Amendment), currently the fungicide manufacturer has only to demonstrate that residues are harmless.

At present levels of fungicide use, with or without zero tolerance, the question is not *whether* foods contain mercury (apples often contain 0.1 ppm, tomatoes 0.1 ppm, potatoes 0.05 ppm, eggs and meat 0.1 ppm), but whether these levels are harmful or may be a cumulative threat in the future.

It should be noted that current tests analyze for total elemental mercury, not the more biologically reactive methylmercury. Also, two different types of analysis are commonly used. The flameless atomic absorption spectrometer method is cheap but less accurate than the neutron activation analysis method, which though quite accurate is expensive. The mercury levels reported are therefore not always accurate, nor do the findings of different research teams using the different methods always agree.

Most European countries have varying tolerance standards for mercury in foodstuffs. Despite the formal zero tolerance level for mercury residues in the United States, few crops are presently being routinely monitored for methylmercury residues by the United States Department of Agriculture (USDA). It might be prudent, in view of uncertainty about the toxicity of the methylmercury in food, to substitute less toxic forms of mercury in widely used fungicides. This would probably reduce, in grain directly and in the food chain indirectly, the mercury residue resulting from the accidental ingestion of treated grain by livestock or birds. The time thus bought might permit the development of fungicides of more selective toxicity, affecting just the target fungus and not all other life as well.

While methylmercury from fungicides has had some spectacular effects on seed-eating birds, the amounts used are too small to explain widespread contamination in other animal species. Analysis of the mercury content of a number of coal sources suggests that at least 3000 tons of mercury are released per year by the burning of coal all over the world, an amount close to that released by industrial processes. However, containment of mercury from the combustion process would be far more difficult than controlling industrial mercury.

LEAD

Between 1954 and 1964, 128 children died of lead poisoning in the United States, but the lead source was neither air pollution nor careless use or storage of a biocide. Until 1958 most interior paints contained lead pigments; since that time lead has been replaced by titanium. However, some old houses were painted countless times with lead paint; as a result, thick chips of leaded paint fall off as the walls and ceilings peel (Figure 11.2). Putty also contains lead and is even more likely to be found in substandard dwellings. At least 20 percent of all children between the ages of one and five, regardless of social or economic status, ingest nonfood particles of all types, partly out of curiosity and partly because of a behavior syndrome called pica. Although the reason for the syndrome is not completely clear, it seems

Fig. 11.2 • As old paint flakes off and falls to the floor the chips are picked up and eaten by children who, as they crawl about casually, have been known to eat or attempt to eat virtually anything they encounter.

to be related to a deficiency of iron in the body. Pica is not limited to children; some adults, particularly pregnant women, eat clay or cornstarch in excessive amounts. One out of fifteen children in ghetto areas suffers from lead poisoning resulting from the ingestion of lead in old paint or putty. Since one chip the size of a fingernail can contain 50 to 100 mg of lead, a few chips could easily amount to 100 times the total adult intake per day.

Lead based paint need not be ingested however, to cause problems. In the past few years several cases of lead poisoning have been reported in houses undergoing remodeling. The sanding and burning of lead based paint places fine dust and fumes in the air which are incorporated into the body when inhaled. Children with no previous history of eating paint chips have been hospitilized, as well as their parents, from the Capitol Hill area of Washington, D.C., a neighborhood undergoing renewal. If you are planning to remodel an old house or apartment by removing prewar paint, use a face mask and arrange for alternative living space until the dust settles and can be vacuumed up.

Although acute lead poisoning has easily recognizable symptoms, chronic or asymptomatic poisoning is either without symptoms or has vague symptoms that could be ascribed to any of a dozen causes. Decreased appetite, clumsiness, irritability, headache, vomiting, and fatigue are the usual nonspecific symptoms of chronic lead poisoning. Consequently, the condition may never be diagnosed at all, or it may be recognized only after irreversible damage has been done.

Lead ions are absorbed from paint chips in the stomach and distributed, initially, throughout the soft tissues of the body. While in this soluble ion form, lead is both mobile and toxic. Damage, of course, depends upon the concentration of lead in the body. Because the soluble lead in the soft tissues is in equilibrium with the blood, blood analysis usually gives a fair indication of the amount of soluble lead in the body at any given time. Most of this soluble lead is excreted in the urine, which also provides an indication of the level of body lead. Some lead is deposited as insoluble salts in bone tissue. If it remained there, treatment for lead poisoning would be simple, but under certain conditions, in children especially, the bone lead can be returned in a soluble form to the bloodstream.

In its most toxic state lead interferes with the biosynthesis of heme, a component of the hemoglobin of red blood cells and of the cytochrome enzymes that are important in energy metabolism. In the central nervous system lead causes the brain capillaries to leak blood serum. Because of the rigidity of the skull, the resultant pressure injures brain cells and causes permanent brain damage. Most of these symptoms, reflecting damage from large doses of lead, are quickly recognizable and it is possible to administer treatment in time to prevent permanent damage. The most frequently used treatment is to bind the soluble lead in the system to special compounds called chelators, which carry the lead out of the body in the urine. But chelators must be used with great care, for if there is a large lead residue in the stomach, chelators may encourage such rapid absorption of lead as to cause death. Chelators may also release lead already stored in the bones, causing further complications. So it is important that the digestive tract of a child with acute lead poisoning be free of all leaded materials before treatment begins. Unfortunately, many children saved from one episode of lead poisoning are returned to the same dangerous environment only to succumb to the same syndrome again.

Asymptomatic Lead Poisoning

It is the asymptomatic aspect of chronic lead poisoning that can be the most dangerous in children. Often several diagnostic tests are required before lead poisoning can be determined and, unless it is looked for, it may go unnoticed. A sample of 1000 children in Chicago ghettos found seventy victims of lead poisoning without symptoms and three with obvious symptoms; and a study of 79,199 New York City children found three times the frequency of excessive lead levels in the blood of black two-year-olds as in Puerto Rican or white children. It has been calculated that as many as 50,000 children in the United States suffer from the asymptomatic form of lead poisoning. Even though physical recovery is possible, brain damage often ensues, for the ages one to five are critical years in the growth and development of the brain. In another study of 425 Chicago children suffering from lead poisoning, 39 percent developed various neural injuries leading to mental retardation.

The solution to this problem is in part a medical one. Ghetto children subject to possible lead poisoning should be examined much more frequently to uncover and treat as promptly as possible any cases of lead poisoning that show up. Leaded paints must be removed from substandard housing in a rehabilitation program designed to upgrade this type of housing wherever possible. Where rehabilitation is impossible, urban redevelopment projects should replace substandard housing. This is the only sure way to eliminate thousands of cases, recognized and unsuspected, of lead poisoning. It is bad enough for ghetto dwellers to endure the deaths of lead-poisoned children, but the insidious effect of sublethal doses in causing mental deterioration or retardation is a totally unnecessary result for which the entire community must share the blame and ultimate cost.

Atmospheric Lead

While dramatic and most certainly tragic, poisoning of children from doses of leaded paint does not represent man's sole exposure to lead in the environment. Lead, like most other elements, is present in the natural environment, albeit in very low concentrations. The industrial revolution, however, followed by the proliferation of the automobile, particularly in the Northern Hemisphere, has gradually increased the background of lead in the environment. Dated samples of snow and ice from the Greenland ice cap near Thule indicate that up to 1750 there were about 20 micrograms (μg) of lead per ton of ice; by 1860 this had increased to 50 μg per ton; in 1940, 80 μg; in 1950, 120 μg; and in 1965, 210 μg. This represents an increase of 400 percent between 1750 and 1940, and a 300 percent increase in background levels since 1940. The highest samples from Antarctica are equal to the lowest from the Arctic. This reflects the great preponderance of land area in the Northern Hemisphere, the presence there of most of the sources of environmental lead, and the general lack of air circulation between the Northern and Southern Hemispheres, the same phenomenon encountered in the concentration of fallout (see Chapter 9). Although some of this lead comes from industrial sources, the striking correlation of atmospheric lead with other automobile emissions suggests that the internal combustion engine has been the major source of lead in the atmosphere. This has been confirmed by Chow and Card, the Scripps Institute of Oceanography, who report that it is possible to distinguish between the lead from coal and gasoline by comparing their isotopic composition.

Lead has been added to most gasolines in the form of tetraethyl lead mixed with ethylene dibromide and ethylene dichloride plus a marker dye, since 1923. The lead is added to increase the effective octane rating of the gasoline without increasing its octane number; that is, without increasing the proportion of the more expensive, more volatile components of gasoline, its performance is improved. Lead also protects the engine exhaust valves from rust accumulation.

In 1923, leaded gasoline was inconsequential; but in the intervening years, 2.6×10^{12} grams of lead have been combusted and distributed over the entire Northern Hemisphere. Much of this lead has run off into the oceans, leading to an increase in the lead content of the surface waters. Runoff from roads following a dry spell can contain lead concentrations of 1 to 14 ppm.

On a typical day the "autopolis" of Los Angeles generates 30,-000 pounds of lead from the 7 million gallons of gasoline that are burned, which helps to explain why the atmospheric lead concentration of Los Angeles is 2.5 μg per cubic meter compared with 1.4 μg in Cincinnati or 1.6 μg in Philadelphia. This urban concentration is about fifty times the concentration of lead in rural air, and 5000 times the natural concentration. At peak traffic periods at congested points, the lead concentration may rise as high as 45 μg per cubic meter. Inhaling city air with a lead concentration of 1 μg per cubic meter, a man in one day would absorb 20 μg of lead. In addition to the lead we breathe in from the air, we ingest about 300 μg of lead a day in foods and liquids on an average.

One curious source of lead in our diet is the lead chromate that until recently was used in the cheerful yellow paint on most pencils. One maker used 12 percent lead, while 1 percent was considered the maximum safe level. More important was the total amount of lead per pencil, about 0.2 mg—dangerously close to the 0.3 mg per day maximum allowed for children. While the Pencil Maker's Association has voluntarily stopped using leaded paints or uses no more than 1 percent lead, there are untold millions of leaded pencils at large. Pencil chewers beware!

A final demonstration of the ubiquity of lead: near-empty toothpaste tubes have been found to contain enough lead to deliver a dose of 1800 ppm of lead to a child who brushes his teeth twice a day. Since the normal childhood dose of lead is 130 ppm per day, a child who enjoyed eating toothpaste because of its candy-like taste could be in trouble. In response to this and related research, most toothpaste manufacturers are shifting to plastic or plastic-aluminum tubes. Since a choice now exists, it would be wise to avoid leaded toothpaste tubes whenever possible.

Since our oral intake of lead every day is fifteen times the amount of lead we breathe, airborne lead is thought by some people not to be a significant hazard. On the other hand, only 5 percent of the lead that passes through the alimentary canal is absorbed, and close to 40 percent of the lead inhaled by the lungs is absorbed. Cigarette smokers, for example, have a slight but consistently higher concentration of lead in their blood than nonsmokers of cigarettes since lead is concentrated in tobacco leaves by lead arsenate residues in the soil from previous spraying or lead aerosols from vehicular traffic.

In either case, little lead is permanently retained by the body, but lead is toxic while it is in its temporary soluble state. When the lead content of the air rises from 1 μg per cubic meter to 15 μg, as it frequently does in large cities, the intake of lead rises from 20 μg to 300 μg a day. This is no longer one-fifteenth of orally consumed lead, but its equal, effectively doubling the intake of lead to 600 μg per day. No city has an *average* of 15 μg per cubic meter of lead in its air today. Nor would any be likely to in the future, if the automobile and oil companies agreed to eliminate tetraethyl lead from gasoline. Elimination would not result from any recognition of the danger from airborne lead in the atmosphere, but rather because leaded gasoline fouls antipollution devices required by federal and state governments.

Future Effects

To assess the effect on man of lead from industrial and automotive sources, we must consider whether the rate of environmental accumulation of lead has increased over the last few years, whether such an increase is mirrored by an increase in the body lead, and whether there is a threshold below which atmospheric lead causes no physiological damage.

Observations from the Arctic, Antarctic, and the world ocean indicate that lead is increasing in the environment. Reports of increasing body burden, however, are conflicting; one survey indicates a significantly higher lead content in American lungs (51 ppm) than in African lungs (26 ppm). But another report analyzing the lead content in the blood of various populations found as much lead in remote tribes in New Guinea as in people living in some large American cities. Unfortunately we have no base line with which to compare contemporary populations. Because of the ubiquity of lead in the environment for hundreds of years, *all* human populations have probably been exposed and contaminated. Since the sea surface has an abnormally high lead content compared with its depths, a result of atmospheric fallout, there is no reason to suspect that any human population, however remote from civilization, can be used as a control. However, a report by workers in California concludes that "further increases in atmospheric lead will result in higher blood lead levels in the population in a predictable relationship."[3]

We know that normal levels of lead in the blood range between 0.05 and 0.4 ppm. The range between 0.4 and 0.6 represents occupational exposure and 0.6 to 0.8 is abnormally high. Above 0.8 ppm recognizable lead poisoning usually results. The average level of lead in the American population is 0.25 ppm. The question is whether this average is "normal" or whether it already represents a serious increase from past levels, thereby reflecting the documented increase in environmental lead. If that is so, then any further increase might well be dangerous. We have the example from the past of chronic lead poisoning in Rome. The use of lead water pipes and, more seriously, lead cooking vessels in ancient Rome by the upper class probably caused the virtual disappearance of this class by increasing the rate

[3] Goldsmith, J. R. and A. C. Hexter, 1967. "Respiratory exposure to lead: epidemiological and experimental dose-response relationships." *Science* **158**, pp. 132–134.

of stillbirth. All of which quite probably contributed to the ultimate fall of the Roman Empire.

Whereas we know that 0.8 ppm in the blood can cause lead poisoning, we cannot be sure that 0.7, 0.6, or 0.5 ppm are innocuous. Indeed, recent research with animals indicated that body burdens of lead comparable to man's reduced the animals' resistance to infectious diseases. While this relationship has not been established in man, it is a possibility. But perhaps we are still too clumsy in our perception of correlations between environmental stress factors and abnormal physiology or psychology to recognize low-level effects when we see them.

CADMIUM

Closely related to zinc and mercury, cadmium is a soft, silvery-white metal used in large quantities in electroplating and in nickel-cadmium batteries, but found as a contaminant in a host of other materials—motor oil, rubber tires, plastics, scrap metal, metal ores, coal and oil, and fungicides, to name just a few. When these products are heated, burned, or used, cadmium is released to the atmosphere as an oxide; almost 5 million pounds of cadmium entered the atmosphere in 1968.

Cadmium, which is used in twice the quantity of mercury, made its debut as a potential threat to man's health about the same time and place—the 1950s, in Japan. Since that time over fifty people in mining areas in northern Japan have died of a degenerative bone disease called itai-itai (literally "ouch-ouch" in Japanese). Many other Japanese are severely crippled as a result of excessive exposure to cadmium. Apparently, low levels of calcium and vitamin D in the diet sensitized the Japanese to the excess cadmium in their environment.

Although no one in the United States has yet died of exposure to cadmium, there is a growing suspicion that airborne cadmium is related to hypertension and associated heart diseases. Unlike mercury, cadmium's toxicity does not depend on its chemical form; any cadmium ion will bind to the sulfhydryl groups in enzymes, halting oxidative phosphorylation in the respiratory cycle. Since cadmium is concentrated in some blue-green algae, which are responsible for poisoning man and other animals, many cases of cadmium poisoning may go undetected. A strange association between hypertension and soft water may relate to cadmium in the following fashion: soft water, low in magnesium and calcium, is more acidic than hard water, acidic enough perhaps to leach cadmium from water pipes that in hard-water regions are insulated by layers of precipitated salts or scale.

Like lead and mercury, cadmium accumulates in the body. The

average ingestion of 70 to 80 μg per day in the United States is mostly excreted in the urine, but the 2 μg that are retained lead to a body burden of 30 mg by middle age. Smoking a pack of cigarettes a day can increase uptake of cadmium by 4 μg per day. Since cadmium binds to blood protein as well as enzymes, 75 percent of the body burden is found in the liver and kidneys. Fortunately the placenta seems to form an effective barrier to cadmium movement during pregnancy.

The cadmium content of food is generally quite low, less than 0.05 ppm. But rice and wheat may contain as much as 1 ppm, oysters 3 ppm, and animal organs 12 ppm. The relatively high level of cadmium in oysters, which selectively accumulate the metal, has delayed the setting of a federal guideline on permissible levels of cadmium. A level that has been suggested by some, 2 ppm, would rule out our eating most oysters harvested in the United States.

BERYLLIUM

Because of its long-standing familiarity and use, the toxicity of lead was recognized quite early. The toxicity of beryllium was much less obvious.

In 1933 some German workers employed in a factory extracting beryllium metal from its ore became ill with severe respiratory symptoms including bronchitis. At the time it was thought that fluorides associated with beryllium were the cause. But in the mid-1940s, with the growth of the fluorescent lamp industry, a number of cases of similar description were reported. In 1946, seventeen workers in a fluorescent lamp factory in Salem, Massachusetts became seriously ill and some died. Although the evidence seemed to point to beryllium compounds used to make the lamp phosphor (beryllium carbonate and zinc beryllium nitrate), the United States Public Health Service concluded that beryllium was not toxic. The toxicity of beryllium was finally established a few years later when similar symptoms were observed in foundry, lamp, and neon sign workers, all of whom worked with beryllium metal or its oxide.

Beryllium evokes a variety of symptoms, which makes diagnosis difficult at times; victims usually complain of shortness of breath, aches and pains, and a dry cough. This is often followed by severe weight loss and kidney damage. Cortisone or other steroid hormones are sometimes useful in alleviation of symptoms, but because beryllium is difficult to remove from the body, there is no known cure for beryllium disease. Often, by the time of diagnosis, irreversible damage has taken place.

Beryllium is indeed quite toxic; whereas 150 μg of lead per cubic meter of air is considered a safe exposure, as little as 5 μg of beryllium per cubic meter of air can cause chronic beryllium poisoning. Further, beryllium behaves in a most peculiar manner. First, there is the neighborhood effect; despite large amounts of toxic arsenic, mercury, and lead released into the atmosphere by industry, no case of poisoning from these metals has been recorded in the neighborhood of the polluting factory. Beryllium poisoning, however, has been recognized in people living as far as three-quarters of a mile from a beryllium processing plant. One such community experienced sixteen cases, including five fatalities. Then there is the oddity that in the neighborhood of the plant, while people living closest to the beryllium source had a greater tendency to develop beryllium disease than those living farther away, within the plant itself this dose relationship did not apply. The beryllium concentration in the air of the neighborhood reporting the beryllium poisoning ranged from 0.01 to 0.1 μg per cubic meter, while that in the plant was 1000 times higher. Yet of 1700 workers only nine developed beryllium disease (0.5 percent) whereas neighborhood people had a beryllium disease rate twice as high. Stranger still, the nine plant workers who became ill were not those who had been exposed longest to the largest amount of beryllium, but people who had been exposed for only a few months or had worked only a short time at the factory. In some instances fifteen years elapsed between exposure and illness. Finally, postmortems on victims showed no relationship between the amount of beryllium concentrated in the lungs and the severity of the disease. Facts like these give many sleepless nights to epidemiologists, who study the nature and spread of disease. The best conclusion that can be drawn from these observations is that people vary in their sensitivity to beryllium: some are almost immune and others extremely sensitive to very small amounts.

As a result of these demonstrations of beryllium toxicity, the use of beryllium-containing phosphors in fluorescent lamps was discontinued in mid-1949, and, although cases of beryllium poisoning from broken lamps were described for several years thereafter, once the old tubes had been replaced there were no further complications either in manufacture or use of fluorescent lamps. Most industries that use or process beryllium today exercise elaborate precautions involving high-speed ventilators and shields to eliminate or sharply reduce the amount of beryllium particles or oxide in the air. Standards have been established, although on rather flimsy experimental evidence, that recommend not more than 2 μg per cubic meter of beryllium over an eight-hour day in the atmosphere of a factory. Further,

Control

no one should be exposed to more than 25 μg per cubic meter at any time. Finally, neighborhood effects may be eliminated if the atmospheric concentration of beryllium is less than 0.01 μg per cubic meter.

Today the incidence of beryllium poisoning in large plants maintaining responsible programs of industrial hygiene is quite low, certainly comparable with other types of job hazards. No fatalities from beryllium have been recorded since 1950.

But as beryllium finds new uses, the probability of its release into the environment increases. Beryllium has entered the space age with the discovery that suspensions of beryllium fibers provide a magnificent source of power when burned in a rocket engine. When research had progressed to the test firing stage, it was calculated that so much beryllium would be released in the firing of a rocket that someone eighteen miles from the launching site would be exposed to the same dose as a beryllium worker if he were on the job twenty-four hours a day for thirty consecutive days. Fortunately, little has been heard since about beryllium-fueled rockets.

ASBESTOS

A few years ago a study was made of the Insulation Workers Union of New York starting with the 632 members enrolled on December 31, 1942. Ten years later the cause of death of union members was compared with a statistical sample of men in comparable circumstances except for their exposure to asbestos. It was found that 255 union men had died compared with 203 men in the comparison group; forty-five workers died of lung cancer or disease, compared with six or seven in the comparison group; twenty-nine died of stomach cancer, compared with nine or ten in the comparison group; and twelve died of asbestos disease, while there were no deaths from this cause in the comparison group.

Some 100,000 to 125,000 people work directly with asbestos, and up to 3.5 million or 5 percent of the total work force of the United States work with asbestos at some time in some capacity in the construction and shipbuilding industries. With the widespread use of asbestos, there has been a distinct increase in the number of mesotheliomas. A diffuse cancer that develops on the mesothelium or lining of the chest or abdominal cavity, mesothelioma was extremely rare until fairly recently. Fourteen of the 120 asbestos workers who died most recently in New York died of mesothelioma. This was to be expected since the connection was made between high concentrations of asbestos fibers in the air and mesothelioma.

Somewhat less expected was the report of forty-seven cases of mesothelioma from a sample in Cape Province in South Africa. Although this area is rich in asbestos deposits, very few of the victims had worked with asbestos directly. Still more surprising was a report in 1960 that of 6312 people X-rayed in a Finnish county that had asbestos mines, 499 had a condition previously associated only with asbestos workers. None of these 499 ever worked in a asbestos mine, but most lived near one. A survey in a nearby county without asbestos mines found no diseased condition in a population of 7101.

Even more remotely circumstantial but still suggestive is the case of

a fourteen-year-old boy with mesothelioma, [who] had no occupational contact with asbestos nor did any of his family. On questioning, the boy's father told the authors that his boy . . . had helped him while he had replaced most of the plaster board during extensive remodeling of his house.[4]

While plasterboard itself does not contain any asbestos (indeed the gypsum used is covered with recycled paper), the compound used to seal the joints between wall board sheets contains asbestos fibers, which can be released to the air if the joints are sanded before painting.

Apparently, when asbestos fibers are inhaled into the lungs they become coated with a material containing iron and an asbestos body is formed (Figure 11.3). While large amounts of asbestos bodies were found in the lungs of asbestos workers, careful autopsies of people supposedly remote from asbestos exposure showed a surprising incidence of asbestos bodies: 26 percent of 500 consecutive autopsies in Capetown, South Africa; 50 percent of 1975 autopsies in New York; 30 percent of housewives; 40 percent of white collar workers; 50 percent of blue collar workers; and 70 percent of construction or shipyard workers. Of forty-two cases of mesothelioma investigated in Pennsylvania, twenty represented people who worked with asbestos, eight lived in the vicinity of an asbestos plant, and three had a familial contact with someone who worked with asbestos. These pat-

[4] Lieben, J. and H. Pistawka, 1967. "Mesothelioma and asbestos exposure." *Arch. Env. Health* **14,** pp. 559–563.

Fig. 11.3 • The rod-shaped particles in these photomicrographs are asbestos bodies in lung tissue. (*Professor I. J. Selikoff, Mt. Sinai School of Medicine, New York*)

terns of neighborhood and familial exposure resemble those of beryllium and suggest that asbestos, like beryllium, may affect people differentially rather than on a linear dose scale.

Fiber Inhalation

The major concern about asbestos is not for the safety of present asbestos workers, for once the danger of asbestos fiber inhalation was recognized several years ago, the way was open for responsible industry to provide, and for responsible unions to demand, a margin of safety by proper ventilation, respirators, and shielding. The chief concern is that asbestos can be extremely damaging to the health of many individuals who are removed from direct contact with the stimulus. Unfortunately, the potential toxicity of asbestos is not at all widely recognized and it is handled quite casually by people who assume it is as harmless as cement or plaster. There are many case histories of people who died some years after such a casual exposure. If this is true of asbestos, what about fiber glass, which is being used more in houses, or various other synthetic fibers? We simply do not know what the long-range effects of these materials are.

Fiber Ingestion

For some time it was considered a medical curiosity that Japanese men had an incidence of stomach cancer seven times that of United States males. Examination of the talc-coated rice that is favored in the Japanese diet indicated that there were several million asbestos-form fibers per gram. In Japan the incidence of stomach cancer is lower in areas where rice is less frequently eaten; it is also less among overseas Japanese who follow more westernized diets. The evidence implicating asbestos fibers is circumstantial but suggestive. Since asbestos fibers are used in the filtering of beer and have been isolated from the finished product in numbers well over a million per liter, public water supplies have been investigated and found contaminated with asbestos fibers as well. Other popular products such as sherry, port wine, vermouth, and soft drinks, were found to contain substantial numbers of asbestos fibers. The hazard of orally ingested asbestos fibers is not fully known but studies are under way.

SOLUTIONS

The basic problem with all five inorganics mentioned in this chapter has been the almost total lack of knowledge about either the threshold at which human health is adversely affected or the cumulative effect of low levels of inorganics already present in the body and which are likely to increase as atmospheric pollution worsens. Mercury, lead, and cadmium seem to follow a linear curve of dose-effect; the more lead, cadmium, or mercury, the greater the likelihood of certain symptoms developing in the exposed population. But beryllium and asbestos follow no rules, making them by far the most dangerous to handle and be exposed to. Although you can assign maximum permissible doses of mercury, lead, and cadmium (subject to revision as more information becomes available about low-level effects), the apparent random effect of beryllium and asbestos vitiates the permissible dose approach.

Perhaps inorganic materials are best treated like poison ivy. This is a plant to which people are differentially sensitive. Some people break out with a rash from head to toe by simply walking downwind from an ivy patch, while others can roll in it without ill effect. The only advice botanists or physicians can give to the general public about poison ivy at our present crude level of understanding is to avoid it wherever possible. Maybe this is the best advice for the general public to follow for the present with beryllium or asbestos. But, of course, the public is not always in a position voluntarily to avoid certain potentially dangerous inorganic materials. It then becomes the responsibility of the industry producing these materials and finally the government or agencies controlling the environmental relations of the industries to protect the general public. This can be done by either strictly controlling the emissions of dangerous by-products, dusts, aerosols, fibers, or whatever, to a point well below that demonstrated to have deleterious effects on nonworkers, or by eliminating such emissions altogether.

It is possible that a small amount of these inorganic materials is indeed harmless and that our concern over a few isolated cases represents an alarmist position. But until much more experimental data is available to provide rational scientific guidelines about the quantities of various inorganic materials that can be tolerated by the environment and ourselves, discretion had better play a greater role in our value system than it has previously.

FURTHER READING

Chisolm, J. J., Jr., 1971. "Lead Poisoning." *Scientific American* **224** (2), pp. 15–23. An excellent review of the effect of lead on human physiology.

D'Itri, F. M., 1972. *The environmental mercury problem.* CRC Press, Cleveland. Good background material.

Friberg, L., 1971. *Cadmium in the environment.* CRC Press, Cleveland. A systematic overview of what is known about the distribution and effect of cadmium.

**MAN MAKES NEW
TRAUMAS**

Gilfillian, S. C., 1965. "Lead poisoning and the fall of Rome." *Jour. Occup. Med.* **7,** pp. 53–60. Intriguing account of the effect of lead upon history.

Hyman, M. H., 1971. "Timetable for lead." *Env.* **13**(5), pp. 14–23. A good discussion about lead as a gasoline additive.

McCaull, J., 1971. "Building a shorter life." *Env.* **13**(7), pp. 2–15, 38–41. An easily read treatment of the cadmium problem.

Selikoff, I. J., 1969. "Asbestos." *Env.* **11**(2), pp. 2–7. An excellent review of the asbestos problem by the world's foremost authority on the medical aspects of asbestos.

Sherrill, R., 1973. "Asbestos, the saver of lives, has a deadly side." *The New York Times Magazine,* 21 January, p. 12.

Stokinger, H. E. (ed.), 1966. *Beryllium: its industrial hygiene aspects.* Academic Press, New York.

Tepper, L. B., H. L. Hardy and R. I. Chamberlin, 1961. *Toxicity of beryllium compounds.* Elsevier, New York. Stokinger and Tepper present a comprehensive picture of beryllium as an environmental hazard.

Wood, J. M., 1972. "A progress report on mercury." *Env.* **14**(1), pp. 33–39. The prospects for cleanup of mercury-poisoned environments are considered.

12

ORGANICS AND THE ENVIRONMENT: EAT, DRINK, BUT BE WARY

It's a very odd thing—
As sad as can be—
That whatever Miss T. eats
Turns into Miss T.

WALTER DE LA MARE

AMERICAN SUPERMARKETS are filled with the greatest abundance and variety of food ever available to man. But the list that sweeps from proletarian potato to gourmet truffle includes more and more foods that have been grown or processed in some way that has included contaminants or additives. Contaminants are compounds, organic or inorganic, that are retained accidentally in marketed foods—biocide residues on plants and antibiotics or hormones in meat. Additives, on the other hand, are substances purposely added to foods during their preparation or processing to assure longer shelf life or greater attractiveness, consistency, flavor, or ease of preparation.

THE FEDERAL FOOD AND DRUG ADMINISTRATION (FDA)

Historically, the concern of the public and hence of government regulating agencies has been with adulterants—the "chalk in milk" or "water in beer" kind of thing that was common until the turn of the century. Public indignation reached a peak in 1906 with the publication of Upton Sinclair's muckraking novel *The Jungle,* exposing conditions in the meatpacking industry (Figure 12.1). As a result of the widespread demand for standards, the administration of

MAN MAKES NEW TRAUMAS

Theodore Roosevelt set up a Food and Drug Administration to control the practices of the food and drug industry and to protect the public. Its first director, Harvey W. Wiley, interpreted the Food and Drug Act quite broadly and within six years had incurred the wrath of his sponsor by attacking Coca-Cola for its caffeine content and questioning the safety of saccharin. Wiley resigned under pressure in 1912 and the Food and Drug Administration has since interpreted its mandate to protect the public somewhat more conservatively. However, the scope of the FDA's activities has of necessity broadened as the food industry adapted increasingly modern technology to the preparation and processing of food for the expanding market. Because of the incredible complexity of pest control, as will be seen in Chapter 13, one of the chief functions of the FDA today is to license the use of biocides, set standards for residues, and monitor food products for their residue content. But of equal importance to the FDA and the public is the increasing flood of additives used by the food industry in processing food for public consumption.

Fig. 12.1 • Upton Sinclair, one of the classic muckraking writers of the twentieth century. (*UPI*)

THE ADDITIVE ARRAY

In the "good old days" products were simple, the range limited, and the quality, by today's standards, poor; but processing and packaging were relatively direct (Figure 12.2). Today few foods

Fig. 12.2 • This re-creation of an old general store is a marked contrast to today's discount emporium, where most of the items come prepackaged in boxes or plastic. (*New York State Historical Association, Cooperstown*)

escape some treatment between farm and market. On your next trip to the supermarket, read some of the labels which, by law, must indicate at least in general terms the contents of all food items (Figure 12.3). A surprising number of labels fairly bristle with such words

1. *Tang Instant Breakfast Drink*
2. *Special Morning Vanilla Instant Breakfast*
3. *Betty Crocker Ready-to-Spread Frosting*
4. *Cremora Instant Non-Dairy Creamer*
5. *Royal Instant Butterscotch Flavor Pudding and Pie Filling*
6. *Weight Watchers Dietary Frozen Dessert*
7. *Lucky Whip Dessert Topping*
8. *Pillsbury Creamy Vanilla Frosting Mix*

A. sugar, shortening (with freshness preserver), water, cocoa processed with alkali, corn syrup, wheat and corn starch, mono and diglycerides, nonfat dry milk, salt, polysorbate 60, artificial and natural flavors, potassium sorbate, soy lecithin, sodium phosphate, citric acid, pectin, dextrose, sodium citrate

B. pasteurized blend of water, hydrogenated vegetable oil, sugar, starch, sodium phosphate derivatives of mono and diglycerides, sodium caseinate, polyoxyethylene (20) sorbitan tristearate, mono and diglycerides, salt, cellulose gum, calcium chloride, vanilla and artificial flavor and color, charged with nitrous oxide and carbon dioxide

C. nonfat dry milk, corn syrup solids, sodium caseinate, lactose, vanilla extract, dibasic calcium phosphate, magnesium hydroxide, sodium ascorbate, lecithin, artificial flavors, ammonium carrageenan, tetrasodium pyrophosphate, ferric orthophosphate, vitamin A, vitamin E acetate, niacinamide, calcium phantothenate, manganese sulphate, basic copper carbonate, pyridoxine hydrochloride, cyanacobalamin, thiamine mononitrate, potassium iodide

D. corn syrup solids, vegetable fat, sodium caseinate, dipotassium phosphate, monoglycerides, sodium silico aluminate, calcium gluconate, beta carotene, and riboflavin (artificial colors)

E. sugar, citric acid, calcium phosphates, gum arabic, natural flavor, potassium citrate, vitamin C, cellulose gum, hydrogenated coconut oil, artificial flavor, artificial color, vitamin A, BHA

F. sugar and dextrose, modified food starch, sodium and calcium phosphates, salt, vegetable shortening, algin, artificial flavor and color, BHA and citric acid in corn oil

G. sugar, corn syrup, vegetable shortening, wheat starch, nonfat dry milk, modified food starch, salt, mono- and di-glycerides, polysorbate 80, artificial flavoring, artificial coloring

H. concentrated skim milk, water, sugar, polyglycerol esters of fatty acids (emulsifier), micro-cyrstalline cellulose, fructose, cellulose gum, artificial flavor, dibasic potassium phosphate, guar gum, carrageenan, ascorbic acid, ferrous sulfate, artificial color, vitamin A palmitate, niacinamide, calcium pantothenate, pyridoxine hydrochloride, thiamin hydrochloride, riboflavin, and calciferol

Fig. 12.3 • Here is a scrambled list of common foods found in most supermarkets. Above are the products, below are the contents listed on the labels. Can you match them?

Tang is a trademark of General Foods Corp.; Special Morning, of Carnation Corp.; Betty Crocker, of General Mills, Inc.; Cremora, of Borden, Inc.; Royal, of Standard Brands, Inc.; Weight Watchers, of Weight Watchers International, Inc.; and Lucky Whip, of Lever Brothers Co.

as acid, alkali, anticaking agent, antioxidant, bleach, buffer, disinfectant, drying agent, emulsifier, extender, artificial flavor, fortifier, moistener, neutralizer, preservative, sweetener, thickener—to mention just a few. Table salt, for example, may contain sodium silico aluminate, dextrose, tricalcium phosphate, potassium iodide, and polysorbate, in addition to sodium chloride. Some food additives are common inorganic salts, but many are compounds whose effect may extend well beyond their intended sphere of activity, much like the detergents discussed in Chapter 6, which kept sudsing and sudsing and caused environmental problems well beyond their manufacturers' intent. With additives, this larger sphere of unintended action is the human body.

The Pros and Cons

From the very beginning, the use of additives by the food industry polarized the consuming public. The larger group accepts whatever the industry wishes to include in its processing, assuming that the industry knows best and that if the ultimate product is "bettered" somehow, then it is worth the cost. A smaller but much more vociferous group equates the word chemical with poison and assumes that *any* chemical, organic or inorganic, added to food is poisonous. Unfortunately many faddists have joined the cause, so that the issue is no longer whether certain additives are harmful or not; the desire for wholesomeness and the yearning for the natural blend into a great crusade against all food additives.

The food-additive protest, however multifarious its background and nature, is by no means as irrational as its critics suggest. Some additives *are* poisonous when taken in large enough doses (Figure 12.4), although few people have been literally "poisoned" by an additive. Most problems have arisen from the implication through animal testing that large quantities of certain organic compounds may cause cancer in a great variety of body sites and organs. Since the threat of cancer is often brandished as a side-effect of many additives, we should take a closer look at the disease and its relationship to chemical additives.

The nature and cause of cancer is one of the great medical puzzles. Something apparently goes wrong in a cell, which causes it to divide without differentiation. That "something" might be a virus, a particle of beryllium oxide, an asbestos body, or some organic compound, perhaps innocuous in itself, that is metabolized by the body into a form that interferes with cell metabolism and causes cancer to develop. The proliferation of cancerous cells ultimately becomes so demanding of the body's resources that organ failure and death ensue. We cannot examine the structure of a prospective additive and predict if, where, and when it may cause cancer. We cannot predict how such a compound will be metabolized by the body,

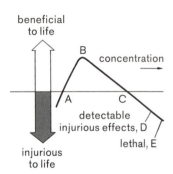

Fig. 12.4 • For many substances the difference between being beneficial or poisonous is merely a matter of concentration. At point A there may be no demonstrable effect; at point B the effect may be beneficial; at point C a substance may again be neither beneficial nor harmful; at point D harmful effects can be demonstrated; and at point E death may occur. (*From a letter to the editor by R. A. Horne, Science 177, p. 1152. Copyright 1972 by the American Association for the Advancement of Science*)

whether the metabolite will be rendered harmless or whether it will be far more destructive than the original material ingested. Today we can only infer from animal testing on a purely trial and error basis that if a compound causes cancer in a rat or monkey then it might do likewise in man. When cancer has been indisputably demonstrated in animals, the offending material is assumed to be a carcinogen in man and is either not used or is withdrawn from use.

It would be extremely convenient if all we had to do to assure our complete safety in using additives was to demonstrate that they did not cause cancer or other debilities in a variety of test animals. But it is not that simple. Some compounds do not cause cancer in all test animals; a dye, 2-naphthylamine, for example, causes cancer in dogs and man, but not rats, rabbits, or cats, while sodium arsenite causes cancer in man but not in experimental animals. In addition there are the problems of deciding on the number of experimental animals that will assure a fair test, the diet on which to place them during testing, the stage in the life cycle to test (young, adult, pregnant, old), the duration of the experiment (some cancers develop in the offspring of the adults exposed to the test material), and whether administration should be oral, by injection, or by application to the skin surface. While many of these problems can be handled by standard experimental procedures, the testing of new drugs or additives is both time-consuming and extremely expensive.

Further uncertainties arise from the latent period between the time of incipient damage and its expression in the form of a detectable cancer. It is not the young man of twenty-one smoking five packs of cigarettes a day, inhaling asbestos fibers, or ingesting a coal tar food dye who develops lung cancer, but the prematurely aging man of fifty or sixty. Who can remember what he might have eaten or breathed twenty or thirty years ago that could explain his present stomach, kidney, or lung cancer? Human metabolism varies widely; even in the best of health, no two people are likely to metabolize a given compound in exactly the same way. During illness or organ malfunction a compound may be metabolized differently than during a normal state. Finally, additives may interact; an additive that normally is not absorbed in the intestine may be readily absorbed with ill effect if an emulsifier is present. So the whole subject of the impact of food additives on human health is far more complex than the "purists" and "adulterers" would have us believe.

The position of the food-processing industry is concisely stated by R. Blackwell Smith, Jr. of the Medical College of Virginia:

Substantially every substance, including pure water and table salt, may be harmful if a sufficient quantity is swallowed or otherwise introduced into the body; and conversely, it is a generally accepted fact that there is no substance sure to kill or harm if swallowed or taken otherwise, provided the amount taken be sufficiently small . . . an additive or other chemical is not and cannot be of itself either poisonous or non-poisonous, hazardous or non-hazardous, harmful or safe; but every additive or other chemical may be safe at some level or mode of intake and hazardous at another level or mode of intake. The question, then, is not one of *whether* but essentially one of *how much*.[1]

This argument is typically rebutted by William Longgood:

Poison is harmful to the human organism. . . . When it is ingested by a human being there is damage. The more poison, the more damage. . . . The fact that the dose may be reduced until damage no longer can be seen or measured by man's instruments does not mean that the damage no longer exists; it merely means that it no longer can be seen.

The vested interests that profit from the sale and use of chemicals in foods are scornful of this attitude. They say it is not scientific. As "scientific" proof of the alleged harmlessness of eating small amounts of poisons in foods, they solemnly point out that it is possible for a person to cram enough salt or water down his throat to kill himself. This is supposed to prove that everything is harmful and even fatal if taken in large enough jolts.

What this strange logic claims is that because a little salt is innocent and a lot is harmful it follows that all other substances that are harmful in large amounts are safe in small amounts.

Of course salt or water and the hundreds of poisons used in food cannot be equated in this way. Salt and water are necessary to life, while virtually none of the food chemicals is necessary or even useful to life; with only rare exceptions, these chemicals are antagonistic to living tissue. The question is not whether they harm those who consume them, but the extent of the harm.[2]

Realizing that we may be faced with a value judgment, perhaps the question to ask is not whether additives harm or how much, but why use them at all. Suppose, because of an atypical quirk of metabolism, one person in a million developed cancer from benzoate of soda, used to prevent spoilage of cider. The additive allows this delicious and healthful beverage to be available and enjoyed year round by the 999,999 people unaffected. Would the death of this person be justified by the pleasure or nutrition of the survivors? How about 1 in 100,000, or 1 in 1000? At what point is the cost balanced by the benefit?

CONTAMINANTS

Biocides as contaminants will be discussed in Chapter 13. But there are three other groups of contaminants that can and have caused problems by not being metabolized as was supposed and so being carried onto the consumer's table.

[1] *Science and food: today and tomorrow.* 1961, NAS-NRS Publ., 887. Washington, D.C.

[2] Longgood, W. F., 1960. *The poisons in your food.* Simon & Schuster, New York.

Antibiotics

The discovery of "wonder drugs" led to their widespread application and use for everything from a minor infection to a major disease. This tendency to over-prescribe led to extreme sensitivity of some people to certain antibiotics, penicillin especially. There was also an increasing resistance of disease organisms to antibiotics, a situation quite parallel to insects and biocides.

When applied to animals, antibiotics gave effective control, often for the first time, of various poultry and stock diseases. But like the overanxious farmers with biocides, stock raisers began to treat their animals frequently, first as a preventive, then as a growth inducer. For it was found that when antibiotics were routinely given to chickens, gut bacteria, which normally slow the rate of growth, were sufficiently inhibited to allow the chicken to be marketed sooner at a greater weight and profit, though perhaps contaminated with the antibiotics. Mastitis, an udder disease in cows, can be treated with penicillin, but the penicillin is passed into the milk. In this instance, milk should be discarded for three days following treatment and antibiotic use should be halted for several days before marketing either the animal or the product to prevent contamination. *But this is not always done.*

The danger is twofold. People who are highly sensitive to penicillin or other antibiotics may be exposed without their knowledge or control. In addition, if the population at large is constantly and unknowingly exposed to antibiotics, resistant forms of organisms may develop, making the drugs useless in a time of real need. Furthermore, recent evidence indicates that drug resistance is a genetic factor that can be transferred from one organism to another, even one species to another. Thus, gastrointestinal bacteria such as *E. coli* found in the large intestine of man *and* animals could acquire and pass on the resistance factor to any of several pathogenic bacteria normally responsive to the effects of antibiotic drugs. Routine feeding of almost 3 million pounds of antibiotics to livestock each year enhances the real possibility of the development and spread of drug resistance not only in livestock and man, but in pathogens as well. Antibiotics are also used to inhibit bacterial growth on dressed chickens in stores. Although most of the antibiotic is destroyed in the cooking process, the danger lies in the false confidence that *all* disease organisms are thereby inhibited. *Salmonella,* a food-poisoning organism, is not inhibited and may increase to a dangerous state in a few days. Also, the illusion is created for the consumer that the food is fresher than it is. The consumer can only purchase by appearance, depending on the honesty of the producer or distributor for quality. Often one suspects that this dependence is abused.

Antibiotics, if they are to live up to their name, should never be used routinely, for the short-term profit is never worth the long-term risk to the livestock or the consumer, either during the life of the stock or upon its processing. In their enthusiasm for progressive techniques, food handlers do not always follow precautions, like the English butcher who sprayed his beef with DDT, then proudly said that he spent eighteen shillings a week on "modern sprays." One can safely assume that antibiotics are occasionally misused with similar exuberance.

Hormones

With increased interest in and research on the development of hormonal contraceptives, a compound called diethylstilbestrol (DES) was synthesized that had biological effects similar to those of the female sex hormone, estrogen. It was found that when a pellet of DES was inserted under the skin of a chicken's neck, the animal rapidly gained weight and became plump and fat; a further useful effect was its ability to emasculate cocks into capons chemically, without castration. Poultry raisers could produce more chickens at lower cost and, of course, higher profit. The pellets, however, did not always dissolve before marketing, which must have led at times to a biologically effective dose of sex hormones for the unlucky consumer of a soup made with chicken necks.

The balance between male and female is a fine one and a very small amount of estrogen-simulating hormone (hormones are unaffected by cooking) can prove disastrous. Then too, some chickens were marketed containing as many as twelve pellets, some of which had migrated to other parts of the chicken. Furthermore, it was discovered after a few years of use that although a DES-treated chicken did indeed look better and weigh more than a normal chicken, it was because of increased fat and water content, not increased protein. Even the fat proved worthless, forming a gooey mess when fried.

In 1960 the FDA declared DES off-limits for chickens. DES was used in beef and lamb production, however, until 1973. While sixteen cents' worth of DES implanted behind a heifer's ear supposedly produced $12 of extra beef, or 15 percent faster growth on 12 percent less feed, no such dramatic response was found in pigs, which thus remain untreated. While the amount of synthetic hormone ingested by eating beef treated according to directions is well below the level necessary to affect the delicate physiology of sex hormones, there are suggestions that synthetic estrogens may cause cancer in men under certain conditions. Certainly, improperly applied amounts with an insufficient period allowed for full absorption and metabolic breakdown present a danger to the beef-eating public. With growing evidence of side-effects from synthetic hormones used as contraceptives (see Chapter 23), the risk of contaminating virtually all beef produced in the United States with synthetic hormones for the short-term gain of $12 a head seemed to outweigh the benefits.

In response to increasing concern about the potential carcinogenic effect of residual amounts of DES in meat, the FDA banned the use of DES in cattle food in 1972. In 1973, when small amounts of DES continued to turn up in liver, the last remaining use of DES in the form of an implanted pellet was also discontinued amid threats of increased beef prices as a result. At no time, however, has any study demonstrated that growth hormones make it possible for the consumer to get more protein for his money; only the grower benefits by selling a heavier steer sooner.

ADDITIVES

One of the chief problems with food is its relatively short storage life. Since earliest times man has tried to preserve his perishable foods by drying or adding salt, sugar, or spices. Today, nitrates and nitrites are commonly used to preserve meat products, especially cold cuts or frankfurters, while benzoic acid or sodium benzoate is often used to preserve liquids which might otherwise ferment. Sulfur dioxide is generally used to preserve dried fruit, and sorbic acid is used for various other products. At customary levels of usage these preservatives are effective, but it has been suspected that some, like nitrite, are converted in the stomach to nitrosamine, a potent carcinogen. If this suspicion is borne out and the use of nitrite in foods curtailed, substitute preservatives can be used: the hot dog will surely survive.

Besides the danger of the preservative, the motivation of the preserver might be questioned. It is one thing to preserve foods safely to allow broad distribution and reasonable shelf life. It is quite another to embalm food that is too far gone to be acceptable as fresh. Such a practice leads to the adding of artificial colors, then flavors—compounds that might be more dangerous to health than the original preservatives.

Antioxidants

An adjunct to the preserving process is the inhibition of the natural tendency of fatty acids, especially those which are unsaturated, to oxidize or become rancid (Figure 12.5). Most prepared foods with unsaturated fatty acids contain an antioxidant to preserve the original fresh flavor of the food and extend shelf or storage life. Some commonly used antioxidants are butylated hydroxyanisole (BHA), propyl gallate, and butylated hydroxytoluene (BHT). The antioxidant properties of these compounds are often enhanced by ascorbic, citric, and phosphoric acid. Once again, as with certain emulsifiers and detergents, the effect of some antioxidants, especially BHT, is suspected of extending beyond the food in which it is ingested to the body itself, inhibiting the uptake of oxygen by hemoglobin in the red blood cells.

Fig. 12.5 • An unsaturated fatty acid has many double bonds between carbon atoms. When these double bonds are replaced by hydrogen atoms the fatty acid is said to be "saturated." "Polyunsaturated" simply means that there are many double bonds present. These double bonds must be protected from oxidation by additives called antioxidants.

Acids

Most acid additives are perfectly safe and have been used for many years without problems. Baking powder or cream of tartar (tartaric acid) reacts with baking soda and produces carbon dioxide, a leavening agent that makes bread and cake light. Phosphoric, citric, and malic acids are extensively used to counteract the otherwise excessive sweetness of most soft drinks. It is not the acids that are a problem to health, but the overindulgence in pastry and soda to the exclusion of more nutritious food.

Emulsifiers

In general, an emulsifier breaks up fats and oils into tiny particles; when used in bread it acts as a softening agent. Homemade bread hardens quickly as it grows stale; commercial bread with emulsifiers retains a soft texture for a much longer period. This gives the shopper the illusion of freshness and reinforces the now erroneous conclusion that soft bread is fresh bread. Emulsifiers also allow the water content of bread to be increased, resulting in the characteristic soft spongy texture of commercial white bread. Pound for pound there is likely to be more air and water in commercial loaves than in homemade bread—air and water that the consumer pays for in lieu of flour. When used for these purposes no emulsifier is worth risking its possible synergistic effect. One emulsifier, polyoxyethylene, no longer used in the United States, was found to increase greatly the rate of iron absorption in some animals. Excessive deposition in the liver can lead to cirrhosis and cancer. One effect in man has been an increase in the absorption of vitamin A. Too much of a vitamin can be as much of a problem as too little (see Chapter 1).

Other emulsifiers such as glycerides, lecithin, sorbatin, and alginate are used in baked goods, ice cream, and confections. In baked goods, emulsifiers improve the fineness, uniformity of texture, and softness. Smooth consistency in ice cream depends upon the uniform size of ice crystals, which is controlled by alginate or emulsifiers and stabilizers.

Artificial Sweeteners

Even a food purist would have to agree that natural sugar poses no direct problem to health. But some substitute sweeteners, compounds that taste sweet but have no food value, may have adverse effects upon metabolism. Saccharin, discovered in 1879, has been used for many years as a sugar substitute, especially by diabetics, who must control their sugar intake. In 1950 a second group of nonnutritive sweeteners, the cyclamates, was introduced. While saccharin has over 300 times the sweetening power of sucrose (cane sugar), cyclamates were only thirty times as sweet.

Since saccharin was not thought to be metabolized by the body, it was considered safe within recommended dose limits (1 g per day). But recent research seems to indicate that saccharin, like cyclamates, may be carcinogenic, and the drug is likely to be restricted in its use or outlawed for all but medical uses. However, many countries restrict or prohibit its use, not because of ill effects but to avoid misleading people into substituting a nonfood for a food of high caloric value. Cyclamates were apparently metabolized, at least by some people; some 20 percent of cyclamate users generated a breakdown product that, when applied to rat cells, caused chromosomes to break up. Further possible danger was suggested by the linking of cyclamates to birth defects in chickens. For this reason cyclamates were banned from foods in 1970.

While saccharin might be excused as a useful sugar substitute for diabetics, cyclamates were totally unnecessary. They entered the market in response to the recent fetish for losing weight by every possible means except reduction of food intake. Rather than give up soft drinks or even cut down consumption, the weight-conscious public was only too happy to have its soft drinks without their calories. But deprived of energy-yielding sugar, a soft drink becomes a nonfood, an expensive way to restore body fluids compared with water.

Food Dyes

Of all the food additives, nonvegetable or coal tar dyes are at once the most hazardous, based on past experience, and the most unnecessary. No brief can be made for their nutritional or preservative value, only the rather weak rationale that the consumer demands vivid colors in his food—oranges must be orange, hot dogs pink, and cherries scarlet. Many dyes have been demonstrated to be carcinogenic; therefore, many closely related compounds are suspect. But which to suspect is an apparent source of confusion. England allows thirty dyes, the United States fifteen; but the English ban nine of our dyes and we ban most of their thirty. This reinforces real doubt about the wisdom of ingesting *any* coal tar dyes.

Many of the additives mentioned are, of course, combined in certain products; indeed, a few of these products contain little but additives. Maraschino cherries, for example, are preserved with sodium benzoate, firmed with calcium hydroxide, bleached with sulfur dioxide, artificially flavored, then given their bright green or red color with a coal tar dye.

A few years ago two firms marketed a blueberry pancake mix depicting luscious blueberries prominently on the packages. The "blueberries" in one mix, according to the FDA, were made chiefly of sugars, nonfat dry milk, starch, coconut pulp, artificial coal tar coloring, artificial flavoring, and a very small amount of blueberry pulp. The second firm's "blueberries" were of an equally startling composition: sugar, gum acacia, citric acid, starch, artificial coloring and flavoring, and some blueberry pulp.[3] Since these purple pellets were hardly blueberries, the FDA forced the manufacturers to change the labels.

This is not a unique example of exaggeration in the food processing industry. Another product that has reached the point of processing to the absurd is the nonfood synthetic cream. One especially acerbic housewife reacted as follows:

Dear Sirs:

Under separate cover I am returning the free sample of your product, "Dreammix" which was placed in my mailbox. I am a consumer who has learned to read labels of foodstuffs. When I read "Hydrogenated Vegetable Oil, Sugar, Sodium Saseinate, Emulsifiers, Preserving Stabilizer, Artificial Flavour, Colour" as the ingredients of this concoction, I decided that your product is foodless trash, not worthy of being ingested into the human body. However, I was faced with a dilemma. Pouring the contents of the box down the kitchen sink might clog the plumbing. Throwing it into the woodstove might lead to an explosion. Placing it on the compost pile might interfere with proper bacterial action. Tossing it on the town dump might be deleterious to scavenging birds and wildlife.

[3] Longgood, *Poisons.*

Burying it in the ground might harm the soil. So, after due deliberation, I am returning the stuff to you, its creator. May I suggest that your copy writers might have named this product "Nightmare Mix" rather than "Dreammix". On those occasions when I desire whipped cream, I prefer to go straight to the cow.[4]

While doctored cherries, fake blueberries, and make-believe whipped cream can easily be eliminated from the diet, other more basic foods have a depressingly long list of additives on their labels. Assuming that the ideal food is transported from the field to the dinner table with the minimum amount of tampering and maximum retention of plain good taste and food value, bread has fallen rather far from the ideal. The wheat grain as harvested is mostly endosperm, a storage tissue of starch and protein attached to the small embryo or wheat germ. Surrounding the grain are the seed coats. They contain vitamins E and B, certain minerals, and amino acids, while the wheat germ contains unsaturated fatty acids.

Over-processed bread

Previously in flour processing, the grain was ground into flour including both the germ and the seed coats or bran. Thus all of the vitamins and minerals were preserved and passed on through the baked bread to the consumer. But to many, the bread was coarse and heavy, and the flour quickly spoiled by the activities of insects attracted to the rich grain and by the fatty acids in the wheat germ becoming rancid. In the late nineteenth century a new process was developed to roll the grain instead of grinding it. In this process, the starchy endosperm powdered while the oily germ rolled flat. When shaken through sieves, the germ could be separated from the flour. This allowed the flour to be kept longer without spoiling. But flour must be matured or aged to reach its maximum workability; during this period of storage there is an opportunity for insects to attack the flour.

Then it was discovered that nitrogen trichloride or agene (today dyox or chlorine dioxide is used because of potential danger from agene) could mature the flour instantly. But at the same time it bleached the flour white, decoloring the pale yellow fragments of the seed coats and endosperm left from the milling process. While this produced a flour of indefinite keeping power, it was practically devoid of its vitamins and minerals—vitamins A, B, and E, calcium, and iron. What was left was mostly starch and 7 to 11 percent protein.

In previous years, when bread was an integral part of every meal and fresh fruit and vegetables were not available year round, many people depended upon bread for vitamins. During World War II, white bread was finally enriched by adding vitamins. Bread producers continue to advertise the fantastic energy and vitamin values of their ultrasoft and moist bread, which must often be toasted to attain sufficient substance to retain the contents of a sandwich. Few

[4] Grant, D., 1961. *Your bread and your life.* Faber & Faber, London.

appreciate the irony of a situation where the flour processor mills out some twenty vitamins and minerals, puts four back, then touts its enrichment!

Of course this need not be, for flour can be milled to preserve much of the nutrition without sacrificing the storage life of the flour, and matured without resorting to vitamin-destroying bleaches. But the public seems more concerned with the aesthetics of the product than its nutritional value. This faulty sense of priorities is unfortunately characteristic of the consuming public—an overconcern with appearance coupled with a lack of awareness or concern for the real purpose of food. If bread is merely a convenient way to sop up gravy, then we might as well use our spoons. If it is a source of vitamins and minerals necessary for maintaining good health, then it is senseless to mill out these vitamins for whatever reason and replace a few at the consumer's additional expense in the name of good nutrition. The alternative is not a whole wheat bread, which some faddists tout as a cure-all, but a flour milled to retain most or all of its native nutrients, obviating the expense of adding them in the baking process.

Perhaps as a result of the recent furor over rats dying of malnutrition on a diet of white bread or the "empty" calories of many breakfast cereals, a wide range of specialty breads, breakfast cereals, and, for the "scratch" cook, unbleached flour, has become available. So the public at last has a choice.

Because of the decreased importance of bread in our diet, its nutritional value is perhaps a moot point. But there remains a tremendous list of foods with additives that are regularly prepared and consumed. One estimate claims that additives totaled three of the 1400 pounds of food consumed per capita by the American public per year. As the demand for convenience foods continues to increase, the consumption of an even greater variety of additives will follow. For additives have become indispensable in assuring the palatability of the instant heat 'n eat, Shake 'n Bake, brown 'n serve, or Whip 'n Chill products.[5]

Fortunately much processing with its concomitant additives can be avoided by old-fashioned cooking from scratch, mashing your own potatoes, whipping your own cream. While the time lost is inconsequential, the gain in flavor is readily apparent. But for the majority of consumers who apparently wish for even more highly processed foods, much more careful screening and testing is required. A careful re-examination of the "safe" category maintained by the FDA, the so-called GRAS (Generally Regarded As Safe) list, is certainly due. After all, cyclamates were marketed for fifteen years before their possible harmfulness was announced.

[5] Shake 'n Bake and Whip 'n Chill are trademarks of General Foods Corp.

Considering the ubiquitous contamination by biocides, hormones, and the spiraling use of additives, we may never again have the luxury of eating simple, nutritious food, direct from the field, orchard, or pasture. But we have every right to be protected from materials that might prove injurious to our health.

FURTHER READING

Bicknell, F., 1961. *Chemicals in your food, and in farm produce: their harmful effects.* Emerson Books, New York. A fairly low-key presentation by an M.D.

Grant, D., 1961. *Your bread and your life.* Faber & Faber, London, England. Grant gives an interesting, somewhat emotional account of bread and its role in nutrition. There is also a fine recipe for whole wheat bread.

Kohlmeier, L. M., Jr., 1969. *The regulators: watchdog agencies and the public interest.* Harper & Row, New York. The ground rules between government and the consumer.

Longgood, W. F., 1960. *The poisons in your food.* Simon & Schuster, New York. As the title suggests, this book is in the muckraking tradition.

National Research Council Food Protection Committee, 1961. *Science and food, today and tomorrow.* NAS–NRS. Publ. 877, Washington, D.C. A somewhat stiff, but revealing, presentation.

Winter, R., 1972. *A consumer's dictionary of food additives.* Crown, New York. Extremely useful guide to food product labels.

13

BIOCIDES

IF IT WERE POSSIBLE to spray a cotton field with some miraculous compound and kill all the boll weevils without harming any other organism in the community, we would have a pesticide, a material capable of selectively killing a pest. Unfortunately, no compound is known which has this selectivity; if the boll weevils are killed, so are many other animal species, both destructive and beneficial—spiders, mites, and occasionally fish, amphibians, birds, and even mammals. So the term pesticide is somewhat general, even misleading; insecticide, while more accurate, gives no hint of potential injury beyond insects. The most general term which avoids these overlapping difficulties is biocide. If this has a grim ring to it, like genocide, perhaps it is fitting. For many people still feel that the earth would be better off without insects, mites, ticks, spiders, and all manner of "creepy-crawly" things.

Although odious in extreme usage, biocides *do* have a use and there is no intention here of insisting otherwise. We should, however, have some idea of the role that these materials play in the systems into which we introduce them; we should have a rational basis for determining at what point the cost of biocide use becomes prohibitive in economic and environmental terms. Since most of what we know about the environmental effects of biocides has come from twenty-five years of experience with DDT, many examples will deal with this biocide. But keep in mind that many environmental effects of DDT are potentially possible with *any* biologically active compound thoughtlessly introduced into the environment.

PANDORA'S PANACEA

Almost a hundred years ago, in 1874, a German Ph.D. candidate, Othmar Zeidlar, synthesized an organic compound for his dissertation, as generations of students have since done. Young Zeidlar published his work as a short note in a professional journal, took his degree, and dropped from sight.

The compound Zeidlar synthesized was not unusual, even for 1874; a series of substitutions had transformed the original ethane (C_2H_6) into a *dichloro-diphenyl-trichloro-ethane* (Figure 13.1). So dichlorodiphenyltrichloroethane (DDT) sat on the shelf with a thousand other organic chemicals synthesized over the years until the late 1930s, when Paul Müller, a Swiss entomologist looking at the insect-killing properties of various compounds, found that Zeidlar's material was an extremely effective insect killer. DDT was "discovered" and for his contribution Müller was awarded a Nobel Prize in 1944.

ethane   1, 1, 1, trichloroethane

Fig. 13.1 • In a series of reactions substituting three atoms of chlorine for three of hydrogen, then two chlorinated rings for two more hydrogens, dichlorodiphenyltrichloroethane (DDT) can be synthesized from the simple compound ethane.

In 1942, some DDT was brought to the United States, where the military quickly appreciated its potential usefulness. As a result of DDT, World War II was the first war where more men died of battle wounds than of typhus and other communicable diseases spread by insects. After the war, DDT was most effectively used to combat a number of insects which were carriers of thirty diseases, including malaria, yellow fever, and plague; within ten years of its first widespread use, DDT had saved at least 5 million lives. Millions of houses and tens of millions of people were sprayed with DDT in campaigns against fleas, houseflies, and mosquitoes and the diseases these carriers transmit.

Clearly a new age of chemical control of pests was underway. DDT was not the first chemical used against insects, of course. Paris green (copper acetoarsenite), first used in 1867 against the potato beetle, holds that distinction. But DDT represented the first time a totally new synthetic material was introduced into the environment on a large scale.

BIOCIDE SPECTRUM

Spurred by the success of DDT, the chemical industry explored the biocide possibilities of other biologically active chemicals.

Today there are seven main groups of biocides available to control a broad variety of plants and animals: arsenicals, botanicals, organophosphates, carbamates, organochlorines, rodenticides, and herbicides. But the three most important and dangerous from the point of view of environmental contamination are the organophosphates, organochlorines, and herbicides.

Organophosphates

Among the first organophosphates to be marketed were parathion and malathion, which are still in use today. Their mode of action on terrestrial organisms is quite specific.

Nerve impulses are conducted across the gap between adjoining nerve fibers by a compound called acetylcholine. As soon as the impulse has bridged the gap an enzyme, cholinesterase, destroys the acetylcholine present, preventing further impulse transmission. Organophosphates deactivate the cholinesterase, allowing a stream of impulses to flow uninterruptedly along the nervous system, resulting in spastic uncoordination, convulsions, paralysis, and death in short order. While quite toxic, organophosphates are not persistent and have much less effect on aquatic than terrestrial organisms.

Also called chlorinated hydrocarbons, organochlorines include the best known of all the synthetic poisons: endrin, heptachlor, aldrin, toxaphene, dieldrin, lindane, DDT, chlordane, and methoxychlor.

Organochlorines

Unfortunately, there is no clear understanding of just how the organochlorines work. Apparently the central nervous system is affected, for typical symptoms of acute poisoning are tremors and convulsions. Chronic levels have various effects. In aquatic organisms, which are especially sensitive to this class of compounds, the uptake of oxygen through the gills is disrupted and death is associated with suffocation rather than nervous disorder.

Another group of chlorinated hydrocarbons closely related to DDT, the polychlorinated biphenyls (PCBs), have recently emerged as an environmental threat; ironically these compounds were never intended to be used as biocides, but because of thermal stability, resistance to oxidation and hydrolysis, and solubility in organic compounds, they have found many industrial uses. Because the structure of PCBs is basically two phenol rings joined together, with a variety of possible chlorine substitutions (Figure 13.2), PCBs are a mixture of many compounds with varying characteristics. But their resistance to oxidation and hydrolysis makes them more stable and resistant in the environment than DDT.

Fig. 13.2 • The PCB molecule is made up of two benzene rings joined together. Chlorine atoms may become attached at any of the points marked by Xs. Because of the large number of possible chlorine substitutions, there is no one formula for PCB, but rather a mixture of several of the possibilities.

The widespread distribution of PCBs closely parallels that of DDT although in lower concentrations. This had been considered something of a coincidence until it was discovered that the environment contains only a small fraction of the DDT that has been distributed since 1940. Although DDT is quite resistant to breakdown it is degraded slowly into DDE by microorganisms. Since DDE is even more stable than DDT the discrepancy between DDT dispersal and DDT and DDE accumulation in the biosphere became even more puzzling. This gap now seems due to the conversion, aided by ultraviolet light, of DDT and DDE into other chemicals such as

PCBs. The possibility of conversion was long discounted because PCBs are themselves an industrial contaminant; if DDT does indeed break down into PCBs in nature, much of the environmental destruction previously attributed to DDT must be reexamined and its possible interaction with PCBs and other degradation products considered. Of course, if the DDT had not been so widely distributed by man the distribution of PCBs would have been far more limited. Despite this new development, which suggests that industry is not the sole source of PCBs in the environment, industry in both the United States and abroad has voluntarily limited the use of PCBs in products whose disposal is likely to introduce these compounds into the air, soil, or water.

Plant killers are not new. Indeed, some arsenicals have been used for years; but shortly after World War II a new group of herbicides was developed that involve plant hormones or auxins. Auxins normally are produced by the leaves of a plant and in the proper concentration serve to keep the leaves attached to the stalk or stem. In the fall, auxin levels normally drop and a layer of large, thin-walled cells forms where the leaf connects to the plant. When these cells rupture, the leaf falls. By applying a compound which lowers the auxin content of a leaf it is possible to defoliate a plant prematurely at will. This is commonly done before harvesting cotton to avoid plugging the mechanical cotton harvester with cotton leaves.

Herbicides

Conversely, by adding auxin at the right concentration, leaf fall and fruit drop can be inhibited, thereby decreasing loss from preharvest drop of fruit. If an excess of auxin is applied, however, plants respond by increasing their respiratory activities considerably beyond their ability to produce food, causing the plants literally to grow themselves to death. Strangely, the commonly used herbicides with high auxin activity, 2,4-D and 2,4,5-T, affect only broad-leafed plants. This makes them useful tools for keeping lawns free of dandelions and other broad-leafed weeds without damaging the grass. Evidence indicates that 2,4,5-T often contains an impurity, dioxin, which can elicit birth defects similar to those caused by thalidomide; as a result, 2,4,5-T is no longer recommended for areas near human habitation in the United States.

Development of Biocides

To develop a commercial biocide, chemical companies must invest a great deal of time, research, and money. First, a large variety of materials must be screened for biological activity; toxicity is not always obvious from a structural formula. Then tests in the laboratory on a wide spectrum of pests determine toxicity levels of likely compounds. Those that are strongly toxic to one or more pests are tested for the effect of multiple doses and long-term effects, up to a month (a considerable part of the life cycle of many insects).

Small plot studies by cooperating state agricultural research stations then work out problems of formulation and field dosage. Next the cooperation of farmers is enlisted for large-scale trials on a number of crop plants in different soils, climates, and regions. All of the data from these preliminary steps are then presented to the proper government regulatory agency. If approved, the compound is registered for use and marketed. Long-term toxicity tests continue as the product is used commercially. Ultimately, under this system, the consumer is the final test of any unforeseen hazard. Since the cost of marketing a biocide may run to several million dollars, none of which is returned until the product is sold, chemical companies are understandably anxious after years of development to recover their investment by promoting the use of their product. Fortunately with most large chemical companies, biocides are usually a small part of their total business, which lessens somewhat the pressure to rush a product onto the market prematurely. However in 1970 1 billion pounds of biocides were used in the United States.

Despite testing by industry and regulation by government, biocides have caused problems, some resulting from the very characteristics that made them useful pest killers. These problems can be conveniently grouped into four categories: resistance of the target organism; accumulation in food chains; persistence in the environment; and nonselectivity. Let's look at each of these problem areas in some detail.

Resistance

For a short while, DDT seemed about to push the mosquito and housefly into extinction. But by 1947, trouble developed. Italian researchers reported that the housefly was becoming resistant to DDT; corroboration soon followed from California. Soon other insects became resistant too: the malaria-carrying mosquito, then lice, fleas, and a number of other disease vectors that had formerly been controlled by DDT. Twelve insects were found to be resistant by 1948, 25 in 1954, 76 in 1957, 137 in 1960, and 165 in 1967.

Ironically, resistant pests have appeared everywhere man has sprayed, simply because the development of resistance depends upon killing as many pests as can be exposed to the selected poison. Until recently, this has been the goal of pest control.

When an insect population is exposed to a new biocide, up to 99 percent of the population is eliminated. The remaining 1 percent is by random coincidence immune or resistant to the chemical. A 99 percent wipe-out sounds almost as good as extinction. But the few resistant individuals are now able to reproduce almost without competition, a result of the elimination of both their peers and their predators. Breeding at a rate of ten generations a season, a resistant housefly population could become the most abundant strain in about five years.

Simple resistance to biocides is bad enough for the farmer, but in some instances even more serious problems were elicited. After several years' exposure to cyclodiene, cabbage maggot strains were selected that were not only resistant to the biocide, but lived twice as long and produced twice as many eggs as nonresistant forms.

A classic example of unanticipated selection for pest resistance is seen in the lygus bug on California cotton (Figure 13.3). The lygus bug sucks plant juices from buds, blossoms, and young cotton bolls. Each one of these stages that fails to develop means one less cotton boll. In 1967, a large chemical company marketed azodrin, a new organophosphate; after much advertising, over a million acres were sprayed in California alone. While the new biocide was not persistent, it was toxic to a broad spectrum of insects and eliminated almost all insects in the sprayed fields. Its lack of persistence allowed migration of lygus bugs into the sprayed fields where their natural enemies had been eliminated (predators are always much less numerous than their prey, and therefore recover much more slowly from biocide applications). Then the lygus bug population increased and another spraying was required—then another and another. After a few seasons, it can be expected that the lygus bug will be resistant to azodrin.

Fig. 13.3 • The lygus bug found on cotton in California may not be the pest it has been assumed to be. (*U.S. Department of Agriculture*)

The killing of beneficial insects along with the lygus bug increased the population of, and damage from, the boll worm, with the result that application of azodrin may well have decreased the yield of cotton through the increased activities of the boll worm. In addition, studies on the cotton plants showed that many buds do not form bolls and all bolls do not necessarily ripen even under the best of conditions. In fact, many buds and bolls that were attacked by lygus bugs would not have ripened anyway. In an experiment set up by University of California entomologists, the yield of cotton was not significantly different between sprayed and unsprayed plots. Hence it was quite probable that the lygus bug was not even a pest!

This type of situation can easily be repeated, because the grower, often an absentee landholder or corporation, wants someone to sample his fields occasionally, advise on the presence of potentially dangerous pests, and suggest some control. Very often the "someone" is a field salesman for a chemical company. In the cotton-growing areas of California, there are a handful of extension entomologists and over 200 salesmen representing 100 companies. Should anyone be surprised that fields are sprayed unnecessarily?

Accumulation

Although DDT is only slightly soluble in water, it is quite soluble in fat. Therefore algae accumulate DDT in their cellular fat bodies on the order of parts per million. An organism that eats algae may accumulate DDT in tens of parts per million. This accumulation and concentration of biocides in a food chain is called biological magnification. By means of this process, an extremely small quantity of a persistent biocide in water, in equilibrium with a much larger amount in the bottom mud, has unexpected effects in an ecosystem—far beyond the original intent of those who first introduced these materials into the environment. For example, bottom mud in Green Bay, Wisconsin contained 0.014 ppm DDT. But small crustaceans in the same environment had accumulated 0.41 ppm, fish 3 to 6 ppm, and herring gulls at the top of the food chain, 99 ppm—enough to interfere with their reproduction (Figure 13.4).

Of course, biological magnification is not limited to aquatic organisms. One early example of this process developed from the attempt to control Dutch elm disease with DDT. The Dutch elm disease is caused by a fungus that plugs the water-conducting vessels

Fig. 13.4 • Estuarine food chains have been contaminated with DDT. The primary producers—phytoplankton, algae, and marsh plants—absorb DDT from mud or the water and store it in their cellular fat bodies. By passing this DDT on to the consumer, the level may gradually accumulate to large doses. (*Brookhaven National Laboratory*)

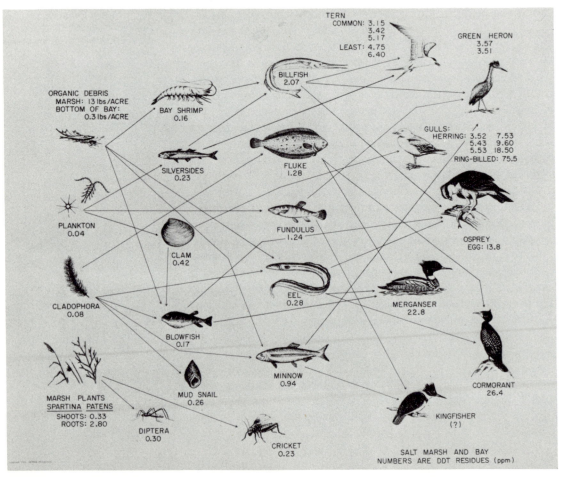

of the elm, leading to the death of the tree (Figure 13.5). Since the fungus is spread by two species of bark beetles, attempts to control the disease have focused, quite naturally, on the insects. The program of spraying elms with DDT began in 1947 and until recently was still being practiced by some communities. While DDT effectively kills the bark beetles and slows the spread of the disease,

Fig. 13.5 • (a) At one time the streets of most northeastern cities and towns were lined with these airy and graceful American elms; (b) but the Dutch elm disease slowly began to kill American elms. (c) Today elms are no longer found in most towns and the stark remains haunt hedgerows and country lanes.

it seemed in some instances to kill birds too, especially robins. The robin population on the campus of Michigan State University in East Lansing, Michigan, was reported to have dropped from 370 to 4 birds in four years, and almost no nests produced young during these years. In another instance the robin population of Hanover, New Hampshire, a town which regularly sprayed its elms with DDT, fell considerably below that of a nearby town which did not spray. In both Michigan and New Hampshire, large numbers of dead birds were found to contain DDT in excess of the 30 ppm that has been observed to be lethal in robins.

In both instances, spraying seems to have been rather exuberantly carried out, applying DDT in doses much higher than necessary, thus killing more robins than usual. If DDT is applied to trees either in leaf or during dormancy, it drips to the ground beneath the trees and is accumulated by earthworms. The soil beneath a sprayed elm tree may contain 5 to 10 ppm DDT; an earthworm that ingests this soil for its organic content may contain 30 to 160 ppm. A robin feeding almost exclusively on earthworms during the

Birds of prey and DDT

early spring, when elms are usually sprayed, might easily receive a lethal dose of DDT through this chain. Many apparently have. In this way a food chain can be poisoned and increasing levels of a noxious substance passed upward through the links.

The most serious and best documented example of biological magnification can be seen in the reproductive failure of certain birds of prey, especially the peregrine falcon, bald eagle, and osprey, which have undergone disastrous population decreases since 1945 (Figure 13.6).

Fig. 13.6 • (a) The peregrine falcon has almost completely disappeared from the U.S. because DDT has interfered with its reproductive cycle. (b) Ironically, the national symbol, the bald eagle, is also in grave danger. (c) Even the once abundant osprey or fish hawk is rarely seen as chlorinated hydrocarbons are concentrated in the food supply of each of these species. (*Photos from U.S. Dept. of Interior, Sport Fisheries and Wildlife*)

The peregrine falcon has completely disappeared from the eastern United States, the bald eagle has become rare, and the osprey uncommon. Charles Broley, who studied the bald eagle in Florida for a number of years, watched the population of eagles in one area fall from 125 nests producing about 150 birds a year in 1940 to 43 nests producing 8 young in 1957. In 1958 only 10 nests and 1 young were found. Similar decreases have been reported in osprey populations in Maine and Connecticut. In all three species eggs fail to hatch or are easily broken and eaten by the brooding bird. DDT concentrations in both the eggs and adult birds were found to be high. Falcons from an area as remote as the Northwest Territories of Canada were reported to contain 369 ppm of DDT in their fat.

How is DDT implicated in the decline of these birds? The high levels of DDT found in the fat of many birds of prey results from the same concentration process seen in the earthworm-robin chain. Birds of prey are carnivores; they eat other birds and animals which may have substantial quantities of DDT in *their* fat picked up from *their* habitats. DDT seems to affect calcium metabolism; it is, of course, the calcium deposited around the egg which produces a strong shell. This metabolic disturbance results in eggs with much thinner and weaker shells than normal, eggs that are easily broken (Figure 13.7). This was suggested by a study which compared the weight of bird eggs in museum collections before and after the introduction of DDT in

Fig. 13.7 • (a) DDT interferes with calcium metabolism in many birds, resulting in delicate eggshells which break easily, as in this pelican egg. (*Photo by W. Gordon Menzie*) (b) Here are two cross sections of Japanese quail eggshells, magnified 400 times. The bird producing the egg on the left was fed a diet containing DDT; the diet of the bird that produced the normal egg on the right did not contain DDT. The structural weaknesses make the former egg quite fragile. (*Wide World Photos*)

1947. Those birds with declining populations had substantially lighter (hence, thinner) shells after 1947 than those birds with stable populations (Table 13.1). It should be pointed out that the declining species tend to feed on species which themselves are removed by several steps from the primary consumer. The birds of prey with stable populations usually feed on the primary consumer directly. Hence they are exposed to much lower amounts of DDT. While DDT is clearly implicated in the decline of certain birds of prey, PCB levels are often as high or higher and may in some cases be more harmful.

Further evidence pointing to a DDT interference in eggshell

Table 13.1

Relationship of Eggshell Thickness to Population Decline of Certain Birds of Prey*

POPULATIONS	WEIGHT OF EGGSHELL (% CHANGE)
DECLINING	
Bald eagle	− 18
Osprey	− 25.1
Peregrine falcon	− 18.8
STABLE	
Red tailed hawk	+ 2.7
Golden eagle	+ 2.9
Great horned owl	+ 2.4

* Hickey, J. J. and D. W. Anderson, 1968. "Chlorinated hydrocarbons and eggshell changes in raptorial and fish-eating birds." *Science 162*, pp. 271–273.

characteristics was shown in a study where sparrow hawks, which were fed a mixture of 3 ppm dieldrin and 15 ppm DDT, laid eggs 8 to 10 percent thinner than control birds. The offspring of these birds laid eggs 15 to 17 percent thinner than the controls. Clearly DDT interferes with calcium metabolism.

Birds tend to accumulate DDT because they cannot excrete liquids as rapidly as mammals. Some mammals also accumulate DDT, however. A study of mice and shrews in a spruce-fir forest in Maine after one pound of DDT per acre was applied to control an outbreak of spruce budworm showed that shrews had 10 to 40 times the amount of DDT residue of mice. This was due to the carnivorous diet of shrews and led to a maximum of 41 ppm the year the forest was sprayed. After nine years, while the DDT residues of mice seemed to be approaching normal, the level in shrews remained unusually high, 2 to 6 ppm. Even though the level probably had little effect on the shrews themselves, what about the predators feeding on shrews? Reproductive failure has been noted in ospreys feeding on organisms with an equivalent amount of DDT residue to that found in the shrews. One can only wonder what DDT load foxes and bobcats are carrying around in forests, years after the initial introduction of DDT in their environment.

Contamination of the Ocean

It is not surprising that local ecosystems that have been sprayed with various persistent biocides pass these materials on in ever higher concentration through food chains. It was something of a shock, however, to learn that Adelie penguins on the Ross Ice Shelf

DDT in mammals

of Antarctica had traces of DDT in their body fat. Although the amounts were small, 13 to 115 parts per billion (ppb) in liver tissue and 24 to 152 ppb in fat, the fact that there was *any* DDT at all, thousands of miles from the nearest possible point of contamination, was alarming. At first it was thought that the presence of several thousand scientific and support personnel nearby might be contaminating the environment in some way through discarded material, clothing, garbage, or fecal wastes, for DDT was not used for any purpose in Antarctica. But subsequent investigation in still more remote areas of Antarctica indicated that local penguins and cormorants contained 0.001 to 0.48 ppm DDE (a degradation product of DDT) and 0.011 to 0.140 ppm DDE respectively. The wide-ranging skua (a gull-like sea bird) contained 0.89 to 26.0 ppm DDE. Furthermore, pelagic or open ocean birds like the albatross, which touch land only to breed, were found to contain traces of various biocides.

Even more alarming than food chain contamination has been the report that concentrations of DDT and PCBs as low as a few parts per billion have reduced the rate of photosynthesis in four species of marine algae grown under laboratory conditions. While the present level of DDT in the world ocean is in the parts per trillion range, should it ever increase to parts per billion, photosynthesis by the oceanic algae might be curtailed. Considering that 90 percent of the earth's photosynthetic organisms which produce oxygen are found in the ocean, the oxygen content of the air could conceivably be affected with both predictable and unpredictable results.

The conclusion was unmistakably clear: contamination of food chains was no longer a local affair in direct relation to biocide application at some specific point; the oceanic food chain, remote from land-based biocide usage, was contaminated. How could the enormous world ocean, stretching thousands of miles around the earth, become contaminated, whatever the scale of man's activities? Some hint might be taken from the Bravo series of H-bomb explosions in the Pacific (see Chapter 9), which resulted in the catching of radioactive tuna in Japanese waters thousands of miles from the blast site. Tuna, the last link (before man) in their particular food chain, accumulated fallout to a dangerous degree. Likewise, DDT has collected in atmospheric dust hundreds and thousands of miles from dusting or spraying activities. One study showed that only 26 percent of the DDT spray intended to reach corn plants at tassel height actually made it; the rest was wafted away by wind currents. When washed out of the air by rain, the DDT can enter any ecosystem anywhere in the world. Since oceans occupy 75 percent of the earth's surface, contamination was just a matter of time. The pathway of contamination seems to be, then, land to air to ocean and perhaps finally burial in the ocean depths.

Biocides in Man

If animals can accumulate and concentrate biocides from their food, one wonders how much DDT man has accumulated from his food chain and if there is any evidence of potential harm. Man has, along with just about every other organism, accumulated DDT in his body fat. A sample of several hundred people in the United States averaged 8 to 10 ppm DDT in body fat. Whether this is or will be harmful is the subject of a sharp and continuing debate. Until 1965 there were no medically documented cases of sickness or death in man that could be traced to proper use of biocides. Of course any number of children and agricultural workers have died through careless handling of biocides, for often people do not read labels or follow instructions. Careless handling of virtually any substance, biocides not excepted, can be disastrous at times. In 1962, for example, seven infants in a hospital nursery died when salt was accidentally added to their formula instead of sugar. If, then, there is no evidence that disease or death in humans has resulted *directly* from the proper use of biocides, what are the long-term prospects?

Delayed Effects?

Apparently no one need fear inordinate residues of DDT on fruits and vegetables, for the Food and Drug Administration has set up maximum allowable levels of biocides on food products and keeps careful watch on food shipped through interstate channels. The fear of dying from one poisoned apple is as unreal as the Snow White fairy tale. But, the fact that we are all carrying a supply of DDT, regularly replenished from our environment, leads us to wonder about long-term problems caused by a body burden of DDT. There are three logical categories of DDT effects: no effects, effects that are too subtle to be correlated with DDT levels; and effects that will appear in the future. One recalls with some uneasiness the case involving a group of women who worked in a watch factory in the 1920s painting the dials with a luminescent paste containing radium. At the time, radium was regarded as harmless, so the women habitually pointed the tips of their brushes by wetting them with their tongues. Thirty years later, many of these women began dying of mouth and tongue cancer. Today we can be appalled at the naïveté of allowing people to "eat" radium, but at the time there were no known adverse effects.

In tests where men were fed 3.5 and 35 mg of DDT a day for a period of time, no ill effects were seen. DDT accumulated in proportion to the amount consumed, then reached an equilibrium point between storage and excretion. When the dose was discontinued the body burden began to fall. Apparently the body stores DDT in response to intake rather than independent accumulation, so a person in a high DDT environment will store more DDT than a person in an environment with lower DDT levels. Although no one has died directly because of DDT properly used, many claims have been made and are being made that associate high DDT body burdens with various organ failures and malfunctions. But whether DDT directly or indirectly causes these problems or is just an innocent bystander has not been clearly demonstrated.

As long as there is any doubt about the role of DDT in human metabolism there is risk—an unnecessary risk, for we are not obliged to use DDT or any biocide. It is this doubt that led to the banning of DDT in several foreign countries and for most purposes in the United States, and will probably force restrictions on the use of other persistent chlorinated hydrocarbons as well. While there has been much obfuscation on both sides of the persistent chlorinated hydrocarbon controversy, anyone who presents an either-or choice, elms or robins, is mistaken because, as we shall see later, we can have both.

PERSISTENCE

We tend to regard the soil as some kind of biological incinerator; whatever we dump is supposed to be decomposed. But some of the organic molecules like DDT are unknown in a natural environment and enzymes that can degrade them are simply not available. The half-life (the time required for half the original quantity of a substance in the environment to decompose) of DDT is around 15 years, toxaphene 11, aldrin 9, dieldrin 7, chlordane 6, heptachlor 2 to 4, and lindane 2. These are maximum values; in some environments biocides decompose much more quickly, but the figures give some idea of the time it can take for the environment to cope with these materials. Decomposition, fortunately, does not depend solely on microbial attack, or we might never get rid of some persistent organochlorines. Volatilization into the air, photodecomposition by various wavelengths of solar energy, mechanical removal in crops, and leaching are also involved in reducing levels of persistent biocides; but volatilization and microbial degradation are the most important.

The danger of persistence is best seen in the organic biocide with the longest half-life, DDT. Though only slightly soluble in water (ppb), DDT is readily absorbed by the bottom mud of aquatic systems. As DDT in the water is degraded or removed by organisms, fresh supplies from the bottom mud are released. Because of its relatively long half-life, DDT can be released continuously into an aquatic system for years.

Residues

Persistence of biocides is also found in terrestrial environments. When spray schedules are maintained over a long period of time, residue levels build up. The soil beneath some orchards in Oregon was estimated to have retained over 40 percent of the DDT applied over a seventeen-year period. Some well-sprayed orchards accumulate 30 to 40 pounds of DDT per acre per year; others have totaled, despite degradation processes, as much as 113 pounds per acre after six years of spraying. Persistence is not limited to the newer organochlorines, though; arsenic trioxide has accumulated in some parts of the Pacific Northwest up to 1400 pounds per acre—a concentration that often poisons the crops themselves.

Some years ago the watershed around Lake George, New York was sprayed with 10,000 pounds of DDT per year for several years to control a gypsy moth outbreak. A decline in the population of lake trout followed. Although DDT residues ranged from 8 to 835 ppm in the adult trout and 3 to 355 ppm in the eggs, the adult fish were unaffected and the eggs hatched normally. Further research showed that the young fry were very sensitive to DDT, just at the time they were about to begin feeding. At DDT levels over 5 ppm there was 100 percent mortality.

Herbicide residues have also had some odd unanticipated effects. Coccinellids, the ladybird beetles which are voracious eaters of aphids, may become sluggish and far less effective predators and their larvae may grow 60 percent more slowly when affected by low levels of 2,4-D. This plant killer has also been observed to cause rice stem borers to grow 45 percent larger in treated rice plants and it may reduce the natural resistance of fish to parasitic microorganisms.

In a sublethal dosage, 2,4-D has increased the potassium nitrate content of sugar beets, which are frequently used for cattle food, from 0.22 to 4.5 percent dry weight, a nitrate level that is quite toxic to cattle. In another instance a low exposure to 2,4-D increased the sugar content of ragwort, a plant toxic to cattle but usually unpalatable to them. With an increased sugar content their avoidance was overcome and the cattle were attracted to a still poisonous plant.

While 2,4-D is quickly metabolized and 2,4,5-T lingers only a few weeks, picloram is quite persistent. Coupled with its biological activity (100 times 2,4-D) and solubility, the persistence of picloram means that its percolation into the water table and transfer to unsprayed areas is quite likely. Its persistence and potency were strikingly demonstrated in a tobacco field with an unusual distribution of stunted plants. With much effort, this pattern was traced to droppings left by mules used to plow the field. The mules had been pastured in a field treated with picloram, which had passed unchanged through the mules' guts and leached out of the droppings, affecting the tobacco.

Another unsuspected problem with herbicides was their potential to be transformed by soil microorganisms into a variety of other compounds, some harmless, some dangerous. Propanil (3,4-dichloro-proprionanilide) is commonly applied to the soil for postemergence control of weeds. Propanil is particularly useful in controlling barnyard grass in rice fields, so it is widely used. In the soil propanil breaks down to proprionic acid, which is quickly metabolized by soil microorganisms, and DCA (3,4-dichloroaniline). Logically the DCA molecule should be split in two and metabolized; instead it is polymerized or stuck together with other DCA molecules to form TCAB, which is stable, persistent, and closely related to carcinogenic azo compounds. Although TCAB does not seem dangerous, its mere presence indicates that we don't really know and cannot always predict what happens to biocides when they enter the soil and are transformed by microorganisms.

Perhaps the most difficult aspect of persistent biocides is the impossibility of their containment. About 15 percent of cultivated land in the United States receives biocides; 3 percent of grasslands and less than 0.3 percent of forests are treated each year. Altogether, 75 percent of the total United States land area remains untreated, and yet virtually all animals, man included, carry traces of DDT and other persistent biocides.

In response to mounting evidence from all sides indicting DDT as an environmental hazard, the Environmental Protection Agency, despite the continuing resistance of the USDA, has virtually eliminated the domestic use of DDT. The ban does not affect the export of 30 million pounds of DDT per year and a few minor, geographically limited uses may be excepted. But widespread use of DDT in the United States is apparently over.

NONSELECTIVITY

Not all of the problems caused by biocides in the environment derive from resistance, biological magnification, or persistence. Killing the wrong organism at the wrong time can be just as great a problem. One example having unexpected effects occurred a few years ago in the Near East. Jackals had been a problem for years in the settlements along the Mediterranean coast of Israel. Attracted by garbage, they stayed on to eat crops and increased greatly in number. In 1965 the government set out bait poisoned with "1080" (sodium fluoroacetate). The bait was widely distributed and no attempt was made to avoid poisoning other animals. As a result of this program, the jackals were eliminated but so were the mongoose, the wild cat, and the fox. With their predators eliminated, hares increased enormously, causing greater damage to crops than the jackals had before. Moreover, Palestine vipers, usually controlled by the mongoose, appeared in large numbers around settlements, to the consternation of the inhabitants.

Even more unpredictable were the results of spraying thatched-roofed huts in Borneo with DDT to control a malaria-carrying mosquito. The cockroaches and lizards that are normally quite common in these rather simple structures were injured or killed by the biocide and eaten by cats, many of which died as a result. Without cats to keep their population in check, rats increased greatly, giving rise to the fear of plague and other rat-transmitted diseases. As if this weren't enough to bear, the houses themselves began to collapse: the population of moth larvae, which normally feed on the thatch roofs, increased considerably, for while the larvae avoided DDT-sprayed materials and survived, their predators and parasites were killed. Although the British Royal Air Force came to the rescue with a massive cat replacement effort called Operation Cat Drop, which alleviated the rat problem, the villagers not surprisingly were reluctant to endure another round of house spraying, whatever the benefits.

Intensively cultivated land is highly sensitive to the disruptive effects of nonselective biocides. Rice fields which are flooded during the growing season in Louisiana also produce red crayfish, a local delicacy which nets the rice farmer an addition to his income beyond the income from the rice crop. However, to control the rice stinkbug, toxaphene often mixed with DDT or dieldrin was used. Because of an extreme sensitivity to toxaphene, fish were sometimes killed in large numbers. Gallinules and tree ducks were also killed by eating sprayed rice. When the switch was made to malathion and parathion to avoid damage to fish and wildlife, it was found that crayfish were quite sensitive to parathion and substantial crayfish kills took place before a combination of chemicals could be worked out that endangered neither fish, birds, nor crayfish and still protected the rice.

Natural ecosystems can also be strongly affected by the application of broad-spectrum biocides, which kill insects indiscriminately. The red spider mite normally is kept under good control by a variety of organisms, especially a predatory mite. When spruce forests along the Yellowstone River in Montana were sprayed to control the Englemann spruce beetle, the spider mites, which lived in web-like structures on the undersides of the needles, were protected from the spray, but the predatory mites were killed. The following year, although the Englemann spruce beetle was controlled, there was a huge wave of spider mite damage that was worse than that of the spruce beetle.

Finally, there are examples in the recent past where whole regions, not just fields or forests, were sprayed unnecessarily and indiscriminately. During the early enthusiasm over the innate superiority of chemicals to all other approaches, the concept of mass control arose. By means of various biocides the fire ant was to be eliminated and the Japanese beetle contained.

The fire ant is a minor pest introduced from Argentina into the southeastern United States. It builds mounds one to two feet high (Figure 13.8), inflicts painful stings, and occasionally damages crops and livestock. The USDA with its technological muscle came rushing to the rescue and began spraying large areas of the Southeast with dieldrin or heptachlor at the rate of two pounds per acre. While this was reduced in later years, over 2.5 million acres of land were treated by air in this program. Although the fire ant did not destroy livestock, wildlife, or crops to any important extent, the biocides in some instances did. After $15 million was spent on mass control, the Southeast still had to live with the fire ant—plus heptachlor residues in the environment. Here was one operation whose failure was attested to by the continued life of the pest. Unfortunately the USDA has continued pressuring the states to spray aerially, now with mirex, a persistent, broad-spectrum, chlorinated hydrocarbon. Even more unfortunate is the insistence that the fire ant could, much less should, be eradicated. Perhaps the USDA's concern is real, but its purview is narrow since it deals only with crop plants and their problems. Mirex in the environment shows no such discrimination and kills native ants of great ecological importance in their environments as well as the introduced fire ant.

The fire ant

Fig. 13.8 • Fire ant mounds in an Alabama pasture seem to be of little concern to the livestock. (*U.S. Dept. of Agriculture*)

Another example of overenthusiastic mass control was seen in the Japanese beetle eradication program. In the first few decades after its introduction in the United States from Japan, this pest was widely distributed in the Northeast and ate everything in its path. But with the introduction of biological controls (see Chapter 14), particularly a bacterium (milky disease) that attacks the beetle grub, populations have stabilized so that the dense clusters that covered grape vines, rose bushes, and most garden plants during the 1940s (Figure 13.9) are no longer a severe problem. Beetles are occasion-

The Japanese beetle

Fig. 13.9 • During the 1940s the Japanese beetle ranged unchecked. Whole trees were defoliated by swarms of beetles. (*U.S. Dept. of Agriculture*)

Fig. 13.10 • Widespread distribution of a bacterium causing the death of the beetle grubs has not eradicated the Japanese beetle, but has reduced its number to a tolerable level.

ally seen during the summer (Figure 13.10) but are nowhere as abundant as formerly.

At first, however, the response by government agencies to the continual westward spread of the Japanese beetle was a compulsive drive to eradicate the beetle in the new areas colonized. Consequently, close to 100,000 acres in Missouri, Kentucky, Illinois, Indiana, Iowa, and Michigan were broadly treated with two to three pounds of aldrin, dieldrin, or heptachlor per acre. As might be expected, chlorinated hydrocarbons at this dose rate were dangerous to vertebrates as well as beetles, not only in initial contact but through stable, persistent residues. A recent attempt in Michigan by an environmentally oriented pressure group to prevent mass control techniques from being used against a small Japanese beetle population was not successful, but the area to be sprayed and the dose rate were decreased.

The aim in both instances, to eliminate or control potentially damaging pests, was certainly in the public interest, but the technique of mass indiscriminate aerial spraying was not the best solution. Instead of thinking, "If one aspirin relieves a headache in thirty minutes, thirty aspirin should relieve it in one minute," we should be thinking, "If aspirin cures the headache but causes an upset stomach, maybe we ought to try another remedy."

But the most ominous of all the problems generated by biocides has been the degree to which farmers have become overdependent on them, perhaps seen at its most extreme in orchards across the country.

Spray Schedules

A hundred years ago orchards were small, containing perhaps a few dozen or at most a few hundred trees. There were enough adjoining fields and forest to allow a considerable amount of natural control. But as the orchards became larger, introduced pests could easily move from orchard to orchard, and because of the huge concentrations of trees they could inflict considerable damage. With the advent of organic sprays of low selectivity, whatever natural controls existed were reduced or eliminated, letting more pests survive and requiring more sprays. Today a typical commercial apple orchard requires a minimum of nine sprays of various biocides (Figure 13.11):

Fig. 13.11 • These peach trees were being sprayed with parathion at a time (1960) when the potential ill effects of broad spectrum biocides were unrecognized. The tractor driver, without protection, was doubtless being exposed to far more parathion than the peach trees. (*U.S. Dept. of Agriculture*)

1. In early spring when the trees are still dormant a spray is applied to eliminate aphid eggs that have overwintered.
2. When the fruit buds begin to show green, trees are sprayed to control mildew, scab, mites, and red bug.
3. As flower buds begin to turn pink, a spray is needed every seven days to attack apple scab. There is no spraying during full flower to allow for the pollinating activities of bees, without which there would be no apples.
4. After petals fall, a spray is applied for codling moth, curculio, redbanded leaf roller, cankerworms, mites, mildew, and scab.
5. A cover spray is added a week after petal fall to get codling moth, leaf rollers, cankerworms, curculio, and scale.
6. After ten more days a second cover spray is added to combat curculio, codling moth, leaf roller, fruit spot, and scab.
7. In ten to twelve days a third cover spray is added.
8. A fourth cover spray is added twelve to fourteen days later, to

catch codling moth, sooty blotch, scab, and fruit spot.
9. Two weeks later, a fifth cover spray is added to kill apple maggot, leaf roller, fruit spot, sooty blotch, scab, and codling moth.

By mid-July, a second batch of codling moths is likely; then, of course, there is black rot, white rot, fire blight, and so on endlessly.

The farmer is caught squarely in the middle. If he stops spraying he is left with either no crop at all or gnarled, wormy crabapples that no one will buy. If he continues spraying, new pests will appear, old ones will become resistant, and he will sooner or later be forced out of business, unable to afford spraying twenty-four hours a day.

A similar dilemma was faced by farmers in the Cañete Valley in Peru. Modern organic-synthetic biocides were introduced between 1949 and 1956 to control seven major pests on the chief crop, cotton. At first the yield of cotton increased, from 406 pounds per acre to 649 pounds per acre. But by 1965 the yield had dropped to 296 pounds per acre and major pests increased from 7 to 13, several of them highly resistant to the biocides. Worst of all, the lowered yield of cotton plus the increasing frequency and expense of biocide application was driving the farmers into bankruptcy.

Integrated Control

The solution in Peru was a break with total dependence upon organic, synthetic biocides and adaptation of a more sensible blend of control measures called integrated control. The heart of this program is decreased use of biocides and the toleration of a few pests in return for freedom from biocide-resistant hordes. There are alternatives to the exclusive use of biocides: development of resistant crops; biological control (see Chapter 14); and cultural methods like destruction of crop refuse, deep plowing, timed planting, and rotation of crops, which collectively can eliminate overwintering stages or upset the often delicate timing that synchronizes pest and crop.

Knowing the problem and trying to work with it can be more successful in the long run than a blind attack with a broad-spectrum biocide. The California spotted alfalfa aphid becomes a problem when large tracts of alfalfa are cut. By harvesting the fields in strips, refuges are left in the unmown strips for the natural enemies of the aphid—fifty-six predators and parasites per square foot in the specially mown fields, versus fourteen per square foot in fields harvested normally.

The California grape leafhopper developed resistance to the organophosphates ordinarily used for control. Workers noted that the wasp *Anagrus* actively attacked the leafhopper but only during the latter part of the growing season, which left the young grape leaves open to leafhopper attack early in the season. When wasps were absent from the vineyards they were attacking another leafhopper on blackberry bushes some distance away. Blackberry bushes had grown in the vineyards, but were regarded as weeds and killed with herbicides. When blackberries were reintroduced to vineyards, the predatory wasp had an alternate prey within easy reach, which helped to maintain the wasp population at a high enough level to control effectively the grape leafhopper.

But pest control need not always involve killing the pest. In many parts of the West, reforestation of cut or burned areas by aerial seeding is ineffective because of extensive populations of white-footed mice which devour the seeds almost at once. If mice are poisoned, new mice from surrounding areas quickly repopulate the area and eat the remainder of the tree seeds. One ingenious approach pioneered by the mammalogist Lloyd Tevis was to condition the mice to avoid all tree seeds by poisoning some with sublethal doses. Experiments indicated that one sickened mouse was worth a dozen dead ones, for once recovered, the mouse not only avoided all tree seed, poisoned or not, but taught its offspring to do likewise. Unfortunately the variations in individual mouse behavior prevented widescale use of this approach in the situation we have described, but the idea is one that might be studied further.

Integrated control, however, is not just alternatives to organic chemicals but moderate use of all controls, including chemicals. A rational, balanced approach that considers ecological factors rather than ignoring them might use the guidelines that follow.

Before any control is attempted it must be determined that the supposed pest is actually causing the damage attributed to it. Millions of dollars worth of spray chemicals have been wasted on false assumptions. Then a survey should be made of the natural control agents that normally work to limit the population of the pest. If the pest is introduced and has no apparent enemies, locate the predators, parasites, and pathogens from its place of origin (no organism is without these associates!), screen them carefully to avoid making the situation worse by adding more pests, and attempt controlled introduction. Make every effort to preserve and increase the number of natural control organisms, native or introduced. Such biological controls have been much more effective than chemicals in many instances, for there is no possibility of the pest developing resistance. Do not panic at the first appearance of a pest. Give the natural controls some time to operate. If chemicals prove necessary to achieve control, use the least amount that will be effective, instead of trying to kill every pest in the field. Then spray only when necessary, avoiding routine spray schedules. Remember that overkill and routine spraying quickly lead to pest resistance and elimination of natural controls, requiring still more sprays, which is what we are trying to avoid. Wherever possible, avoid broad-spectrum biocides, particularly if they are persistent. Persistence tends over the long period to eliminate more beneficial organisms than pests.

If these guidelines are followed, most of the disadvantages of biocides can be avoided. In fact, many of these materials can take their place as useful tools in the continuing war against the great variety of potential pests that man has inadvertently been nurturing through his ecosystem disruption and simplification. If mass indiscriminate spraying continues, problems seen by the late Rachel Carson will seem like bedtime stories compared with the grim realities ahead.

FURTHER READING

Carson, R., 1962. *Silent spring.* Houghton Mifflin, Boston, Massachusetts. The pioneering book on the environmental effects of biocides.

Graham, F., Jr., 1970. *Since silent spring.* Houghton Mifflin, Boston, Massachusetts. A sequel to Carson's book, updating many of the incidents reported in the earlier book.

Gustafson, C. G., 1970. "PCBs—prevalent and persistent." *Env. Sci. & Tech.* **4**(10), pp. 814–819. Fairly comprehensive review of a widespread problem.

Peakall, D. B. and J. L. Lincer, 1970. "Polychlorinated biphenyls, another long-life widespread chemical in the environment." *BioSci* **20**(17), pp. 958–964.

Pimentel, D., 1971. "Ecological effects of pesticides on nontarget species." Office of Science and Technology, Washington, D.C. Concise review of some of the problems of broad-spectrum biocides.

———, 1973. "Realities of a pesticide ban." *Env.* **15**(2), pp. 18–20, 25–29. While a total pesticide ban is most unlikely and not necessarily desirable, its impact is carefully evaluated here.

Rudd, R. L., 1964. *Pesticides and the living landscape.* University of Wisconsin Press, Madison. A systematic survey of the various types of chemical pest controls and their environmental effects.

Woodwell, G. M., C. F. Wurster, Jr., and D. A. Isaacson, 1967. "DDT residues in an east coast estuary: a case of biological concentration of a persistent insecticide." *Science* **156**, pp. 821–824. Poisoning of a salt marsh food chain by DDT.

Woodwell, G. M., P. P. Craig, and H. A. Johnson, 1971. "DDT in the biosphere: where does it go?" *Science* **174**, pp. 1101–1107.

4

The Biotic World and Man

In natural environments nothing goes to waste. When this cheetah finishes his meal, the vultures will pick the remains clean. No one is sure what will happen to the vultures or the prey species when the last cheetah is processed into a fur coat. All of these species may have to give way to growing demands for space by the developing countries of Africa. (*Wide World Photos*)

14

INTRODUCTION OF
EXOTICS
AND THEIR CONTROL

IN THE GEOLOGIC PAST the continental land masses were much closer together than they are now, allowing a much broader natural distribution of plants and animals than is possible today. At that time there were fewer climatic or geographic barriers to isolate species. Even as late as the Tertiary period, which began 65 million years ago, much of the land area of the earth was covered with broad belts of tropic, subtropic, and temperate types of vegetation and inhabited by animals of a more general and wide-ranging sort than we have today.

About 100 million years ago, South America and Africa, Australia and Antarctica, Europe and North America, began to be pulled apart by the forces of continental drift—slow-moving but enormous convectional cells whose circulation deep within the earth moved the shallowly rooted continents about like the film on scalding milk (Figure 14.1). Continuous land masses began to be broken by straits that became channels, then seas, and finally oceans. Because of continental motion, long quiescent land areas underwent mountain building episodes that further divided the inhabiting species from one another and limited their distribution. Climatic alteration, related both to mountain building and other global phenomena, ultimately led to continental glaciation that ended the Tertiary period and further isolated the distribution of organisms. The stresses produced by these changes resulted in extinction for some groups and reduction in range for others; but the same stresses also stimulated rapid evolution in the plants and animals that were genetically able to adapt to changing conditions.

NATURAL INTRODUCTIONS

The slow emergence of savannah, grassland, and then desert encouraged the rapid evolution of the host of grazing and browsing

Fig. 14.1 • A hypothetical land mass called Gondwanaland included the present continents of South America, Africa, Antarctica, and Australia before being broken up by continental drift. (*Modified from A. C. Smith and A. Hallam,* Nature, *1970.*)

Fig. 14.2 • Cattle egrets are often found in pastures, where they feed on insects stirred up by the movement of livestock. (*Walter Dawn from National Audubon Society*)

animals that can be seen today in east Africa. These groups reached their greatest geographic isolation towards the middle of the Tertiary period. The later formation of a land connection between North and South America (the Isthmus of Panama) and between Siberia and Alaska (the Bering land bridge) allowed the movement of animals and plants from their isolated centers of specialization and speciation, movements that are still taking place today. Glaciation further mixed the once-isolated stocks of plants and animals.

South America was invaded by tapirs, llamas, peccaries, foxes, dogs, cats, otter, and deer from North America, driving many of the primitive marsupials into either extinction or migration. One marsupial, the opossum, invaded North America from South America and is still extending its range northward. The porcupine and armadillo are also fairly recent invaders of North America. The Bering land bridge allowed elephants and deer to move into North America from Asia, while it also allowed horses, camels, and tapirs to move from North America to Asia.

Usually this process of range extension is extremely slow, but it may be quite rapid. Both the cardinal and mockingbird, formerly limited to the South, were considered rare visitors in Pennsylvania or Connecticut thirty years ago. Today, both birds breed in these northern states and seem to be extending their range farther north. Shortly after World War II, cattle egrets (Figure 14.2) were moved, with the help of storm winds, to South America from Africa. They then worked their way to Florida and now regularly breed along the southern coast from North Carolina to Texas.

By and large, the post-Tertiary plant and animal redistribution had stabilized by the time man arrived on the scene. But once again these relative isolations are being disturbed, not by continental drift, climate, or land bridges, but by the unique mobility of man. The wanderings of man have encouraged the introduction of plants and animals on a scale never before possible. For thousands of years these introductions were limited to crop plants and weeds, domesticated animals and their pests; but more recently, a greater and more varied number of plants and animals has been shifted about, accidentally or purposefully—sometimes with catastrophic results.

MAN THE INTRODUCER

Despite their proximity to other continents during the early phase of the breakup of Gondwanaland, for the last 70 million years New Zealand and Australia have been isolated from the floral and faunal developments that swept the rest of the continents. These continents have a unique fauna and flora containing many species found nowhere else in the world, while the plants and animals characteristic of the rest of the world, especially Europe, were, until recently, altogether lacking. New Zealand, for example, had only two native species of mammals, both bats. The Maoris, early Polynesian settlers of New Zealand, brought in dogs and rats, but it was not until white settlement that widespread introduction of animals began.

Because New Zealand was so far from their home in the British Isles, halfway around the world, the early white settlers were wretchedly homesick. Not only were there no animals to remind them of home, but the shrubs, trees, and flowers bore no resemblance to anything known in England. So great was this feeling of isolation and homesickness that the strange plants and animals were attacked as an evil to be replaced with something familiar. (The same attitude might have taken root in North America had the plants and animals not resembled those of Europe so closely.) Plants and animals began to be introduced at a rapid rate so that New Zealand would look like England as quickly as possible. What followed was an orgy of introduction. In the 120 years since white settlement, 207 species of vertebrates have been introduced. Of these, ninety-one became established, a fairly high rate of success due almost certainly to the lack of competition from native mammals. With such an auspicious start one might picture New Zealand today as a kind of zoo without bars like that famous primitive American painting, *The Peaceable Kingdom* by Edward Hicks (Figure 14.3). But of the ninety-one successful introductions, twenty-nine became problems of one sort or another.

Originally native to Europe, red deer were introduced early to New Zealand. With no serious grazing competition and no predators, there were soon many more deer than could be controlled by hunting. They began to compete with sheep, destroy crops, inhibit forest regeneration, and finally, by overgrazing, to cause serious soil erosion. Today there is no limit or closed season on red deer. In fact, the government offers free ammunition to anyone who will kill them, but

Fig. 14.3 • This painting, *The Peaceable Kingdom,* by the American artist Edward Hicks, depicts the peaceful coexistence of man and animals. Unfortunately the introduction of a variety of disparate animals is never peaceable. *(Courtesy of the Brooklyn Museum, Dick S. Ramsay Fund)*

the red deer are still flourishing.

A similar story can be told of Australia. Unlike New Zealand, Australia had many vertebrates, but because of Australia's long isolation its vertebrates were mostly the primitive marsupials, possum, kangaroos, wallabies, and the like, which were generally outcompeted when faced with the more advanced mammalian stock introduced by settlers. Like the New Zealanders, new Australians missed the familiar animals of home and soon introduced the rabbit. To their credit, they introduced the fox as well, as a predator. But in Australia foxes and rabbits seemed to go their own ways; the fox never amounted to much as a rabbit predator while the rabbit became one of the greatest pests of all time, as will be discussed later in this chapter.

The Biological Context

What went wrong? Why did perfectly well-behaved European animals suddenly become pests? No animal or plant exists in a world limited solely by physical and chemical factors. With few exceptions, for every plant or animal there is a potential consumer or predator that serves to check its population growth. There are also parasites, and parasites of parasites. A few years ago, John Tilden took a long careful look at a plant that grows on the sand dunes of California and Oregon. "Associated with it he found, by two years of systematic observation, 257 species of arthropods of which 221 were insects. Of the latter 65 were parasites and that was not complete. There were 53 species of primary herbivores—leaf-nibblers, leaf miners, stemborers, leaf-suckers, root feeders, gall-makers; 23 species of predators; 55 species of primary parasites, 9 of secondary parasites; and even one tertiary parasite."[1] All of these organisms interact so that there are rarely too many or too few objects of their predation. This is why in a stable ecosystem there are usually relatively constant numbers of plants and animals from year to year. But when you pluck such a plant or animal from its ecological context and fling it into another environment it will either disappear almost immediately or proliferate with unbelievable fecundity. This is a direct measure of the importance of ecological associations in maintaining stable populations.

Even today, we know far too little about the ecological relationships of plants or animals to be able to predict which course the introduced plants or animals will take in a new environment. Obviously elephants would be short-lived in Labrador, but what about zebras in New Mexico or bamboo in Louisiana? Of course not all introductions have been disastrous, nor have they all been intentional; let us examine the record.

Types of Introductions

Few insects have been intentionally introduced except in a desperate attempt to combat proven pests biologically, but accidental introductions have been rife. The Japanese beetle has already been discussed in Chapter 13. Of even greater impact has been the gypsy moth. Leopold Trouvelot, a French astronomer employed by the Harvard Observatory, tried to crossbreed several silk-producing moths with the aim of finding a strain resistant to a disease threatening the French silk industry. One of his subjects was the gypsy moth, notorious in Europe for its destructive effect on the leaves of shade trees. In 1869 Trouvelot accidentally allowed some gypsy moths in his collection to escape near his home in Medford, Massachusetts. Seemingly, they disappeared into the blue. But twenty years later in 1889, Medford was overwhelmed by an incredible swarm of caterpillars.

Insects

The numbers were so enormous that the trees were completely stripped of their leaves, the crawling caterpillars covered the sidewalks, the trunks of the shade trees, the fences and the sides of the houses, entering the houses and getting into the food and into the beds. . . . The numbers were so great that in the still, summer nights the sound of their feeding could plainly be heard, while the pattering of their excremental pellets on the ground sounded like rain. Valuable fruit and shade trees were killed in numbers by their work, and the value of real estate was very considerably reduced. . . . To read the testimony of the older inhabitants of the town, which was collected and published by a committee, reminds one vividly of one of the plagues of Egypt as described in the Bible.[2]

[1] Elton, C. S., 1958. *The ecology of invasions by animals and plants.* Methuen, London.

[2] Howard, L. O., 1930. *A history of applied entomology (somewhat anecdotal).* Smithson. Misc. Coll., 84.

In response to this scourge, the state government was persuaded to employ some control measures which were moderately successful until 1901, when Massachusetts decided enough money had been spent killing caterpillars and canceled the appropriation. By 1905 the gypsy moth infestation was not only out of control again but beyond the resources of state action; the federal government stepped in and has been attempting to prevent further spread of the gypsy moth ever since (Figure 14.4). For many years, spread of the infestation was checked by the Hudson River and by annual DDT aerial spraying programs. Then a series of hurricanes carried moths to the west, embroiling the federal government in a controversial mass spraying program over much of the Northeast. Taxpayer suits finally dissuaded use of the shotgun approach and today other approaches to control are being explored.

Fig. 14.4 • (a) This young maple was attacked by gypsy moth caterpillars early in the summer. The green leaf blades were soon skeletonized to the tough veins and petioles, which are still attached to the twigs. (*Photo by Dan Jacobs*) (b) Two months later this same tree has put out a second set of leaves, completely obscuring the earlier attack.

The problem seems to have been a classic example of negative feedback. Populations of gypsy moths built up to peak levels, which were then sprayed with DDT or sevin. While the moth population was thinned, its natural predators were wiped out. Unfortunately, the moth was able to build up its population before the predators and another population cycle took place, eliciting another aerial spraying. The net effect of spraying, despite the gratification of seeing larvae die, seems to be setting up population outbreak after outbreak. Spraying of individual trees in one's backyard poses few problems, but aerial spraying of whole forests is misguided at best.

Practically every crop in the United States has at least one introduced pest of economic importance. Introductions have not been one-way, however. Two examples of a United States native causing problems abroad are the Colorado potato beetle (Figure 14.5) and the grape phylloxera. The potato beetle, when originally discovered, was feeding quite contentedly on a wild relative of the potato in the Rocky Mountains. In 1859 potato cultivation came to Colorado. Sometime that year the beetle discovered the domestic potato. By 1874 the beetle had reached the Atlantic coast and a short time later it was inadvertently carried to Europe. In 1930 most of western, central, and southern Europe had become infested.

Fig. 14.5 • The infamous Colorado potato beetles with their eggs. (*Courtesy of the American Museum of Natural History*)

During the Korean War, North Korea and China accused the United States of dumping canisters of potato beetles on their potato fields. But in view of the expansion of its range in the hundred years since its discovery, the Colorado potato beetle hardly needs the assistance of the CIA to manage its dispersal!

The phylloxera is an insect that sucks juices from the rootstock of grape vines. Native to the grapes of eastern North America, the phylloxera usually does no particular damage. When introduced into France on the rootstocks of American grapes just before the Franco-Prussian war in 1870, the result was catastrophic. At a time when France could least afford further calamity, vines began dying throughout her world-famous wine districts. For a few years grape juice for wine production was limited to that wrung from dried grapes shipped in from uninfected areas abroad, surely a low point for the French spirit. Within ten years the insect was killing vines in Germany and Italy as well. But the wild American grape came to the rescue by providing resistant rootstocks to which the superior European grapes could be grafted. In another few years, production was back to normal with the curious result that today all French grapes have "American" roots.

Fish introduction *seems* to have caused few problems. But we have no way of being sure since displacement and even extinction of anything but the most popular game fish have until quite recently gone unnoticed by most people.

Fish

Probably the most widely introduced fish is the mosquitofish (*Gambusia*) which is native to the southeastern United States. The mosquitofish has been introduced into subtropical and mild climates all over the world because of the avidity with which it feeds on mosquito larvae, hence the effectiveness with which it controls the incidence of malaria. Although its total ecological effect beyond mosquito control has never been assessed in any of the areas into which it has been introduced, the mosquitofish has apparently caused remarkably few problems.

Introduction of *Gambusia* is not without effects, however. Experimental cultures of mosquitofish have shown that the fish feed heavily on zooplankton. This feeding pressure allows the algae that the zooplankton normally eat to multiply, causing dense algae blooms to develop. These greatly reduce light penetration and increase water temperature. This work suggests that eutrophication may sometimes be due to manipulation of fish populations by man as well as to direct increases in the nutrient load (Chapter 6).

Trout and salmon have also been widely introduced into areas where native sport fish are absent or poorly regarded. Again, the impact of these exotics on native fish has never been investigated, but no one is apt to complain about the abundance of trout or salmon in a stream, and the side-effects have not yet been noticed.

While fears of released piranha fish tearing Florida swimmers to ribbons have not yet been realized, a catfish introduced from Asia has caused some concern in southern Florida. Able to breathe out of water for up to twelve hours and wriggle cross-country, these catfish can live under conditions lethal to other species. Naturally the Florida Fish Commission is concerned that with its voracious appetite the "walking" catfish will outcompete more desirable freshwater fish.

Another fish widely and thoughtlessly introduced by people is the goldfish. Most people have compunctions about flushing unwanted goldfish down the toilet, so they release them in nearby ponds or streams where they often reproduce vigorously to the exclusion of more desirable fish—goldfish may be pretty, but tasty they are not.

Birds   For centuries, birds as pets have been moved about the earth at the whim of man. During the Victorian age this effort reached an almost frenzied peak. In city after city acclimatization societies were formed to introduce the avian glories of Europe and the East to the urban areas of the northeastern United States. Birds such as the nightingale, skylark, and English robin were introduced again and again, but without success; even the English sparrow (house sparrow), which had long been regarded as a pest in Europe, was repeatedly introduced. Some measure of the extent to which this craze gripped otherwise conservative institutions can be seen in the purchase and release of 1000 sparrows by the city government of Philadelphia in 1869.

Unfortunately, the sparrow succeeded where the nightingale and skylark did not. Soon the noisy, dirty, little birds were everywhere, chasing away native birds, nesting in great piles of trash and under eaves and eating large quantities of grain. By 1883, Philadelphia thought better of its humanitarian action and passed a law making it lawful to kill or destroy in any way possible the bird it had so enthusiastically introduced.

Before long, however, the house sparrow had extended its range all over the United States and Canada. But like the house fly, the Norway rat, and the common pigeon or rock dove, the sparrow was never able to sever its connection with man. It was unwilling or unable to wander too far from the largesse provided by man's disturbance of the environment—the true measure of a pest or weed. In a few decades, house sparrow populations gradually declined, perhaps because of the decline of the horse when the automobile came into widespread use. Horse droppings undoubtedly provided a nutritional margin that helped carry sparrow populations through severe winters. Today house sparrows remain common and seem to have made a secure niche for themselves as an accompaniment of human civilization, but they are nowhere overly abundant. The starling has had a similar history except that its numbers have not yet stabilized.

Another bird associated with man and his cities is the rock dove or city pigeon (Figure 14.6). Kept alive and thriving almost exclusively by what they can glean from abundant city garbage or crumbs generously scattered about by well-meaning urbanites eager for the sight of any bird, pigeons have increased into enormous flocks in many cities. Ledges, windowsills, and the exuberant architectural ornaments on older buildings are used as nest sites and the pigeon's copious excrement has become one of the vicissitudes of life in most cities. Pigeons have persisted despite occasional efforts to reduce their numbers. It has become clear, however, that pigeons are capable of transmitting several diseases to man—parrot fever (psittacosis or ornithosis), cryptococcal meningitis, histoplasmosis, and encephalitis.

Fig. 14.6 • Pigeons in the city survive by scavenging to some extent, but large populations are possible only through the largesse of people who feed them. (*Frank Puza, courtesy of the American Museum of Natural History*)

301

Fig. 14.7 • The monk parakeet, a recent accidental introduction to the avifauna of the eastern United States, may very well reoccupy the niche vacated by the native but now extinct Carolina parakeet. This monk parakeet is attacking figs, but a wide variety of less exotic fruit is also attractive to this species. (O. S. Pettingill, Jr. from National Audubon Society)

Mammals

Today, efforts to control urban pigeon populations are taken more seriously. By generously spreading pigeon feed treated with a contraceptive drug in suitable gathering places, large numbers of pigeons are rendered infertile, allowing the ravages of age, alley cats, and taxicabs gradually to reduce the pigeon population to manageable terms. There are still pigeons in New York City, but not the huge flocks that there once were.

In the late 1960s the monk parakeet (Figure 14.7) became one of the most recent additions to the fauna of the United States. Imported by the thousands for the caged bird market, dozens were released or escaped in the New York area and soon were building nests and laying eggs. At first these efforts were unsuccessful because the monk parakeet's native habitat is Argentina and Bolivia, south of the equator; hence the birds began nesting activities in their spring, which is our fall. This should have limited their spread, but after a year the birds became acclimated to the Northern Hemisphere and began rearing young. By the early 1970s monk parakeets were being spotted in twenty-five states, from Maine to Alabama and west to North Dakota. In view of the large size of the bird, ironically about the same as the Carolina parakeet (see Chapter 15), and its voracious appetite for fruit, many people are concerned that another starling or house sparrow has been unleashed. Consequently, attempts are being made to locate and eradicate colonies of the monk parakeet before the situation gets out of hand.

In 1859 Thomas Austin imported a dozen pairs of European rabbits to Australia and released them on his ranch. Six years later, after killing 20,000 of them, Austin still had 10,000 left. This fecundity is quickly explained by noting that a female rabbit can breed at four months, then average six litters of six young (called kittens) a year thereafter. Theoretically, one pair of rabbits can generate about 13.7 million offspring in three years! Fortunately, mortality is around 80 percent and adults rarely live more than one year. Nonetheless, in a few short years Australia was faced with a national problem because five rabbits ate as much grass as one sheep. Trying to capitalize on the grave mistake of introducing the rabbit, Australia soon became the world's major supplier of rabbit meat and rabbit fur. But the loss of sheep range more than balanced any economic gain from the rabbit. Besides direct grazing competition, the rabbit burrows extensively undermined buildings, initiating soil erosion and further ruining range land.

All means of control were attempted—trapping, shooting, netting, poisoning, fumigation of burrows, and even erecting a 1500-mile fence designed to keep the rabbits from moving into western Australia, but these measures were of little avail. The combination of ideal habitat and minimal natural predation allowed the rabbit population to increase by 1953, despite control measures, to an estimated *one billion rabbits* over an area of a million square miles (Figure

14.8). Some of the economic burden this rabbit population placed on the Australian economy can be seen by the calculation that if there were no rabbits in Australia the land could support another 100 million sheep. Recently the rabbit population was reduced considerably by biological control. This kind of control will be discussed later in this chapter.

Fig. 14.8 • Rabbits gathering at a water hole in Australia during a drought. At the time this photo was taken there were one billion rabbits in Australia. (*Australian Information Service*)

Australian experience with the rabbit is perhaps an exception, for although there have been problem introductions of muskrat into Europe and nutria into the United States, the most severe problems have usually arisen when various mammals have been introduced to small islands. Occasionally chains of errors have followed an initial introduction. Many years ago rabbits were introduced as a supplemental food source on some rather barren windswept islands southeast of Australia. When the rabbits began to compete with the sheep, the economic mainstay of the islands, cats were brought in to reduce the rabbit population. But the cats preferred the eggs of sea birds nesting on the islands and used by the natives as food. In desperation, the islanders brought in dogs to control the cats; however, while the dogs merely chased the cats, they attacked and ate seals on beaches, depriving the islanders of still another food source. In view of the wide range of problems stemming from introduced animals it is heartening that most introductions are unsuccessful.

Plants

One of the most important forest trees in eastern North America until about 1900 was the chestnut. A singularly useful tree, the chestnut tree grew tall and straight providing good saw timber; the bark and wood had a high tannin content, which was used in the tanning of hides and which preserved the wood from rapid decay in contact with soil, making the chestnut especially useful for fence posts and rails; the nuts were of excellent quality for roasting and provided much food for game and livestock; and finally, it was a fine shade tree, widely planted and as widely admired:

> Under a spreading chestnut-tree
> The village smithy stands;[3]

The village smithy went out with the horse and buggy but unfortunately so did the chestnut. A wealthy New York gentleman imported a number of foreign chestnut trees to plant on his Long Island estate just before the turn of the century. A few years later he noticed that his American chestnuts were dying. In a few more years, chestnuts were dying all over Long Island, then New York, Connecticut, and New Jersey. By 1940 the chestnut no longer existed as a forest tree. The cause of its demise was a fungus introduced on trees brought from China. The Chinese species of chestnut apparently evolved with the fungus and developed a tolerance to it, but the American chestnut had no such immunity; in fact it proved extremely susceptible. The fungus slowly girdles the trunk of the tree, filling the water-conducting vessels of the wood with its growth, and the crown of the tree, deprived of water, slowly dies. Because the roots remain viable the chestnut continues to sprout, but just enough fungus spores remain in the environment to kill the sprouts after a dozen years or so, preventing the regrowth of any sizable tree. Some chestnuts have been vainly sprouting in this manner for over fifty years. Although the loss of such an important tree was staggering, because of the resistance of the wood to decay, dead trees are still being cut in the southern Appalachians, where they remain a declining source of the split-rail fencing so popular in suburbia today.

CONTROL OF INTRODUCTIONS

As the list of disastrous introductions lengthened, man slowly became aware that although he had the unquestioned ability to transport organisms—moose, mice, or measles—wherever he pleased, he could not pull the strings that would make these organisms behave as he wished.

Quarantine

The most obvious solution to the problem of introductions is to avoid them in the first place. Most countries have strict regulations prohibiting random importation of any plant or animal stock which might cause problems or which may be a carrier of potentially harmful pests or diseases. The United States requires that all plants entering the country from abroad be fumigated or carefully inspected to

---

[3] Longfellow, H. W., 1903. *The complete poetical works of Henry Wadsworth Longfellow*, Houghton Mifflin, Boston.

eliminate plant diseases or insect pests. Imported wild animals must spend a period of time in quarantine in Clifton, New Jersey, to be sure the animals are disease free. Even then under federal law they cannot be released in this country without a permit from the Secretary of Agriculture. Such strictures in force in the United States have sharply decreased the number of new introductions since 1900. But despite these measures, unwanted plants and animals still occasionally get through.

Eradication

If the infestation is discovered quickly enough eradication is usually attempted. Hoof and mouth disease in cattle or the Mediterranean fruit fly on oranges in Florida are examples where the unwanted organism was eliminated by quick action before it was well-established or spread too far. Eradication is far more easily envisioned than accomplished. Too many times eradication is initiated when the chance for success does not merit the potential damage to other species; this was the case in the previously discussed campaign against the fire ant in the Southeast, and the Japanese beetle in the Midwest.

Biological Control

Should an organism evade both border quarantine and eradication attempts and become established, there is a third line of defense that does not necessitate an endless series of biocide sprays. Every organism, as we have seen, is surrounded in its natural habitat by a plethora of other organisms with which it has evolved and to which it is related through producer-consumer, predator-prey, or parasite-host bonds. If the new habitat of an introduced species is a reasonable facsimile of the old and if the web of associates is left behind, as it almost always is, there are few checks and the population may "explode."

Why are organisms new to a habitat not immediately eaten and eliminated or at least assimilated? The bonds mentioned are developed over long periods of time and are quite conservative. Interlopers have no place. A bird is as loath to sample a beetle it has never seen before as a shopper is to buy rattlesnake meat at the supermarket. After many years, however, predator-prey relationships slowly form and the species is gradually "absorbed" by the environment as the house sparrow and Japanese beetle have been. But the period of "absorption" may take hundreds of years.

The point of biological control is to speed up this process of natural "absorption" by introducing whatever consumers, predators, or parasites are available from the old habitat that will bring an introduced species under control in the shortest possible time. Biological control is not a new approach, but since it is neither quick, easy, nor necessarily successful, biocides were for a time considered the best solution—that is, until their drawbacks became too glaring to ignore. Today there is renewed interest in the possibilities of biological control.

The first step in applying biological pest control is to know the life cycle of the pest intimately to see where the weak points are, and where controls might be successfully applied. Then the original habitat is carefully studied so that the interrelationships between the control species and the target species are fully understood. Before attempting any introduction of control agents, careful screening must be carried out to be sure that a second pest is not introduced. A convenient rule is that an introduced control species should prefer to starve to death rather than attack some species other than the pest to be controlled. The new habitat must also be generally suitable to the intended control species. Finally, the control species must be reared or cultivated in large enough numbers so that when introduced there is a fair chance of success. Then cultural practices should be modified wherever possible to encourage survival and reproduction of the control species. When these steps are carefully carried out with one or more control species, the target species *may* be effectively controlled.

Despite the time, expense, and uncertainties, many pests have been quite effectively controlled, some spectacularly, using biological methods. Just about any organism from virus to sea cow can be used as a biological control.

Viruses

The European spruce sawfly was introduced into eastern Canada in the 1920s, and before long it had destroyed 3000 square miles of timber. When a virus important to its control in Europe was accidentally introduced, the insect was practically eliminated in three years. Even more impressively, the Australian rabbit problem was finally but quickly brought under control by a virus disease, myxomatosis. Indigenous to rabbits in Brazil, where it causes a minor skin disease, myxomatosis was incredibly virulent when introduced to captive Australian rabbits. After determining that the virus affected only the rabbits, infected rabbits were released, and before the end of a year the virus, spread by mosquitoes, was killing rabbits by the hundreds of thousands. After a couple of years the virulence began to decline, eliminating the possibility of extermination, but the measure still provided better control than could have been achieved by any other method.

The appeal of viruses as a biological control lies in their specificity in contrast to biocides. Of the 280 viruses isolated from various insects, the most promising belong to a group called the nuclear polyhedrosis group. Normally, these rod-shaped viruses are encapsulated in a proteinaceous sheath that insulates them from the environment. This coat is dissolved in the insect gut, and the activated virus penetrates the insect cells, ultimately killing them. When properly diluted, evenly distributed, and applied at the right stage of insect development, viruses can be extremely effective.

Viruses are not yet in commercial production because the FDA is quite properly being cautious about possible effects on other organisms. Tests using the corn earworm virus on a variety of animals and preliminary tests on man have shown no ill effects. As soon as additional tests are completed, viruses will probably be licensed and marketed. If proven safe, as well as successful, other viruses capable of attacking other insects may well replace some of the biocides in use today (see Chapter 13).

For a number of years one bacterium (*Bacillus thuringiensis*) has been commercially produced and used as a means of controlling the cabbage worm and alfalfa caterpillar. The bacterium is extremely virulent and spreads rapidly.

Bacteria

Another related bacterium discovered in New Jersey in 1933 attacks the grubs of Japanese beetles. Because the infected grubs turn white, the infection is called the milky disease. The spores, which are resistant to extremes of temperature and moisture, are easily applied to the soil as a dust. Since the spores pass unharmed through the digestive tract of any animal eating the infected grubs, they remain viable and slowly spread throughout the environment. When large areas are treated, very effective, long-lasting control of Japanese beetles can be obtained without the destructive side-effects of mass aerial spraying (Chapter 13). Between 1939 and 1953 the USDA in cooperation with various state and other federal agencies distributed 230,000 pounds of spore dust at 160,000 sites in 14 eastern states. Much of the reduction in the Japanese beetle population over this period may well be due to the slow but effective spread of this bacterial infection.

The most widespread and certainly the oldest form of biological control involves insects. Successful control of both plant and insect pests by insects has been obtained many times. Two examples illustrate the possibilities.

Insects

The prickly pear cactus was introduced into Australia as early as 1788. It was used then as it still is in Mexico and the southwestern United States as a defensive or sheltering hedge around ranch houses. But by 1870 it began to get out of control. In 1900, 10 million acres were covered with the cactus, and by 1920, 60 million acres were infested. Both mechanical and chemical controls were too costly because of the huge areas affected. As early as 1912 research was begun to survey the natural enemies of the prickly pear. Of the several possibilities a plain brown moth called *Cactoblastis* was selected and reared in the laboratory. Over a ten-year period nearly 3 billion eggs were distributed. The eggs hatch into caterpillars which tunnel inside the flat cactus segments, the stems, and even the roots. When this is followed by bacterial rots, the plants are quickly destroyed. By 1933, prickly pear was under effective control in Australia and although the cactus was not eradicated, there is a large enough resident population of *Cactoblastis* to prevent any widespread regrowth of the prickly pear.

Not long after oranges had been introduced to California on a large scale in the late nineteenth century, the cottony-cushion scale, a sucking insect that weakens the tree and reduces the orange crop, was introduced from Australia. A search in Australia by C. V. Riley, an early pioneer in biological control, turned up a ladybug called the vedalia beetle, which when introduced to California proved extraordinarily successful in reducing damage (Figure 14.9). So successfully did the vedalia beetle control the scale that it began to be taken for granted. When broad-spectrum biocides were first used on orange groves in California seventy years later and the vedalia population declined, cottony-cushion scale again became a serious pest almost overnight.

Fig. 14.9 • These vedalia beetles are feeding on the large white cottony-cushion scale. The populations of this pest are effectively reduced by introducing the vedalia beetle. (*University of California, Department of Entomology, Division of Biological Control, Riverside*)

Insect hormones

Springing from the research of Carroll Williams at Harvard in the 1950s, hormones have great promise as an insect control. Later work by Karel Slama at Prague in the late 1960s seems to assure this: less than a millionth of a gram of the juvenile hormone analog DMF applied to the female linden bug sterilized the insect for the rest of its life; yet doses of 10,000 times the sterilizing dose did not kill it; if males were treated with 100 micrograms of DMF not only were the females sterilized after mating but when these females mated with untreated males, enough DMF was passed on to those males to steril-

ize the next females with which they mated; none of the eggs resulting from these matings hatched.

If these hormones are specific and can be developed for, say, the
twenty most damaging insects, the technique could be a tremendous
substitute for the synthetic chemicals on which agriculture has become dependent. But if they are broad-spectrum, capable of affecting
predators as well as prey, hormones will represent no improvement
over present techniques and only postpone facing the inevitable—that
there is no ideal solution to any problem, just as nothing, despite ap-
pearances, is ever free.

Because of much experience with genetic manipulation in fruit
flies and other species, there is a reasonable prospect of introducing genes into specially-bred populations of noxious species.
Suppose several synthetic strains of blowflies were released in early
spring, each carrying a recessive lethal gene that is expressed only
at the onset of low temperatures. By late fall the lethal gene will have
spread through the population and by the end of winter fewer than
one in a million flies would survive the winter and still be fertile.
This technique would be both more effective and far less time-
consuming and expensive than sterilization techniques that have been
used successfully in the past. While it is not appropriate for all insects, houseflies and fruit flies seem particularly suited to it.

The female gypsy moth is such a weak flier that males must seek
the female out over a great distance. To guide the male to his destiny,
the female exudes a powerful scent which can be smelled by a male
several miles away. By grinding up female moths, a lure was developed that could attract males from miles around to a trap or a
poison target where they could be killed. Today this scent, dispar-
lure, identified and synthesized, has been successful in helping to control the gypsy moth. The possibilities of finding and using other sex
attractants offer another useful approach to the selective control of
pest insects. The insect cannot develop resistance to its own sex
attractant without destroying its normal reproductive behavior, so
the attractant should retain its effectiveness indefinitely.

The object of sterilization is to interfere as unobtrusively as possible with the reproductive cycle of a pest organism so that everything
progresses normally except that eggs will not be fertilized and the
reproductive cycle will be broken. This has proven quite successful
with pigeons, as previously mentioned, and has been used with insects such as linden bugs as well. A chemical sterilant is currently
being developed that is intended to halt sperm reproduction in rats
permanently. One dose is calculated to do this effectively without
adverse effect on the rat's sex drive or general health. A female mated
with a sterile male displays pregnancy symptoms but does not mate
again for eleven days, twice the usual nonreceptive period. This
measure reduces the number of possible litters by half and promises
to be quite effective in lowering the rat population.

Know your organism thoroughly, and then focus on a weak point in its life cycle. This rule was followed successfully by E. F. Knipling, using radiation on the screwworm fly. The adult is a metallic blue fly about three times the size of a housefly that lays eggs in wounds or even in the navels of newborn warm-blooded animals. The feeding maggots cause a sore whose discharge attracts more flies until hundreds or even thousands of maggots may infect the animal. After five days the maggots drop to the ground, pupate, then emerge as adults. In a few days the adults mate and the female lays her eggs, completing the life cycle. In warm regions ten to twelve generations a year are possible. The screwworm is limited in the United States to southern Florida and the southern parts of Texas, New Mexico, Arizona, and California.

Knipling's idea was to expose the screwworm pupae to a dose of radiation that would sterilize the males without other side effects; 2500 r was found to do this nicely. Since the females mate but once, mating with a sterilized male would result in sterile eggs. It was found that a ratio of 9:1 sterile to fertile males would result in 83 percent sterile matings. After some preliminary tests, an eradication program was begun to eliminate the screwworm fly from Florida. Using assembly line methods in an airplane hangar, 50 million screwworm pupae per week were irradiated, both male and female, then released. Over a period of eighteen months 2 billion flies were released over 70,000 square miles of Florida, southern Georgia, and Alabama. The program was begun in January 1958. One year later the screwworm fly was eradicated from the southeastern states. Reinfestation is prevented by the colder Gulf coast between Florida and Texas which eliminates the possibility of reinfection from overwintering flies and by animal inspection stations which check all livestock crossing the Mississippi. The technique is now being used in Mexico, where it is hoped that the screwworm fly can be eliminated north of the Isthmus of Tehuantepec.

After this outstanding success, follow-up studies are investigating the possibility of using the method against other large-scale pests such as the gypsy moth, codling moth, boll weevil, and others. But the most valuable lesson demonstrated by this type of biological control is that introduction of sterile but otherwise normal species into a population has far greater impact in reducing the population than simply removing or killing an equal number of individuals. If successful with other major pest species, this technique of sexual sterilization will make biocides unnecessary.

If plants and animals were introduced with care to include the complex consumer-predator-prey-parasite organisms associated with the introduced species, populations would stay under control. But introductions have always been random, a species here and a species there, which precludes any short-term stabilization. Population regulation requires an enormous input of energy. Too often we have

panicked and caused more problems in our clumsy efforts to exterminate the inexterminatable.

The wisest policy is to avoid introduction until enough is known
about the species both in its native home and in the new habitat to
evaluate the possible environmental repercussions of the introduction.
Once a species is introduced we should learn to live with it, but at
the same time we should exploit every means of natural control at
our disposal, for long-term control by biocides is not only ineffective
but self-defeating.

Every species has in its life cycle some point where natural
control can be effectively exerted. By seeking out these weak points
and exploiting them, man can work *with* natural processes rather
than expend large amounts of energy on trying to overwhelm them.
Someday man may come to the realization that much of his so-called
struggle for existence is unnecessary.

FURTHER READING

Beroza, M., 1971. "Insect sex attractants." *American Scientist*
59(3), pp. 320–325. Brief review with a list of attractants
currently being tested.

Debach, P. (ed.), 1964. *Biological control of insect pests and weeds.*
Reinhold, New York. The standard text.

De Vos, A., R. H. Manville, and G. Van Gelder, 1956. "Introduced
mammals and their influence on native biota." *Zoologica* (N.Y.)
41, pp. 163–94. Thorough survey of the impact that introduced
species have on native species.

Elton, C. S., 1958. *The ecology of invasions by animals and plants.*
Methuen, London, England. Beautifully written classic account
by one of the most respected ecologists of the twentieth century.

Hinckley, A. D., 1972. "The gypsy moth." *Env.* **14**(2), pp. 41–47.
An overview of the gypsy moth problem from an ecological
perspective.

Hurlbert, S. H., J. Zedler, and D. Fairbanks, 1972. "Ecosystem alteration by mosquitofish (*Gambusia affinis*) predation." *Science*
175, 639–641. A fresh look at the role of fish in the eutrophication problem.

Lachner, E. A., C. R. Robins, and W. R. Courtenay, Jr., 1970.
"Exotic fishes and other aquatic organisms introduced into
North America." *Smithsonian Contributions to Zoology* **59.**
Systematic review of the introductions and their impact.

Rolls, E. C., 1969. *They all ran wild.* Angus and Robertson, Sydney,
Australia. Good account of the introduction of the rabbit into
Australia and subsequent problems.

Smith, R. H., and R. C. von Borstel, 1972. "Genetic control of
insect populations." *Science* **178,** pp. 1164–1174. Somewhat
technical, but a fine treatment of the possibilities.

15

EXTINCTION:
THE LORD GIVETH AND
MAN TAKETH AWAY

EXAMINING THE evolutionary record, we see the process of species gradually becoming extinct and being replaced by newly evolved species. The overall numbers of species probably remained constant, for there is only a finite number of available niches. It seems odd that in a relatively short period of geological time, the last 20,000 years, dozens of species of mammals and birds have disappeared from their niches without replacement. Something unusual must have happened; the spacing of the extinctions over several thousand years makes it difficult to ascribe them to climatological or geological events alone. Furthermore, most of the extinct animals were large—the mammals over 100 pounds, and the birds certainly large enough to be conspicuous and edible.

The large North American mammals were hardest hit—70 percent became extinct during the last 10,000 to 15,000 years of the Pleistocene epoch. Horses and camels, which had evolved in the New World, became extinct on this continent, as did mammoths and mastodons, which had migrated into the New World over the once dry land of the Bering Strait. The ground sloth, the saber-toothed tiger, the tapir, the dire wolf, the giant buffalo, an antelope, and the giant beaver also disappeared, and yet there was no concomitant loss of small mammals, plants, or aquatic organisms.

A surprising number of the fossilized bones of these animals and birds were discovered associated with charcoal and stone tools, such as arrowheads or spear points, some still embedded in the bones. Was it possible that primitive man had hunted these animals to extinction? At first this idea was denounced as ridiculous. Who could imagine a Paleo-Indian with his crude weapons exterminating the mighty mastodon or the swift horse? We are not certain what really happened, but in a time of severe environmental stress caused by glacially induced climatological and vegetational changes, certain animal populations may have been pushed beyond the point of survival by man, the superpredator. To envision man's possible role as exterminator, we have to look beyond the spear-carrying savage killing one mastodon or horse. As the Plains Indian of historic times demonstrated again and again, herds of animals could be driven by fire off cliffs or bluffs to their destruction, and, if the animals' numbers had already been reduced by unusual natural stresses, this practice could have led to their extinction.

If man was instrumental in the extermination of so many large mammals, how did any escape? Possibly some, like mountain sheep and goats, were saved by occupying rough country; others, like moose, bear, and deer, found refuge in wooded country. Those escaping extinction on the open prairie may have been both abundant and fecund enough to keep pace with man's effort to kill them for food.

The musk ox might well have been saved by an accident of distribution (Figure 15.1). The range of musk ox was divided into two populations by the continental glacier. Like the woolly mammoth, which shared its range, the southern group was probably exterminated by the Paleo-Indians moving into North America from Asia. The northern population along the unglaciated Arctic Ocean coast seems to have survived because of less contact with these hunters. By the time the ice had melted, the Paleo-Indians had evolved into the present Eskimo stock of the area, with whom the musk oxen were apparently able to coexist.

Fig. 15.1 • Musk ox survive to the present day in the arctic, but to the south only deposits of bones indicate that musk ox were ever present. One possible explanation for their extinction in the south is the migration of Stone-Age hunters into the areas south of the glacier. (*Modified from P. Martin*, Nat. Hist. 76, 1967.)

13,000–12,500 B.P.
no stone age hunters

12,000–10,500 B.P.
Stone Age hunters arrive ⟶

today

☐ glacial ice
■ musk ox range

☐ glacial ice
■ unglaciated refuge
■ musk ox disappear

☐ glacial ice
■ musk ox range
■ musk ox bones

But what of Africa, whose broad plains teemed with the greatest variety of grazing animals in the world? At least 30 percent of this fauna was also lost during the late Pleistocene. The fact that more African animals survived to the present day is explained by the concurrent evolution of man in Africa, giving some of the animals time to develop defenses against man's special predatory activities. In North America, however, man suddenly appeared upon the scene from Asia, occupied the continent, and began hunting animals that had never had a chance to adapt to this special type of predator.

Man's role as exterminator can be inferred from the wave of extinctions that followed his occupation of new continents or islands around the world: the disappearance of the giant kangaroo and other large animals when the Bushmen came to Australia; the loss of twenty-seven species of moa (a large ostrich-like bird) when the Maori reached New Zealand; and the elimination of the dodo from Mauritius (an island in the Indian Ocean) by the Dutch and Portuguese sailors in the sixteenth and seventeenth centuries. Most of the niches left by these extinctions remain unfilled; an interesting exception is the niche occupied by the horse in North America. Ten thousand years after its extinction in North America, the horse was reintroduced by the Spaniards and successfully reoccupied its old niche.

For some years the Spanish horse or mustang was quite common in the Southwest. But with the arrival of American settlement and American-type horses, Spanish-descended mustangs were slowly replaced by the more prosaic wild horse. These animals were slaughtered for years for the dog food market (see Arthur Miller's moving presentation of the subject in the movie *The Misfits* on a TV late show sometime), until an aroused public finally pressured the Bureau of Land Management, on whose lands most of the wild horses are found, to set aside a wild horse range in the Pryor Mountains of Montana.

At times it seems that man is bent on reducing the thousands of species of birds and mammals to a few dozen he considers desirable; the rest are viewed as expendable (Table 15.1). The way man has attacked the organisms of the world has varied from outright assault to insidious nibblings, both of which have had the same destructive result. In this chapter we will look in greater detail at the negative impact that man has had on his fellow organisms.

EXTINCTION BY DIRECT ASSAULT

Many animals have disappeared simply because they were edible. Others have become extinct because they became fashionable in man's eyes. A few examples of more recent extinctions are discussed here.

Table 15.1

Some Endangered Animals in the United States*

ANIMAL	NUMBER IN COTERMINOUS UNITED STATES
Timber wolf	400
Grizzly bear	850
Black-footed ferret	Unknown, but rare
Southern sea otter	500
Florida panther	100–300
Guadalope fur seal	600
Florida manatee or sea cow	Unknown, but rare
Key deer	300
California condor	40
Florida everglade kite	10
Southern bald eagle	500
American peregrine falcon	Extinct south of Canada, 5000 in Alaska and Canada
Attwater's greater prairie chicken	750
Masked bobwhite	Extremely rare
Whooping crane	70
Eskimo curlew	Extremely rare
Puerto Rican parrot	200
Ivory-billed woodpecker	Probably less than 12
Kirtland's-warbler	Less than 1000
Ipswich sparrow	Less than 100
Pine barrens tree frog	Unknown, but rare
Atlantic sturgeon	A few thousand, probably
Black fin cisco	Very rare — last specimen taken in 1955
Atlantic salmon	Rare in rivers where formerly abundant

* Source: *Rare and endangered fish and wildlife of the United States, 1966.* Bureau of Sport Fisheries and Wildlife, Resource Publication No. 34.

Half as large again as the mourning dove and at least as savory, the passenger pigeon was probably the most abundant bird ever encountered by man (Figure 15.2), yet in a few decades it had been hunted to extinction. In 1810 the ornithologist Alexander Wilson saw a flock of pigeons in Kentucky, which he calculated to be 240 miles long and a mile wide. Wilson estimated the flock to contain 2.23 billion birds! Unlike geese or ducks, which flock only during migration and never in numbers approaching this scale, passenger pigeons were gregarious throughout the year. Nesting was erratic, depending upon the availability of food. Usually, when beechnuts were plentiful, the birds nested in Michigan and Pennsylvania; when acorns were abundant, Wisconsin and Minnesota were favored. In 1871 one of the largest nestings ever observed took place in Wisconsin. The pigeons nested in almost every available tree over a strip seventy-five by fifteen miles covering over 850 square miles. Anywhere from five to a hundred nests were built in each tree. Such a nesting naturally attracted hunters from all over the country. One

The passenger pigeon

315

Fig. 15.2 • A popular sport in the early nineteenth century was shooting wild pigeons, here the passenger pigeon. (*The Bettmann Archive, Inc.*)

conservative estimate placed the number of pigeons in that nesting area at about 136 million.

The males and females sat on the nest alternately, the males leaving at daybreak, returning at midmorning to relieve the females, who fed until early afternoon, returning to allow the males a final afternoon feeding. This traffic to and from the nesting area presented a unique opportunity for hunters. One eyewitness gives this account of the proceedings:

And now arose a roar, compared with which all previous noises ever heard, are but lullabys, and which caused more than one of the expectant and excited party to drop their guns, and seek shelter behind and beneath the nearest trees. . . . Imagine a thousand threshing machines running under full headway, accompanied by as many steamboats groaning off steam, with an equal quota of R.R. trains passing through covered bridges—imagine these massed into a single flock, and you possibly have a faint conception of the terrific roar following the monstrous black cloud of pigeons as they passed in rapid flight in the gray light of morning, a few feet before our faces. . . . The unearthly roar continued, and as flock after flock, in almost endless line, succeeded each other, nearly on a level with muzzle of our guns, the contents of a score of double barrels was poured into their dense midst. Hundreds, yes thousands, dropped into the open fields below. Not infrequently a hunter would discharge his piece and load and fire the third and fourth time into the same flock. The slaughter was terrible beyond any description. Our guns became so hot by rapid discharges, we were afraid to load them. Then while waiting for them to cool, lying on the damp leaves, we used, those of us who had [them], pistols, while others threw clubs, seldom if ever, failing to bring down some of the passing flock.[1]

[1] Schorger, A. W., 1937. "The great Wisconsin passenger pigeon nesting of 1871." *Proc. Linn. Soc. N.Y.*, 48, pp. 1–26.

Hunting was not limited to birds on the wing; birds were attacked on the nest with sticks, woods were set afire, and trees were chopped down. After a few more years of this kind of hunting, people began to wonder where the pigeons had gone; some said Canada, others, Australia. But there was no escape for the passenger pigeon from the unremitting pressure of the hunter. Why was the passenger pigeon not saved from extinction? Surely a few thousand birds of all those billions could have been preserved.

Here the bird's social habits worked against him. When the large flocks were broken up into hundreds or even thousands, the breeding instinct was somehow inhibited and nesting was sporadic or not even attempted. With continued hunting and cutting of the hardwood forests which supplied most of their food, nesting, and roosting sites, the pigeons faded into oblivion. Once extinction was apparent and perhaps unavoidable, a great wave of remorse swept the land, but to no avail; the last passenger pigeon, an aged female called Martha, died on September 1, 1914, in the Cincinnati Zoo. We must conclude that the passenger pigeon could not be saved: when reduced to numbers congruent with the space we were willing to share with it, reproduction was apparently impossible and extinction inevitable.

Similar sentiment bathed the last pitiful remnants of the buffalo that once blackened the American prairie. However, we are forced to admit that the buffalo and the passenger pigeon would not be compatible with present day use of either the Great Plains or the eastern deciduous forest. Picture today a herd of even 1 million, much less the original 60 million buffalo, wandering from Texas to Montana and back every year, or a flock of a million passenger pigeons settling on an Ohio county to feed! Some buffalo still remain, but they are found only in zoos or reservations.

However, we cannot evade responsibility by assuming that extinction is inevitable for *all* wild animals, predicated on an essential incompatibility of man and beast. Unlike the passenger pigeon, the great auk was never overly abundant (Figure 15.3). A flightless sea bird resembling the penguin, the great auk once nested on rocky islets around the periphery of the North Atlantic from Maine to Spain. Although a superb swimmer and diver, the great auk was extremely clumsy on land and when ashore was easily killed. Fishermen from northern Europe on extended fishing voyages found the auk a welcome source of otherwise unobtainable fresh meat. By the time Jacques Cartier visited the coast of Newfoundland in 1534, the great auk was already becoming much less widely distributed and locally rare in northern Europe. On Funk Island off Newfoundland, Cartier found huge colonies of auks and soon the slaughter began in the New World as well as the Old. As the Grand Banks off Newfoundland became known to fishermen, the North American population of auks was vigorously reduced by expeditions that herded whole flocks into stone compounds where they were methodically clubbed to death and rendered for their oil or salted for their meat.

The great auk

317

THE BIOTIC WORLD AND MAN

Fig. 15.3 • The penguin-like great auk, now extinct, was once extremely abundant in the North Atlantic region. Today few specimens exist even in museums. (*New York Public Library Picture Collection*)

By the early 1800s, the North American great auks were gone. A few lingered on rocky islets off the coast of Iceland but with the butchering of two birds in 1844, the species was clubbed into extinction. Had today's concept of renewable resource been in existence in the nineteenth century the great auk could well have been saved, for even at that time a brief study could have determined how many birds could be taken for their flesh, oil, or feathers without seriously reducing the population. Unfortunately, animals were not then viewed as a valuable resource. We assumed when one was gone there was always another, but we have seen our last great auk.

Whales

While it would be reassuring to think that times have changed and that nations are now concerned with maintaining their animal populations as a valuable resource, the recent dramatic decrease in the population of several species of whales indicates that this is not yet the case.

The great and romantic age of whaling flowered between 1825 and 1860 (Figure 15.4), then it withered as prosaic but dependable

Fig. 15.4 • In 1840, at the height of the golden age of whaling, when this water color was done by a New Bedford whaler, the odds were fairly even between whale and man. (*The Whaling Museum, New Bedford, Mass.*)

kerosene replaced whale oil. While several species of whale produced good oil, only the sperm whale provided ambergris, a wax-like substance still used as a perfume base, and spermaceti, a fine oil of great value. Unlike the toothed sperm whale, the right, bowhead, and gray whales have a mouth full of horny plates through which they strain sea water containing their shrimp-like food. This was the whalebone used for years in corset stays and buggy whips. Although pursued energetically in the nineteenth century, whaling was inefficient and limited to the most easily caught whales. In the twentieth century, as new uses were found for whale oil and whale meat, interest in whaling revived, and using modern techniques, whaling again became a serious business enterprise with huge factory ships capable of locating, killing, and processing *all* useful species of whale with great efficiency (Figure 15.5).

Fig. 15.5 • Today the big ones no longer get away. This factory ship is about to process a fin whale (shown belly up). (*Popperfoto*)

After World War II, an attempt was made by the International Whaling Commission to establish some rules for whaling, but the nations that signed were only obliged to set their own catch limits and regulations. Scientific advice was offered by the Commission but signatory nations were not bound to follow it. As each nation continued to take what it pleased, whales soon became few and far between. Today only the U.S.S.R. and Japan continue to maintain high-pressure whaling operations, and the focus of whaling efforts has shifted from the Antarctic to the northern Pacific. The sperm whale managed to hold its own until recently when scarcity of other whales increased pressure upon it to the extent that 29,000 were killed in 1964.

Apparently the nations that continue to exploit whales have decided that it is to their economic advantage to use the full capacity of their whaling fleets until all whales are exterminated and then dispose of the whaling fleet in toto rather than cut back operations to the level of indefinite sustained yield. Although this may be of interest as a final expression of the nineteenth-century ethic of short-term gain versus long-term yield, it makes a wretched epitaph for the whales.

Some species like the gray, bowhead, and right whales have been protected, and the humpback whale has been partially protected for several years. But protection isn't worth much on the high seas. In 1962 off Tristan da Cunha in the south Atlantic the largest known herd of the supposedly protected southern right whale was "harvested" to the last pound by a whaling fleet that happened by.

The whale in greatest danger of extinction is the blue whale, the largest animal ever to live. Often more than 100 feet long and weighing up to 130 tons, the blue whale has been reduced to a population of around 600, a level so low that even with complete protection it may slowly slip into oblivion. The only answer is to reduce sharply the annual kill of whales and strictly enforce whatever protective regulations may be necessary to maintain and increase the numbers of the most severely reduced whales. In response to the problem, the U.S. Department of the Interior has placed all great whales (bowhead, right, blue, humpback, gray, finback, sei, and sperm) on its endangered species list, halted all whaling by United States citizens, and banned importation of whale products from the endangered species. Unfortunately the gesture was more symbolic than real, for the remaining whaling nations seem determined to extract the last barrel of oil from the last whale.

Numbers of other animals have been reduced to the extinction point because of their attractive fur or feathers. The sea otter was very nearly exterminated before it was completely protected. Although still limited to only a few points in its former range from northern Japan across the north Pacific to California, the sea otter seems to be slowly increasing in numbers and distribution, allowing the Russians to harvest a small number from their otter population.

Until banned by international agreements, a vigorous trade in millinery feathers at the turn of the century threatened the African ostrich (Figure 15.6), several species of egret, and the birds of paradise found in New Guinea. Today the African leopard, Indian tiger, and Florida alligator are threatened because of the demand for their fashionable skins. Even though changes in fashion have spared species from extinction in the past, time is short and pressure on these species great. Most reputable furriers refuse to handle skins from animals officially listed by the International Union for the Conservation of Nature and Natural Resources, which maintains a Red Book of endangered species of the world. Although the alligator is protected in every state over its range but Texas, it was widely

Fig. 15.6 • This young flapper seems unconcerned about the fate of the ostrich who provided her costume. (*McGraw-Hill Films*)

poached in Florida, and even in the Everglades National Park. A national ban on interstate shipment of hides has done much to assure the continuance of alligators in the South; and a recently negotiated international treaty expected to be ratified by over eighty nations offers some hope that the endangered species of the earth may survive for a few more decades.

NOVEL USES LEAD TO EXTINCTION

Killing wild animals for food or for their skins can perhaps be rationalized, but killing a gnu to make a flyswatter out of its tail, an elephant to convert its feet into waste paper baskets, or a buffalo to make a tobacco pouch with zipper out of its scrotum, is perverse (Figure 15.7). In the United States, mummified baby alligators or baskets made of armadillo "shells" are as tasteless as a cookie jar made out of a human skull.

Fig. 15.7 • Tasteless curios such as these elephant foot wastebaskets can exist only when a thoughtless public provides a market. (*Grzimek/Okapia, Frankfurt/Main*)

While no one claims the imminent extinction of elephants to satisfy the waste paper basket trade, the walrus is definitely threatened. Although there are perhaps 30,000 Pacific walruses left, as many as 7000 are killed each year in Siberia and Alaska for their ivory tusks, which are worth $2 a pound. Since the tusks of a large male may weigh twenty pounds each and the ivory carved into ornaments or figures may be worth up to $100 a pound, walruses are actively hunted. Both the U.S.S.R. and the United States regulate the hunting of walruses within their territory, but walruses are citizens of neither country; without an international treaty to guarantee protection, this species may also be doomed.

Superstition

The wild mountain goat or ibex was exterminated in the Swiss Alps in the seventeenth century because of its supposed benevolent powers:

Almost every part of its body was considered to have some healing virtue or other beneficial property, be it aphrodisiac, talisman, or poison detector. A particularly wide range of medicinal effects was attributed to the hair balls occasionally found in the ibex's fourth stomach compartment. These "bezoar balls" were reputed effective against fainting, melancholy, jaundice, hemorrhoids, hemorrhagic diarrhea, pestilence, cancer, and other ills. The ibex's blood was considered a cure for bladder stones; the heel bone helped combat spleen diseases; the heart yielded a strength-giving tonic; and even the droppings were utilized as medicine against anemia and consumption as well as a rejuvenating agent. The many reputed pharmaceutical properties brought the ibex no benefit; on the contrary, its body was so highly sought after that by the seventeenth century it was already extinct in the Swiss Alps.[2]

Of course today's Swiss are somewhat more sophisticated, but many Chinese still believe that rhinoceros horn in powdered form is a powerful aphrodisiac. The result of this superstition has been the reduction of the three Asian species of rhinoceros to the point of extinction; in 1964 there were 600 Indian rhinos, 150 Sumatran rhinos, and only twenty-one Javan rhinos. Even the African species of rhino has been affected by this "magic." Although both the black and white African rhinos are protected, over 1000 are poached each year which has reduced the black rhino to around 13,000 and the white to 3900 individuals.

Trophy Hunting

The Arabian oryx is a graceful gazelle-like animal with long tapering horns. In recent years its extinction in Arabia was assured by oil-rich Arab notables who took to running down the last fragmented herds in jeeps or airplanes, and machine-gunning them to the last individual. In view of the inevitable results of such "sport," an expedition was mounted a few years ago to capture breeding stock of these animals while they still existed. These were shipped to Phoenix, Arizona, where successful efforts were made to establish a breeding herd to preserve the species and ultimately allow reintroduction to Arabia when the safety of the oryx could be assured.

It is only fair to note, however, that trophy hunting is much more regulated than this example would suggest and few other animals are in immediate danger today from this sport.

[2] Ziswiler, V., 1967. *Extinct and vanishing animals.* Springer-Verlag, New York.

Competition with Man

In the early years of their settlement of South Africa, Boer settlers, eking out a living by running stock or farming, viewed any competition from natural sources as expendable and felt justified in leveling their guns on any wild animal in range. In a few years the Cape lion, blue buck, and two species of zebra, the quagga and subspecies of Burchell's zebra, were extinct, and the Cape mountain zebra, bontebok, and white-tailed gnu were reduced to specimens in game reserves.

Similarly, the only parrot native to the United States, the Carolina parakeet, was eliminated because of its fondness for man's crops. It was easy prey because these parrots circled in flocks again and again around fallen members until all were destroyed. The last of these green and yellow birds, a lone specimen in a zoo, died in 1914.

Occasionally the numbers of desirable species are reduced in vain. The tsetse fly carries a parasite (trypanosome) which causes a disease commonly known as sleeping sickness, varieties of which affect both man and his cattle. As a result of the presence of the fly and the parasite, large areas of land in Rhodesia have been closed to settlement and the raising of livestock has been curtailed. Acting under the belief that large mammals were the only reservoir for the trypanosome, control was sought by killing large numbers of zebra, gazelle, and antelope—500,000 during the control attempt. Later it was found that small mammals and birds also acted as hosts for the trypanosome, making this huge loss of large mammals needless, for the tsetse fly remained unchecked.

Control of predators has long been a controversial subject in regions with a large livestock industry. To be sure, mountain lions and coyotes do take livestock occasionally, a few individuals perhaps consistently. But the effects of massive predator control programs have the same destructive environmental impact as broad-spectrum biocides—the poisons are not confined to the target species and wildlife in general suffer. Some effects of predator control are more subtle. By constant predator control pressure, the normal territoriality of coyotes is disrupted and the survivors develop different habits that may include a dependence on livestock. Widespread indiscriminate poisoning may also kill the normal food species, thus driving the predator to the only food left, the livestock the poison was distributed to protect. In view of the vast sums of money that have been spent annually to control predators it might be cheaper for the government to pay a rancher for stock that can be proved to the government's satisfaction to have been killed by predators. In 1972 a sheepman (not a wild-eyed conservationist) estimated that 1500 of Arizona's 350,000 sheep were lost to coyotes the preceding year. This is 0.43 percent, rather less than the claims of 10 or 20 percent or more that are commonly tossed about. At a value of $25 a head, the coyote damage was only $37,500. The same year over $250,000 was spent by state and Federal agencies to control predators in Arizona. Responding finally to public pressure the federal government acted in 1972 to end the use of poisons to kill predators on public lands and to prohibit interstate transport of certain poisons used to kill predators.

Live Animal Trade

The desire of people to see a variety of animals with minimum discomfort and danger has led to the establishment of zoological gardens or zoos. Some, like the zoos at San Diego, the Bronx, or Lincoln Park, are quite extensive; others are quite small. Altogether there are perhaps 100 large zoos in the world creating a steady demand for wild animals, particularly the rarer ones. A platypus or giant panda will always draw a larger crowd than a lion or zebra and since many public zoos depend on admission fees or attendance figures to maintain or increase their appropriations, there is constant pressure to exhibit the rare and the unusual. This pressure is transmitted to collectors and ultimately to animal populations throughout the world. A good case in point is the orangutan (Figure 15.8).

Zoos

Native to Borneo and Sumatra, this large ape is a favorite exhibit in zoos. But as the orangutan became rarer, the demand increased sharply and irresponsible animal collectors redoubled their efforts. The situation was aggravated by placing a premium on the young apes, usually collected only by killing the mother. In an effort to save the orangutan in its native habitat, most zoos, to their credit, no longer purchase this or other animals in danger of extinction. In some cases where extinction seems imminent, however, a special effort is made to establish a breeding stock in the safety of a zoo, as was done with the Arabian oryx.

Fig. 15.8 • Orangutans were eagerly sought by zoos a few years ago, and their population in the wild was dangerously depleted. Today most zoos are careful not to place a premium on the collection of animals in danger of extinction. (*Zoological Society of London*)

Another source of potential danger to wild animal populations, especially monkeys, is the rising demand for animals for experimental purposes, especially in medical and testing laboratories. It is perhaps ironic that as more and more products of potential danger to *man* are generated, such as drugs, additives, and pollutants, more and more *animals* are required in testing programs. Many of these animals—rabbits, hamsters, guinea pigs, and rats—are bred especially for this purpose, but the monkey possesses the best analogue of man's metabolism and thus it is eagerly hunted in Central and South America by collectors. Although some effort is being made today to set up breeding colonies of the most desirable species to increase the supply and to assure greater uniformity and higher-quality animals, it will be several years before the demand can be satisfied.

Animals used for experimental purposes

Both zoos and medical research, however, have used a small number of animals compared with the pet market. Until 1967, parrots were not permitted in this country under ban of the Public Health Service, for parrots were known to carry parrot fever (psittacosis) which could be passed on to man. Recently a wide array of other birds has been found to carry this virus, whose name has accordingly been broadened to ornithosis. Since human contact is much more likely with pet birds than wild ones, parrots remained on the banned list. Then it was found that by holding the birds in an isolation center, treating them for several weeks and observing them for another period, the disease could be controlled. The ban was lifted and a different kind of "parrot fever" swept the country. The five-and-dime stores, which handle most of the parrots sold in the United States, are unable to supply the demand.

Pets

The parrots are collected by Indians in the rain forests of the Amazon Basin, mostly in Colombia. Considering the amount of handling necessary from tree to household cage, it has been estimated that only one in fifty birds survives the transition. This means that in 1968, 500,000 parrots died to supply the 10,000 that were sold. Yearly removal of half a million parrots may be well within the reproductive potential of the species involved, or it may be seriously threatening their survival. The problem is that no one knows what the effect of the soaring market for pet parrots is having in the source country, for the Amazon Basin is still in many ways *terra incognita,* although this is rapidly changing (see Chapter 20), and none but the broadest ecological generalizations can be made. One effect has been noted though; the Indian collectors, attracted to the easy money offered by traders, neglect their traditional means of livelihood in pursuit of the parrots. After the local supply is exhausted, the trader moves on, leaving the Indian collector economically high and dry.

Until the impact of this increased rate of parrot removal can be ascertained, harvesting of these birds should not be left in the hands of irresponsible entrepreneurs, for it has been just this type of unregulated exploitation that has led to extinction of species in the past.

HABITAT DESTRUCTION; EXTINCTION BY SUBVERSION

Direct attack by man on a species is an obvious form of extermination or extinction, but more subtle processes often have the same fatal results for the organisms involved. Short of destroying the organism itself, the destruction of its habitat is perhaps the most effective indirect means of bringing about extinction.

At one time, the island of Madagascar was completely clothed in a rich forest. Its long isolation from the rest of Africa resulted in the preservation of an unusual group of primitive primates, most of which were tree dwellers. In recent years, development has destroyed over 80 percent of Madagascar's forest, leaving this unique collection of primates clinging, literally and figuratively, to the scattered remnants. Already one form is extinct, four more are quite rare, and twenty-three are threatened. Unless steps are taken to preserve some tracts of undisturbed forest as refuges, this entire group of primates will disappear.

Probably the most publicized species courting extinction is the whooping crane. This five-foot-tall wading bird is completely white except for a bright red crown, black moustache, and black wing tips, a spectacular sight on the ground or in the air (Figure 15.9). The whooping crane originally nested in marshes in the prairie states and wintered on the Gulf Coast. Never especially abundant, the crane was initially reduced by hunting as it made its long migrations from its breeding grounds in Canada and the United States to its wintering ground. The bird was not especially edible, but it presented a striking target to the hunter. Later, as the prairie states came under cultivation, marsh after marsh was drained. Through a combination of habitat destruction and hunting, the whooping crane was reduced to a low of fifteen birds in 1942 despite the establishment in 1937 of the 47,000-acre Aransas Wildlife Refuge designed to protect their wintering range.

Fig. 15.9 • Few birds can equal the seven-foot wing-spread of the whooping crane. These four birds are on their wintering range at the Aransas National Wildlife Refuge in Texas. (*U.S. Dept. of the Interior, Sport Fisheries and Wildlife*)

Since then the whooping crane population has increased to around seventy birds. Some of these birds are in captivity in an effort to increase their numbers artificially and to provide some insurance lest, despite man's best efforts, a catastrophe eliminate the natural population.

Certainly much of our empathy lies with endangered species of animals, but some 2000 species of plants across the world are also endangered. Many of these are endemic species, plants whose very limited ranges or specialized habitats make them very sensitive to environmental manipulations. Trees and shrubs are conspicuous enough to elicit some protection—for example, California long ago set aside the last grove of Torrey pine near San Diego as a state park to preserve the species, and the big tree has received state and federal protection in the Sierras (Figure 15.10). But for every tree or shrub under protection there are dozens of rare herbs, grasses, sedges, and ferns whose passing as species goes virtually unnoticed. Much of the problem is due to our incomplete knowledge of plant distributions; an isolated colony of a grass or fern could well be destroyed by highway construction before anyone knew it was there. But despite these problems steps are being taken to compile lists of endangered plants. The Smithsonian Institution is working on a list of endangered American species and the International Union for the Conservation of Nature (IUCN) has published a list of endangered plants of the world.

Perhaps because of the recognition recently given to endangered species, there is a growing awareness that habitats of these species can be endangered too. How can a rare bog plant be preserved without saving the bog where it must grow, or an orangutan, if its rain forest environment has been reduced to a scrubby savannah or manioc field? What is needed at this point is a list of endangered environments as well as plants and animals. Mere listing does not of course save either a species or an environment, but it does focus public attention on those entities that are rapidly being lost and provide some motivation for setting aside not only adequate habitats for rare species, but habitats that in themselves are rare.

But preservation of a habitat does not necessarily assure protection of its species. Careless introduction of foreign plants and animals can have very harmful effects on natural ecosystems too.

Fig. 15.10 • Because of its dramatic size and tendency to shatter into toothpicks when felled, the California big tree or sequoia survives in parks. Some small, less conspicuous plants have not been so protected. (*Michael Capizzi*)

Animal and Plant Introduction

Despite unusual examples like the dodo, the great auk, and the passenger pigeon, modern man has directly caused the extinction of relatively few animals, particularly when compared with the destruction wrought by animals he has introduced in certain areas. Most striking among these man-introduced animals are rats, goats, pigs, cats, and dogs in that order. Rats attack all ground-breeding animals of appropriate size and have caused the extinction of nine species of rails (a small marsh bird, often flightless on isolated islands). Though goats are vegetarians, they have with their prodigious appetite reduced more vegetated land to near-desert than any other animal known (see Chapter 2). Goats were widely introduced on many islands by ships that left them as a kind of living larder to provide a supply of fresh meat for the next visitor. The occurrence of "desert islands" is as much due to these walking vacuum cleaners as to any accident of nature. With the destruction or gross alteration of vegetation, indigenous species are deprived of breeding sites and shelter from rats as well as from their natural predators; hence the faunas of many of these goat-inhabited islands have suffered impoverishment and extinction.

Another calamitous example of the impact of animal introduction is seen in the sugar cane fields of the West Indies, which were plagued with rats whose depredations destroyed much of the crop. The mongoose, a fierce, weasel-like native of Asia, was introduced to reduce the rat population. The mongoose *did* attack rats for a while but soon began to attack every other ground dweller. Amphibians, reptiles, and birds were preyed upon until several species became extremely rare.

Foreign species are not always intentionally introduced. The original grasslands of the Great Valley of California were composed of perennial bunchgrasses. Centuries of overgrazing opened the way for the invasion of and ultimate takeover by annual grasses whose seeds were brought in from the Mediterranean countries in the wool of sheep and with the seeds of crop plants.

Once introduction has taken place, elimination is very difficult, particularly if the plant is annual, or the animal small. Even if eradication of the introduced species were achieved, it may not be possible or desirable to restock the native organisms (see Chapter 16).

Fire-maintained Niches

Damage to endangered species may result, paradoxically, from efforts to protect their habitat by the elimination or control of supposedly destructive factors such as fire.

The California condor, one of the largest vultures in the Western Hemisphere, has been restricted by man-induced disturbances to a horseshoe-shaped range in the coastal mountains of California from Coalinga south to near Ventura and north again to east of Bakersfield. The center of this range has been protected by its inclusion in the Sespe Wild Area of the Los Padres National Forest. Despite its name, Los Padres National Forest, set up to protect part of Los Angeles watershed, contains more chaparral than trees (Figure 15.11). Large areas of easily burned chaparral, before fire protection, presented a mosaic of burned and open patches surrounded by a dense growth of shrubs. When fire protection was begun with the formation of the National Forest, large tracts became solidly covered with dense chaparral.

The condor and chaparral

A recent estimate by Koford suggests that the condor population is rather precariously stabilized at sixty individuals, for the condor, though protected by law, is a conspicuous target, supporting its twenty pounds in flight with a ten-foot wingspread (Figure 15.12).

Fig. 15.11 • Chaparral-covered mountains like these in the Los Padres National Forest in California are the last refuge of the California condor.

Although condors may live up to forty years, they do not begin to breed until they are five years old, and then produce only one young every other year thereafter. The birds are restricted to their present small range mainly by urbanization, since the condors feed upon carrion, which, as a result of small, well-cared-for cattle ranches and the trend toward grain farming, has become scarce.

Because of its enormous wingspread, the condor needs room to approach its food supply, and, more importantly, to take off after heavy feeding. Thus, open space is vital to the condor's survival. Since fire elimination tends to close chaparral stands, controlled burning helps the condor in some respects. However care must be taken to avoid overburning, since the heavy brush helps to keep people out of nesting areas. Perhaps the best way to provide feeding space *and* protective isolation would be to spot burn unconnected

Fig. 15.12 • One of the sixty California condors left, soaring in search of food in its home range. (*U.S. Dept. of the Interior, Sport Fisheries and Wildlife*)

329

Jack pine and warblers

small areas of the chaparral on a rotation basis. In this way it may be possible to increase the survival of the young condors and so increase the number of condors to a point where no single disaster could exterminate the species.

A much less conspicuous bird, Kirtland's warbler breeds only in an 85-by 100-mile area in the north-central region of the lower peninsula of Michigan (Figure 15.13). The nesting area is limited to dry, porous, sandy barrens covered with patches of jack pine, a scrubby pine of minor commercial value. After the white pine forests originally covering this area were cut or burned, jack pine was one of the few plants able not only to survive but to thrive in the frequently burned barrens. This is in part because of the pine cones of the jack pine, which remain tightly closed and attached to the tree, protecting the seeds, often for years. When seared by fire or dried by the death of the tree the cones pop open, scattering several years' supply of seeds over the ground. These quickly germinate in the fire-cleared mineral soil and a stand of young trees results. Although Kirtland's warbler nests on the ground, it requires the protection of living pine boughs in thickets near the ground, which allow the birds to enter and leave the nest unobserved. This arrangement of branches occurs only when the trees are about six feet tall and disappears when the trees get much over fifteen feet tall. Trees shorter than six feet are too far apart and lack the necessary thick lower branches. In older trees the lower branches are shaded out and die, reducing cover.

With the extensive fires following logging in the 1870s, the number of trees with the proper characteristics increased enormously and the population of Kirtland's warbler was probably at its greatest. When fire was eliminated or restricted in this part of Michigan after

Fig. 15.13 • A female Kirtland's warbler feeding her young. Notice the protection provided by the jack pine whose branches can be seen at the top and bottom of the picture. (*Photo by G. Ronald Ausking from National Audubon Society*)

the turn of the century, and large areas of jack pine began to exceed the size required for warbler habitat, the warbler population started to fall. In 1971 it was estimated that there were less than 200 birds remaining. A program of rotational controlled burning of patches of jack pine forest at least forty acres in extent (the minimum breeding territory for a pair of Kirtland's warblers) has recently begun. Hopefully this will ensure a stable warbler population and reduce the chance of extinction.

Before the advent of white men, the Kenai Peninsula in Alaska was primarily a forest of spruce with lichen-covered open patches of ground (a lichen is a primitive plant that in temperate forests usually grows on rocks or tree trunks). Large herds of caribou depended on lichens for their winter food supply. When man appeared in the late nineteenth century much of the forested area was burned and the lichens destroyed. In a few years, deprived of their winter food, the caribou were gone.

Moose and spruce

The spruce killed by fire was succeeded by willow, alder, and aspen. Since moose require the buds and bark of these successional species as their winter browse, in the same way the caribou needed lichens, the large fire-generated stands of these hardwoods allowed a significant increase in moose survival. In fact, so spectacular was the increase in the moose population that the government reserved a very large portion of the Kenai Peninsula as a National Moose Range. But as the short-lived alder, willow, and aspen were replaced by spruce in the natural sequence of events, the moose population declined again, raising the prospect of a mooseless Moose Range.

But when unauthorized fires, for this was a time when fires were forbidden, swept through the Moose Range, restoring successional browse, the moose population quickly recovered. Now that the lesson has been demonstrated, the size of the moose population in the Kenai is being controlled by planned fires to maintain a certain proportion of the forest in a fire recovery stage, thereby providing moose with the winter food supply necessary for their survival.

SUSCEPTIBILITY TO EXTINCTION

Although no species is immune to extinction, some are extinction-prone. Any species that has a naturally low population level or is subject to violent fluctuations in its numbers is in danger. The population of the California condor, like that of the whooping crane, hovers between forty and sixty, for as we have seen a condor pair produces only one egg every two years. With such a low reproduction rate, population growth is unlikely and susceptibility to catastrophe is greatly enhanced. Inbreeding has sharply limited the gene pool. This reduces the genetic variability, which in turn reduces the species' flexibility in the face of environmental change. But reduction to low

Fig. 15.14 • This koala is feeding on eucalyptus leaves, the only food it will eat. (*Australian Information Service*)

Specialized breeding sites

Island distribution

numbers in itself need not doom a species to extinction. A species that has a high rate of reproduction, when there are no pressures against it, may be able to bounce back from the edge of extinction. For example, the great white heron suffered a 40 percent reduction in the Florida Everglades, as a result of Hurricane Donna in 1960, but by 1963 the population had regained its former level of about 1500 individuals.

Some species feed on only one or a few other species, which automatically limits their distribution and makes them prone to extinction. The Everglades kite, for example, feeds on one species of snail. As land is drained for cultivation or development, the snail is being exterminated and its overspecialized predator with it.

The Australian koala feeds mostly on eucalyptus leaves (Figure 15.14). If the eucalyptus forests disappear, the koala will too. The ivory-billed woodpecker is another species with a limited food source—beetle larvae in *recently* dead trees. Most other woodpeckers, including the closely related pileated woodpecker, accept a broad variety of larvae that successively inhabit dead trees for years after the death of the tree. This need of the ivory-billed woodpecker for recently dead trees requires large tracts of mature timber, which have become rare in the South. This need may doom the ivory-billed woodpecker to extinction.

The wood duck at one time became quite rare. Unlike most ducks, it requires a nesting cavity, and as old hollow trees disappeared, so did the duck. When this need was recognized and suitable nesting boxes were erected in marshes, the wood duck population sprang back.

Kirtland's warbler, as we have just seen, is even more specialized in its nesting requirements, which limit it to a small area in Michigan's pine barrens. Some fish have equally specialized nesting sites. Salmon must have streams with gravel bottoms to receive their eggs, and silted streams are unsuitable because the eggs become buried in the silt and suffocate.

Many of the world's birds and mammals, now extinct, were restricted to islands. This limited both their numbers and their ability to survive the introduction of predators. Because many small islands originally had few predators, birds sometimes lost their ability to fly. When man introduced predators they eliminated these flightless birds.

Both lakes and isolated mountain ranges form ecological analogues of islands, limiting the range of a species and often imposing special environmental conditions which lead to the evolution of unique forms distinct from related populations. A subspecies of bighorn sheep became extinct because its limited range in the Black Hills allowed it to be hunted without respite. The freshwater shark and porpoise of Lake Nicaragua, and the seal of Lake Ladoga in the U.S.S.R., are all species that, although still extant, may in the future become extinct for the same reason.

THE COST OF EXTINCTION

During the history of mankind, more than 150 birds and mammals have become extinct; at least half of these have been full-fledged species, the rest have been subspecies, races, or geographical variants of still existing forms. The question we must ask is what are the implications of this loss? With millions of species of animals and plants still remaining, some as yet undescribed, why should 150 animals be missed?

Man, of all these hosts of organisms, is the only animal capable of completely exterminating any other form of life. All other species are bound in predator-prey relationships that exercise checks and balances upon excessive increase or decrease in population levels. Although these levels may cycle with time, they do so within certain limits.

Sometime in the last few million years, man became dissociated from these checks and balances and became a superpredator. But with this ability to drive any species into extinction at will comes at first a need, then a responsibility for restraint. The need develops from the growing awareness that no species exists in an environmental vacuum. Organisms and their environments are inextricably interrelated. Scores, perhaps hundreds, of species may be dependent upon a key plant or animal such as a redwood, buffalo, or sand dune shrub (see Chapter 14). But why are all these associates so important that we should guard against the loss of any one of them? The answer lies in the relationship of man to this species diversity.

Man has traditionally tried to concentrate the energy from sunlight into a few species which are easily harvested, the monoculture system that was mentioned in Chapter 6. While this practice accomplishes its aim reasonably well, the price paid in crop protection is high: biocides to control weeds, insects, fungi; hormones to control fruit set, preharvest drop and so on. Despite this tremendous input of energy, things still go wrong. Insects and fungi develop resistance and new pests are encouraged by the destruction of their predators. At times, as we saw in the spray schedule for an apple orchard (see Chapter 13) it would seem that we have a tiger by the tail in our attempt to maintain simplified ecosystems.

When species diversity is reduced in natural ecosystems, either willfully or inadvertently, the same kinds of problems seen in monoculture begin to arise. But a natural ecosystem cannot be treated like a cornfield and routinely sprayed to control some beetle or caterpillar. For a biocide that might be tolerated in a cornfield could so scramble the food web in a woodlot that the original problem would be lost in the ensuing chaos. Moreover, while we can afford the energy input to maintain a field of corn because of the value of its yield, we cannot

possibly afford the economic burden of maintaining natural ecosystems that have maintained themselves until disturbed by man.
Consequently, until we know exactly the role of *every* organism in
an ecosystem, how it controls or is controlled by others, we cannot
abandon any species as superfluous, not even that pitiful flock of
seventy whooping cranes. If we do we run the risk of having to cope
with future population explosions of fungi, insects, rodents, or whatever, that might well dwarf any problems seen to date.

However logical our need for all extant species may be, to place
the case for continued survival of endangered species solely upon
our present or future need begs the issue of responsibility. If man
were a passive bystander he could rationalize extinction as a purely
natural phenomenon. But it is certainly fair to say that since the
emergence of man *no* extinction has been a purely natural one. Just
as we bear a certain responsibility toward domesticated species which
have become dependent upon us, we have an obligation toward wild
species which through no fault of their own have become dependent
upon man for their continued existence. Each of these species, gnat
or gnu, represents a pool of genetic information which uniquely
adapts it to its environment. When this information is lost through
the process of extinction a valuable key to survival with all of its
potential economic, scientific, and aesthetic worth is irretrievably
lost as well. It is time that we, as self-aware members of most of the
earth's ecosystems, develop the ecological sensitivity to refrain
from invoking the enormous power of extinction that we have inherited from our ancestors. We must do this for the sake of enlightened self-interest, to be sure, but also because we have come to
feel that all other organisms on earth have a right to exist equal to
our own and that our traditional dominion over beast and fowl is
not a mandate for mindless exploitation but a charge of responsibility.

THE ROAD BACK

Most extinctions need not have happened. With the exception
of a very few species like the passenger pigeon, which seemed unable
to adjust its life style to accommodate man in any degree, most
species presently extinct might be living now had there been a little
care and foresight at the right time. Controlled harvesting, for example, might have saved the great auk, the dodo, Steller's sea cow, or
the Black Hills bighorn sheep. Thus it is imperative that we profit
from the mistakes of the past and accept the fact that there is no
inherent reason why man should hurry any organism presently living
into extinction. Large-scale efforts have already saved some organisms from imminent extinction, and for a few others the future
seems hopeful. Techniques of wildlife management, breeding, re-

introduction, and even resynthesis are becoming more widely used and offer hope for saving species now threatened.

Management

Perhaps the greatest challenge is to institute proper management of a wildlife resource before its numbers become so low that emergency measures are necessary. The concept of management is not new but many mistakes were made before it came to be properly applied. In the early 1900s the mule deer herd on the Kaibab Plateau of Arizona was reduced in numbers; sportsmen blamed the traditional enemy, the predator. The government responded with a predator control campaign, which practically exterminated mountain lions from the plateau. The deer quickly responded, but without natural control they increased so greatly that their browse could no longer support them, and thousands starved to death. The survivors so severely overgrazed what remained of the vegetation that the effects were noticeable for decades afterward. Although the mountain lion population was allowed to recover somewhat after this fiasco, the damage had been done.

Management is now viewed as going beyond mere protection of the "good guys" from the "bad guys." Important management tools are removal of surplus stock by controlled shooting, manipulation of vegetation, and control over multiple use of summer *and* winter range of wildlife. Habitat management using controlled burning is employed to create nesting sites for Kirtland's warbler in Michigan, and to retain a mosaic of open spaces for the use of the California condor in Los Padres National Forest in California.

However, efforts to increase the population of bighorn sheep in the Sierra Nevada between California and Nevada have foundered because of inadequate control of the winter range. The sheep have adequate summer grazing in the meadows of Sequoia and King's Canyon National Parks, but when the sheep move down into the National Forests below to graze during the winter much of the range is already occupied by domestic sheep or mule deer. Few National Parks or game refuges are large enough to be self-regulating. The same problem has occurred in the Serengeti Game Reserve in Tanzania: animals migrate out of the protected area during part of the year for some favored graze that is not included in the reserve. In such situations arrangements must be made to accommodate the year-round needs and activities of the animals, if they are to be maintained in a truly wild state.

Natural Breeding

A few species have survived only in captivity. Both the gingko and dawn redwood (*Metasequoia*) trees (Figure 15.15) were re-

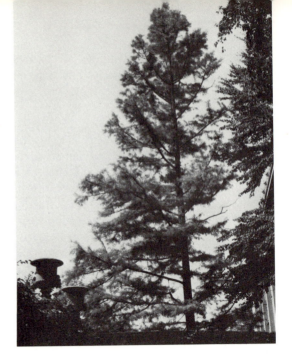

Fig. 15.15 • The *Metasequoia* or dawn redwood, long believed to have been extinct for millions of years, was discovered growing in western China after World War II. This specimen is growing reasonably well on the campus of the Pennsylvania State University.

discovered growing in temple gardens in China. Both had been thought extinct but today they are planted throughout the world. Another Chinese species, Père David's deer, was discovered in Peking's Imperial Garden in 1861, long after all wild individuals had become extinct. Today there are several hundred in zoos and private collections in Europe.

The best known breeding program was initiated in 1967 to save the whooping crane by captive breeding. Experiments with the closely related and more abundant sandhill crane indicated that the whooping cranes would probably breed in captivity and would replace eggs taken from the nest. Accordingly, each season several eggs are taken from whooping crane nests in Wood Buffalo Park in Canada and flown in special incubators to the Federal Endangered Wildlife Research Station at Patuxent, Maryland, where they are hatched. The resulting birds are later bred to build a reserve of captive cranes as a stock for possible reintroduction.

A successful breeding program was carried out with the Hawaiian goose or nene, which had been reduced to less than fifty birds when a few pairs were bred in captivity in England. In a few years there were enough nene to restock their original habitat in Hawaii. Today there are over 500 nene, captive and wild.

Stacking the Deck

Often there isn't time to breed stock, patiently accumulating a population surplus that can be used for reintroduction. Developments in experimental embryology over the past several years have suggested improving on nature by utilizing the techniques of super-ovulation and artificial insemination. Certain drugs can induce super-ovulation—release of many eggs at once instead of the sequence characteristic of the species. Artificial insemination techniques involve sperm that can be frozen for several years and used to fertilize the eggs in the womb of the female. Since all the fertilized eggs cannot be carried to term by the original female, the fertilized eggs can be flushed out and implanted in the womb of closely related, less endangered species, which ultimately give birth to the young of the desired species. Suppose as an example that a female European bison (an endangered species) were superovulated and artificially inseminated with European bison sperm. Suppose 25 fertilized eggs resulted; these could be flushed out and implanted in 25 American bison cows, a closely related but more abundant species. In a few months the world would possess 25 more European bison. Although this process has not yet been performed on an endangered species, there is every reason to expect it to be a valuable, if last-ditch, tool for saving species close to extinction.

Reintroduction

By 1900 the Russian antelope or saiga was almost exterminated. A careful research program by Soviet scientists managed to increase the breeding stock to about 1000 individuals. When more information about its habits, food, and nutrition had been gathered, the saiga was reintroduced to its former range. With over 2.5 million saiga in 1960 a controlled harvest was allowed for the first time. Each year since, over a quarter of a million saiga are harvested, producing 6600 tons of meat, 240,000 square yards of hides, and large amounts of industrial fat. Thus a species in danger of extinction was restored as a resource and sustained productivity was achieved on land unsuitable for cattle grazing.

Similarly, the trumpeter swan of North America was reduced to a few dozen birds in Yellowstone National Park and an adjacent hot spring area at Red Rock Lakes, Montana. This latter area was designated as a wildlife refuge in 1935, and by 1958 enough swans had bred to allow introduction of breeding pairs into refuges in Oregon, Washington, Nevada, Wyoming, and South Dakota where they nested successfully. In 1968 the total population of trumpeter swans was around 5000, with over 800 of them in the United States.

Perhaps of greater interest to hunters was the revival of the wild turkey. This large game animal, which was preferred by Benjamin Franklin over the bald eagle as the national bird to be represented on the Great Seal, was so abundant in colonial days that it was sold for a penny a pound. But by 1920 overhunting, deforestation, and loss of the chestnut, its prime food source, had eliminated the turkey from three-fourths of its natural range. A program initiated by the state game department of Pennsylvania reintroduced wild turkey stock in appropriate gamelands in the state, thereby increasing the number of turkeys in Pennsylvania to nearly 50,000. From this stock, turkeys have been reintroduced to New York, Connecticut, and Massachusetts. As the number of wild turkeys increases in these states they may offer hunters an open season, as has become possible in Pennsylvania.

Fortunately there seems to be enough public concern to continue federal support for programs of the Department of the Interior that are making strenuous efforts to prevent further extinctions of native American species. This support comes somewhat belatedly, though it is nonetheless welcome, considering that the National Audubon Society has been protecting populations of rare birds since the early 1900s. The Boone and Crockett Club and the Izaak Walton League have also been concerned for many years with wildlife values. Internationally the World Wildlife Fund has supplied money for protection of endangered species all over the world (see Appendix 2 for a more complete list of organizations concerned with the preservation of wildlife and its environment). With the concerted efforts of all interested parties there is no reason for further unnatural extinctions to occur.

FURTHER READING

Branch, E. D., 1962. *The hunting of the buffalo.* University of Nebraska Press. Fascinating description of a tragic saga.

Bureau of Sport Fisheries and Wildlife, 1966. *Rare and endangered fish and wildlife of the U.S.* Resource Publ. 34, Sup. Doc., Washington, D.C. This is the best and most current source book available on the problems of American wildlife.

Francoeur, R. T., 1972. "Tomorrow's fossils: experimental embryology and endangered species." *Bull. Atom. Sci.* **28**(2), pp. 11–14. The prospect of using artificial insemination and superovulation to save endangered animals.

Greenway, J. C., Jr., 1967. *Extinct and vanishing birds of the world.* 2nd rev. ed. Dover, New York. A very thorough, updated account of bird extinction.

Koford, C. B., 1966. *The California condor.* Dover, New York. Details of the life history of this slowly vanishing species, with some suggestions of how fire might be used to increase survival rates.

Kroeber, T., 1962. *Ishi in two worlds.* University of California Press, Berkeley. A poignant example of man's deliberate extinction of man.

Matthews, L. H., 1968. *The whale.* Simon & Schuster, New York. Sumptuously illustrated presentation of the whale in time and space.

Mayfield, H., 1960. *The Kirtland's warbler*. Cranbrook Institute Sci. Bull. 40, Bloomfield Hills, Michigan. The relationship between this rare warbler and fire is carefully discussed.

Payne, R. S. and S. McVay, 1971. "Songs of humpback whales." *Science* **173,** pp. 587–597. Fascinating account of a previously unsuspected phenomenon.

Philip, H. R. H., Duke of Edinburgh, and J. Fisher, 1970. *Wildlife crisis*. Cowles, New York. A concise listing as of 1969 of the status of endangered mammals and birds of the world.

Schorger, A. W., 1955. *The passenger pigeon, its natural history and extinction*. University of Wisconsin Press, Madison. *The* account of the extinction of the passenger pigeon.

Small, G. L., 1971. *The blue whale*. Columbia University Press, New York. A definitive study of the blue whale; note particularly its examination of the economics of hunting the species.

Trefethen, J. B., 1970. "The return of the white-tailed deer." *Amer. Heritage*, February, pp. 97–103. Interesting history of the white-tailed deer since white settlement.

16

ENVIRONMENTAL RESTORATION

IF MAN's cultural progress over the last million years of his evolution were graphically portrayed the result would probably be a straight line, but its slope would depend upon the bias of the observer. An optimist would see the line sloping upward, a realist would see a horizontal line, and a pessimist, a line sloping downwards. Any perturbations in the line, and there have been many, would be smoothed over by the great time span. But if we look at just a small time segment we see many peaks and lows, representing the best and worst of man's cultural capabilities. A particularly striking recent example was the intense but short-lived flowering of the arts and sciences in Germany between World Wars I and II. The relaxation of international tension, the sense of a new era that major wars often elicit, the throwing off of an authoritarian government for a more permissive democratic regime—all were part of the reason for it. The sense of that age is beautifully captured in the art of George Grosz and Paul Klee, the music of Alban Berg, Kurt Weill, and Paul Hindemith, the writing of Thomas Mann and Hermann Hesse, and the scientific vision of Albert Einstein and Lise Meitner.

One small but extremely interesting facet of this cultural bubbling up was an idea generated by the brothers Heck. Lutz Heck was director of the Berlin Zoo and his brother Heinz headed the Tierpark Hellabrun, a world-famous zoo in Munich. Perhaps encouraged by being the first to breed the African elephant successfully in captivity, Heinz Heck in 1921 began to assemble in his zoo a widely diverse collection of strange, half-wild cattle: the Hungarian, Podolian steppe, Scottish highland, grey and brown Alpine, Corsican, and piebald Friesian. Heck's goal was to synthesize the long-extinct wild ancestor of his motley herd, the aurochs. While the last aurochs had died hundreds of years before, artifacts remained in the form of bones, skulls, horns, skins, and many pictures. With these fragments as his only guide, Heck began a breeding program to try to pull together the bits and pieces of the aurochs that existed in the genes of the semiwild stock he had collected from all over Europe. Lutz Heck expressed this plan succinctly:

No creature is extinct if the elements of its heritable constitution are still to be found in living descendants. All that needs to be done is to apply the experience of the breeder to the assembling of the inherited elements scattered among these descendants. To this end a sure eye is needed to detect the primitive qualities of the various races of cattle: the nearest approach to the horn of the aurochs, or its combativeness, or its long legs, or the colour of its hide, or the small udder of the wild animal. The next step is to devise a well-considered plan of breeding, various races being crossed in order to combine all the qualities similar to those of the aurochs, and so its heritable constitution, in a single animal. If this is successfully done, the aurochs of ancient times must eventually reappear from its tame descendants.[1]

Much to the surprise of everyone (except perhaps the Heck brothers) the experiment succeeded. An animal whose young closely resembled the parents *was* put together from all the genetic races that resembled strongly the aurochs (Figure 16.1). The result was, of course, not a recreation of the original aurochs, but rather a facsimile much like an artist's conception of a brontosaurus or pterodactyl. The "experience of the breeder" and "a sure eye" are rather different selecting mechanisms than the ability of an individual to survive an unusually cold winter, or a herd to protect itself from a pack of wolves.

But despite these qualifications, the Hecks had accomplished something unique—scattered genes had been rounded up and linked together into a self-reproducing form. If this could be done with an

[1] Street, P., 1964. "Recreating the aurochs and tarpan." *Animals* **5**, London, pp. 250–254.

Fig. 16.1 • A small herd of aurochs bred by Lutz Heck at the Munich Zoo from many breeds of cattle. Although these animals breed true they can only be considered a facsimile of the original aurochs, which became extinct long ago. (*Photo by Lothar Schlawe, courtesy Tierpark Hellabrunn, Munich*)

organism it could be done with an ecosystem. If enough pieces of any natural system still exist to provide some concept of what the original system looked like, there is a reasonable chance that by combining the pieces a self-reproducing system can be established. Like Heck's aurochs, such a reassembled system might differ in subtle but important ways from the original, but given time and a free hand, natural selection could work to bring both organism and system in line with the reality of a natural environment. Heinz Heck was lucky that in 1921 there were still a number of rather primitive cattle stocks available as raw material. As we continue to improve our breeding lines of cattle, corn, and cantaloupe, by reducing their variability, the older varieties, not to mention the original natives, tend to be lost. The recent corn blight dramatically demonstrated how small the genetic pool of most of our staple foods has become through this process. Unfortunately circumstances are similar in natural systems. Most natural systems have been profoundly disturbed by man, particularly in Europe and parts of Asia. The fragments are either widely scattered or, in some cases, lost; so we have no idea how the original ecosystem functioned or even what it looked like. In the remaining fragments, invasion by alien plants and animals has diluted the existing natural elements to the extent of obscuring their original relationships.

Therefore it has become exceedingly important to develop techniques to pull together the scattered pieces of natural systems before further destruction and dilution make any hope of restoration impossible—not for sentimental reasons or as an ecological tour de force, but to preserve from irretrievable loss genetic information that will become invaluable in the next several decades. Fortunately, the importance of environmental restoration has been recognized and restoration is under way in a variety of systems. What we learn from these pioneering projects can then be applied to other endangered systems.

THE HISTORICAL ROOTS

For a very long time man has been managing the environment, or at least the aspects within the grasp of his technology (see Chapter 2), not attempting to imitate nature, but rather to reorder it. Consider the typical farm. It requires a constant input of energy to deter natural processes such as succession, which would increase species diversity of the fields beyond the reduction to the absurd that a farm, after all, is. The same working against natural processes can be seen in the history of gardens. For thousands of years gardens were formal, artificial affairs with an emphasis upon utility, containing fruit trees and often beds of herbs and flowers laid out in intricate geometric patterns (Figure 16.2). This tradition of rigid layouts stretched with-

Fig. 16.2 • The Renaissance garden was a study in geometrical forms illustrating the imposition of man's intellect upon nature.

out a break from the atrium garden of a Roman villa through the medieval gardens of the monastery cloister, reaching its culmination in the geometric (though hardly utilitarian) excesses of seventeenth-century Versailles. The right angle and straight edge reigned supreme.

But in the late eighteenth century the rise of the romantic movement in arts and letters had a strange reverberation in England. Lancelot "Capability" Brown, a gardener and minor architect, almost singlehandedly created the art of landscape gardening and began a revolution in the way people perceived their environment. Capability Brown's genius was to elicit with great care the illusion of being in a natural environment, breaking completely with the ancient gardening tradition of creating an environment totally artificial in its rigid geometry. Sometimes starting with nothing but barren fields, sometimes with an old-fashioned formal garden, Brown would evoke a scene of pastoral beauty, vistas through the forest, a pond nestled in the hills, mounds of greenery swirling in graceful curves, all persuading the viewer that the scene had always been part of the natural landscape (Figure 16.3). It was, of course, an extensive and expensive manipulation involving the damming of streams to form lakes, the construction or leveling of villages, the planting or cutting of trees, and the construction or removal of low hills—all to achieve a desired aesthetic effect. Despite the massive manipulation, the result was intended to appear natural rather than manmade like the formal gardens of the past.

Compared with a formal garden, Brown's landscapes did indeed look natural; but they were not modeled on any natural environment that existed at the time, for even in 1750 truly natural environments had disappeared from much of England. Brown's landscapes were modeled on the highly man-modified environments that then as now characterize most of Europe, that interplay of natural and unnatural that the work of Dutch and English landscape painters has made a

Fig. 16.3 • This typically English landscape was designed and planted by the landscape architect Capability Brown in the eighteenth century.

standard of "natural" beauty for many. Since the aesthetic effect he created was at best only a highly romanticized evocation of nature, Brown had little compunction about using whatever exotic plants were available to fit the needs of his designs.

A Capability Brown landscape, for all its "natural" appearance, is not, nor was it intended to be, a restored environment. But Brown did play a very important role in changing people's view of what remained of their natural environments. From the classical-medieval concept of nature as disorder ultimately to be ordered by man's work, a new view was beginning to be expressed; that nature had its own intricate order and values, which were often aesthetically pleasing and instructive as well.

On the basis of this promising beginning it might be supposed that the British would sooner or later be moved to go several steps further and try to restore one of the natural environments that played such an important role in English history, the oak wood, for example. But thousands of years of management had taken their due and there was little real evidence of what Sherwood Forest looked like to Robin Hood, the New Forest to the Norman kings who hunted there, or what Julius Caesar saw when he stepped ashore in 55 B.C. Some of the pieces probably remain, but as with Humpty Dumpty there are few clues to how they all fit together. But today the Queen's men are trying with increasing success to manage the pieces available to them.

Environmental Restoration in Britain

A classic example of the problems facing restoration of former environments can be found in the Scottish Highlands. Originally, perhaps 2000 years ago, the higher elevations were covered with some kind of pine forest grading through birch and ash to some kind of oak forest below. Today we know most of the species involved, but not very much about how they were organized into communities. Demands over the centuries, first for ship timber, then charcoal, then cropland, gradually reduced the original vegetation to small scraps that may have been cut or replanted, sometimes with alien species; in general the remnants were profoundly affected. But the final destruction of this forest came with the formation of the United Kingdom in 1707, which put an end to the warring between Scots clans and brought social changes culminating in the introduction of sheep in large numbers. Grazing by sheep converted most of the Highlands into an enormous pasture characterized by grasses, heather, and mosses. The rabbit, which had been introduced into Britain by the Normans in the twelfth century, reached the Highlands in the nineteenth century and added to the problem.

This constant grazing pressure, plus the gradual growth of mosses over the mineral soil most tree seeds need to germinate gradually eliminated any chance for reproduction of the original vegetation; and so the moors came into being (Figure 16.4). Since all this happened hundreds of years ago, the Scot has come to regard moors as the natural vegetation of the Highlands rather than a strange result of overgrazing by rabbits and sheep. Much the same process took place in the southwestern United States, although in this case within memory of living men. The original vegetation over much of west Texas

Reforesting the Highlands

Fig. 16.4 • Moorland is characteristic of Scotland and the Appenine Mountains of northern England—this example is Howgill Fells in North Yorkshire, England. If covered with their original forest vegetation, these hills would strongly resemble the Smoky Mountains of North Carolina. (*Peter Stuttard*)

and eastern New Mexico was a desert grassland with occasional shrubs. But overgrazing combined with the elimination of fire allowed a tremendous population explosion of mesquite and juniper, which invaded the grassland. Today the dense clumps of mesquite and juniper seem always to have been there, like the moors of Scotland.

When, after World War II, the British Forestry Commission announced a program of reforestation in the Scottish Highlands, there was quite a public outcry. Since the moor vegetation had come to be regarded as natural, the plan to restore a forest environment was viewed with the concern Americans might express if the Forest Service decided to cut all the windbreaks in Nebraska and replace them with the native prairie grasses. With no emulation whatever of the sensitivity with which Capability Brown placed thousands of trees in his landscapes, the Forestry Commission established plantations of single species in blocks scattered across the rugged Highland landscape like a crazy quilt (Figure 16.5). Since then, the lesson has been learned and reforestation is proceeding in a more aesthetically pleasing manner (Figure 16.6). But the problems are many. Large herds of red deer, maintained for hunting, relish the tender buds and twigs of the young trees; this means the plantations must be fenced. Then the dense accumulation of acid, soggy organic material—peat—has to be scored by heavy machinery (Figure 16.7) and fertilized to get the seedlings planted and off to a good start (Figure 16.8).

Fig. 16.5 • Early efforts to reforest the moorlands of Scotland met with much criticism because of the lack of aesthetic considerations in the planting schemes. Crazy-quilt bands and zigzags persist until the trees are harvested.

Fig. 16.6 • In contrast to Fig. 16.5, these slopes have been reforested in sympathy with the roll of the land.

Fig. 16.7 • To convert moorland turf to forestland, the heavy sod of mosses and grasses must be broken and fertilized before the tree seedlings are set out.

Fig. 16.8 • These young spruce trees are making thrifty growth in former moorland in Scotland.

While this reforestation project can be considered a type of environmental restoration, since trees are being planted in a region that once supported forests, the Forestry Commission is not attempting to restore the original species (about which not enough is known for accurate restoration anyway). At least an effort is being made to return the Highlands to what is probably their most ecologically sound use, the production of timber and wood products. Scots pine, the original Highland pine, is being planted; but other fast-growing species of

347

economic value, native or foreign, are being used in whatever environments seem matched to their needs.

More characteristic of British environmental restoration projects are the attempts to maintain representative examples of the chalk grasslands and the coppice-with-standard forest. This forest type is composed of planted shrub species—hazel, sweet chesnut, hornbeam, or even oak—which are cut or coppiced for poles on a regular basis every eight to twenty-five years. The standards are also planted species, usually oak, grown traditionally for ships' timbers but more recently for lumber (Figure 16.9). Both these types are man-maintained artifacts that have existed for so long that they have gained natural status. Unfortunately, both are highly unstable and revert to other vegetation types without continuing management. Since the tradition of allowing unused land to revert to the self-reproducing or climax vegetation of the region has been lost along with that climax vegetation, some European restoration efforts try to maintain examples of the man-dominated systems. But efforts are also being made to restore some beech woodland on what was recently chalk grassland, and English oak in the coppice-with-standard woodland.

The chalk downs and coppice-with-standard forests

Fig. 16.9 • This young coppice-with-standard woodland in England consists of clumps or coppices of hornbeam interspersed among the thicker boles of the standard species, oak. In the springtime the forest floor is covered with anemones.

The chalk downs are rolling hills in the south of England, which, because of their well-drained and relatively light soil, were cleared of their ash and beech forests and farmed or grazed in neolithic times (Figure 16.10). With the foundation in the Middle Ages of monasteries, which maintained large flocks of sheep, most of the chalk ridges, with their thin soil, became pastures. For nearly 500 years and in some places far longer, the grazing pressure of generation after generation of sheep had a selective effect on the flora of the chalk grasslands, producing a group of species resistant to grazing pressures and forming a community of plants found nowhere else (Figure 16.11).

The moors, once formed, tend to persist even if they are not grazed, partly because of the remoteness of appropriate tree or shrub

Fig. 16.10 • This chalk grassland in southern England is being maintained by grazing. The fine lines along the contours of the hillside on the left are trails left by grazing sheep.

seeds and the unlikelihood of their becoming established in the thick layer of peat even if planted. But the chalk grasslands are quite susceptible to invasion by thorny shrubs once grazing pressure is removed. And that is precisely the problem today over most of the chalk grasslands of southern England. Sheep have been withdrawn from many of the chalk downs for economic reasons, and thousands of acres of downs have either been plowed or are invaded by shrubs. The problem has been aggravated by the severe reduction by myxomatosis (Chapter 14) of the rabbit population that maintained the grassland in the absence of sheep. Why be concerned with the loss of grasslands that aren't natural anyway? Because aesthetically they are a more pleasing landscape than the shrubs that replace them, and because they contain a rich variety of plants and animals that disappear when the shrubs invade.

The heir to this and other problems of vegetation management is the Nature Conservancy (not to be confused with the private American organization of the same name). Set up by the government in 1949, its functions are "to provide scientific advice on the conservation and control of the natural flora and fauna of Great Britain; to establish, maintain, and manage nature reserves in Great Britain, including the maintenance of physical features of scientific interest; and to develop the scientific services related thereto."[2] The Nature Conservancy has been quite active in the campaign to restore and manage representative samples of man-dominated systems, as well as to restore native beech and oak woodland. It maintains an active research program to support its efforts toward these ends.

The most obvious restorative technique for declining chalk grass-

Fig. 16.11 • The old British penny, about the size of an American half dollar, gives a sense of scale to this photo of the dense turf made up of thyme and grasses that characterizes chalk grassland in prime condition.

[2] Blackmore, M. (ed.), *The Nature Conservancy Handbook 1968*, p. 5. Her Majesty's Stationery Office, London.

lands would be to restore the grazers that normally maintained the vegetation. But if this is done in the summer the flower crop, especially orchids, is reduced; if grazing takes place in the winter the livestock refused to eat the coarse grasses and thorny shrubs that tend to invade and replace the desirable plants. If supplemental feed is offered, the hay introduces exotic weeds. In the old days the flocks grazed in spring and summer, but since they were constantly on the move, grazing damage to flowering plants was light. The solution seems to be a short, intense feeding period in May and June, which encourages the eating of coarse grass while it is still green, and the thorny shrubs before the thorns harden. Flowering plants that have been grazed often recover and flower again. But it is not always feasible to run livestock on the relatively small reserves; and so mowing, burning, and local application of herbicides have been used to control the invasion of undesirable species. If managed for insects the grassland should be grazed much less frequently, say every third year. Vegetation management is neither simple nor obvious.

The coppice-with-standard artifact was once quite common in Kent, Sussex, and Hampshire, but is now becoming rather restricted. The extremely slow growth of the oaks coupled with lack of demand for the poles has led to a wholesale replacement of coppice-with-standard woodlands by more productive and valuable species, often exotic conifers. Because conditions during the first seven years after coppicing are quite favorable to many interesting and rare species of insects, the Nature Conservancy has continued to coppice its coppice-with-standard woodlands despite unfavorable economic conditions (Figure 16.12).

Fig. 16.12 • When the coppice species are freshly cut the coppice-with-standard woodland is quite open and park-like. This allows a succession of insect and plant species to develop that is quite characteristic of this vegetation type.

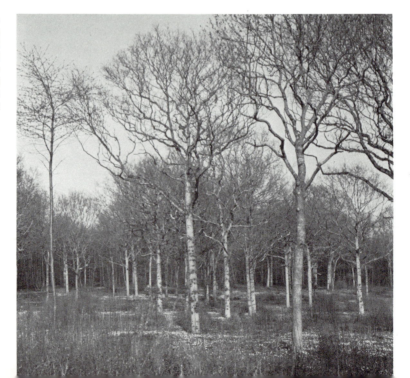

True environmental restoration—an attempt to recreate the original vegetation—is being attempted on the Scottish island of Rhum, an island eight by eight miles in size, originally forested with pine, birch, elm, rowan, and oaks, which has been virtually treeless for several hundred years. Small areas are being fenced and planted with a mixture of native tree species grown locally from seed obtained from the few remaining trees. Much more of the island will be replanted as soon as the herd of red deer is reduced to a sustained-yield level. In another example, Yarner Wood in Devon, a mixed oak and pine forest with some birch, beech, and ash, is being allowed to develop naturally without further intervention by man. But in England that is far more difficult than it might seem. Many woodlands have been grazed or used for winter shelter for livestock or deer and consequently have no understory saplings to replace the trees in the canopy. If the woods are fenced to exclude the larger animals, the ground vegetation becomes so luxuriant that small mammals multiply enough to replace the larger animals as destroyers of seeds and tree seedlings. The situation is exacerbated by the elimination by gamekeepers of most of the larger carnivores and birds of prey that normally keep small mammal populations within bounds. Finally, exotic species such as rhododendron or sycamore (sycamore maple in American usage) may compete with the native species; should they be allowed to remain or be eliminated?

Other efforts

The British experience, then, while quite different from our own in its attempts to maintain systems that would be regarded in the United States as in need of restoration rather than maintenance, does include some experiments designed to allow as close a return as possible to the systems originally present.

The American Experience

Since our continent was so recently settled, compared with Britain, virtually no natural system of any size in the United States has disappeared without a trace. Many systems have been destroyed, but the elements of these systems still exist in large enough associations to allow the possibility of reassembly. However, this will not always be so; hence the urgency in getting research started in as many diverse systems as possible. A major start toward this end has been made by the International Biological Programme, which, although oriented toward productivity of systems and the development of system models, has nevertheless generated a great deal of rather basic information about the major North American ecosystems.

The Ramble

Probably one of the first attempts to restore a natural environment in the United States was undertaken in New York's Central Park over 120 years ago. Around 1850 the area now occupied by the park was covered with squatters' huts and shacks, and was on the outskirts of New York City. Period drawings and woodcuts show almost

no trees or shrubs. In fact, there was little vegetation of any kind
(Figure 16.13). The designer of the park, Frederick Law Olmsted
(Figure 16.15), had to start from scratch. Even knowing that the
original site was treeless, it is difficult to realize that almost every tree
and shrub was planted by hand (Figure 16.14) in the fashion of Capability Brown. For the park as a whole, Olmsted had no intention of
restoring the land to its original appearance. Like Brown, his "professional grandfather," Olmsted wanted to create an informal, aesthetically pleasing landscape; and a great deal of digging, filling, damming,
and pumping was required to achieve this end. But in the midst of all
of the "natural" artificiality, Olmsted intentionally created a small area
he called the Ramble, which was to be as natural a wood as he could
design. So he planted native species that he supposed might have
been there originally: trees, shrubs, and herbs. Today, 120 years later,

Fig. 16.13 • In 1859, squatters' huts occupied the land
that later became Central Park
in New York. (*Museum of the
City of New York*)

Fig. 16.14 • Virtually every
tree and shrub in Central Park
was planted by hand. (*Thomas
Airviews*)

the trees have matured and probably reproduced several times, filling in the spaces, creating their own patterns. Walking through the Ramble early in the morning when the park is deserted, the effect is quite magical—that of being out in the country miles away, surrounded by a very natural-looking landscape (Figure 16.16), yet knowing that the sound of white water (or is it surf?) in the distance is really rush-hour traffic on Fifth Avenue and Central Park West.

If some sort of natural system can be created in the very navel of megalopolis, then why not recreate prairies, forests, or marshes where they have been badly misused or destroyed? The thought has occurred to many. A. S. Leopold, a rather perceptive zoologist at Berkeley, put together in 1963 a four-step procedure for reconstructing ecosystems:

1. Historical research to determine what the area looked like before man began to alter things.
2. Ecological research to find out how the plants and animals are related to each other and to the ecosystem, and which environmental factors are most important in maintaining the area as we wish to see it.
3. Pilot experiments, based on the information gained from ecological research, to establish the feasibility of proposals.
4. Full-scale management of large ecosystems based on the more workable of the pilot studies.

Fig. 16.15 • Frederick Law Olmsted, eminent nineteenth-century landscape architect, designed many of the famous urban parks in the United States, including Golden Gate Park in San Francisco and Central and Prospect Parks in New York. (*New York Public Library Picture Collection*)

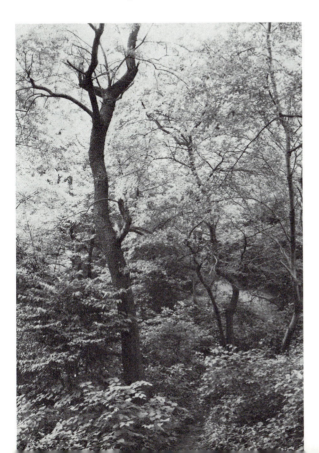

Fig. 16.16 • Planned as a natural area by Olmsted, who used mostly native species, Central Park's Ramble today appears like a bit of forest that has wandered in from Westchester County.

The first three of these steps had already been carried out on the tall grass prairie system by several groups in the Midwest several years before Leopold published his paper. The first such project was undertaken by H. C. Greene, a botanist on the staff of the University of Wisconsin. In 1945, Greene realized that the tall grass prairie, which had played such an important role in the history of the Midwest, had virtually disappeared as a system, plowed into cornfields over the preceding 100 years. All that remained, at least in Wisconsin, were a few small remnants scattered along railroad right-of-ways, in old cemeteries, and around abandoned schoolhouses. Even these odd scraps were disappearing one by one. Consequently Greene took over the management of thirty-five acres of a former sand prairie with small groves of bur oak at the University of Wisconsin Arboretum in Madison and began his restoration program. Many approaches were used at first since there were few precedents to follow: sowing seed of various prairie grasses and herbs, planting seedlings grown in the greenhouse, setting out blocks of sod taken from nearby prairie fragments, and transplanting mature prairie plants from remnants about to be developed. This introduction phase lasted for several years, proceeding as fast as time, funds, and the supply of plant materials allowed. By 1952, 214 species characteristic of the prairie in Wisconsin had been reintroduced. Except for some very persistent introduced grasses such as Kentucky blue grass, the prairie species had become sixteen times more abundant than in 1945 and the introduced weeds, which had formerly dominated the plot, had declined by the same factor.

In 1964 the Morton Arboretum in Lisle, Illinois, twenty-five miles southwest of Chicago, having just lost a small piece of prairie remnant to utility company construction, decided to replace a twenty-five-acre weedy meadow with a tall grass prairie. Since earlier restoration projects had been criticized for using commercial seed from remote sources and transplants that potentially introduced soil organisms foreign to the local area, it was decided to use only mixtures of seed gathered from local sources.

At one time the prairie stretched from horizon to horizon, Saskatchewan to Texas and Colorado to Illinois. While there was ample opportunity for slow genetic exchange among populations over this vast territory, there was a tendency also for local adaptations to photoperiod, soil factors, and temperature extremes. So virtually all widely distributed species differentiated into local adaptation units or ecotypes. The Morton Arboretum wanted to be sure that those ecotypes well adapted to conditions in northern Illinois be used in stocking their prairie. Introduced commercial strains of former components of the prairie such as switch grass can often be so successful that the nongrasses or forbs such as shooting star, sunflower, and compass plant, which are as important a part of the tall grass prairie as the grasses, are crowded out. Within ten years the synthesized prairie at

Tall grass prairie

the Morton Arboretum had accommodated 125 prairie species on six acres, and the acreage is being increased as materials become available.

All the prairie restorations attempted so far have been pilot experiments designed to explore the best techniques for restoring grassland systems. These techniques may become extremely important if and when a prairie national park is established. Over forty years ago, at a time when many splendid sites were still available, Victor Shellford, an ecologist at the University of Illinois, recommended that national parks should preserve not only scenic wonders, but viably large tracts of representative biomes. Many of these ecosystems have been preserved: the eastern deciduous forest in Great Smoky Mountains National Park; tropical forests in the hammocks of Everglades National Park; the coastal and lake dunes in several national seashore or lakeshore areas; the redwoods and big trees in Redwood and Sequoia National Parks; the superb conifer rain forests in Olympic National Park; and many desert systems in several parks and national monuments. But the gorgeous sea of tall grasses that so impressed the early settlers in the Midwest is totally unrepresented. Ideally, such a park should include up to 1 million acres and be fenced to allow introduction of bison, antelope, elk, and their predators, wolves and coyotes. While the ideal is perhaps unlikely because of the numbers of cattle that 1 million acres of good grassland could support, a much smaller tract of 57,000 acres in the Flint Hills of eastern Kansas has been proposed but bitterly opposed by ranchers who see any withdrawal of range land as a threat to the entire cattle industry. Even if such a modest prairie lands national park is finally established, restoration work will be needed to bring the whole tract back to its former productivity, for overgrazing is pandemic in the grasslands of most of North America.

However, the goal of environmental restoration needn't be regarded merely as the establishment of some kind of open-air museum. It is possible to apply the ecological principles inherent in environmental restoration to the rehabilitation of working ranchland with the aim of producing superior beef. An excellent example is provided by the Lasiter Ranch.

After World War II, rising land prices in south Texas obliged Tom Lasiter to relocate to Colorado, where he purchased 25,000 rather battered acres of shortgrass prairie rangeland near Matheson. Departing sharply from contemporary ranching methods, Lasiter began by stocking his ranch lightly to allow recovery of the long-overgrazed grassland. The results were so gratifying—tripling of the yield of forage per acre—that Lasiter determined to keep his stocking rate low. By carefully rotating the stock on his pastures, he prevented overgrazing, weakening of the grass cover, and soil erosion. A no-poisoning, -trapping, or -shooting policy has resulted in a balanced population of native range species; the natural buildup of coyotes and other predators was accompanied by a corresponding decline in jackrabbit and cottontail populations. Poisonous weeds and plants have been controlled by the light stocking rate, which, by maintaining a good cover of grasses and forbs, eliminates the weeds that characteristically invade overgrazed lands. By keeping pastures large the stock have more freedom to choose their food and thus behave more like the original grazers in the system, the bison. In short, Lasiter has required that his cattle adjust to a natural environment complete with the native species of coyotes, gophers, rabbits, and hawks; a natural environment that he has supported by maintaining intact the complex of vegetation on a stabilized soil. A simple application of basic ecological principles perhaps, but these principles are so rarely applied that Lasiter's experiment appears like an oasis in a desert of misuse.

One of the most endangered ecosystems in North America is the coastal wetland type characterized by cord grass, *Spartina alterniflora*. Once this habitat was extremely abundant along the Atlantic coast from New Hampshire to northern Florida. But years of sanitary landfilling and dredging for marinas and seashore developments have gobbled up vast acreages. While there are still huge tracts remaining along the more remote stretches of coast, such as the eastern shore of Virginia and the Georgia coast, the more heavily populated states like Connecticut have about run through their wetland capital.

Wetlands

Because wetlands were long regarded as wasteland—property that could not be conveniently drained for farming, did not provide particularly good grazing, flooded every spring tide, and produced hordes of mosquitoes—their development was viewed as a public benefit. With the pioneering research of Eugene Odum and his students at the University of Georgia we have finally begun to understand how important coastal wetlands really are as primary producers in the food web that reaches offshore into the estuaries and shallow continental shelf (see Chapter 7). Still another important function of coastal wetlands has been discovered recently: their ability to take up phosphates and nitrates from estuarine water and release oxygen. By so doing, wetlands have a significant effect in reducing the nutrient and BOD burden of coastal water. This latter function makes wetlands extremely valuable as natural purification systems in areas of high population density, where increasing amounts of pollutants are finding their way into coastal waters. Unfortunately, this discovery came at a time when most wetlands in the vicinity of urban complexes had long been filled with the city's wastes. After a marsh has been filled with thirty feet of trash there is little chance for restoration. But if new marshes—even if they are simplified compared with the original vegetation—can be established in appropriate sites, then the trend of wetland destruction can be at least slowed or even reversed. Considering the importance of wetlands in the productivity of estuaries and coastal waters and their role in pollution control, it might well repay a coastal state to subsidize establishment of new marshes wherever feasible. Should that enlightenment come to pass, the knowhow will be available. At least two groups have successfully established tidal marshes on the Atlantic coast: one, headed by Ernest Seneca, at North Carolina State University at Raleigh; and a private research group, Environmental Concern, Inc., at St. Michaels, Maryland.

Several years ago Seneca and his associates planted a sand flat near Oregon Inlet on North Carolina's outer banks with plugs of cordgrass. Using hand labor at first, only a few acres were planted. The second season a tobacco planter was adapted to accept the cordgrass plants and a larger area was planted. Although snow geese attracted to the new planting the first winter caused much damage, the plants recovered and the marsh flourished.

The plan of Environmental Concern was somewhat more ambitious. An island in Chesapeake Bay that was being eroded into fragments (Figure 16.17) was selected and purchased by the Nature Conservancy (this time the private American group), with whom the project was initially associated. Bank sand was barged out to the gap between two of the fragments and dumped to provide the proper elevation for cordgrass establishment. In the spring of 1972 planting began of cordgrass plus several other species of plants capable of tolerating the brackish bay water. Instead of using vegetative shoots dug from existing marshes, plants were grown from seed and peat pots of the seedlings were then placed in a two- to three-foot grid on the flat. Growth was quite vigorous the first summer and many plants flowered (Figure 16.18). During the first winter Canada geese fed heavily on the plants but enough plants survived, together with a plentiful germination of seeds produced the preceding summer, to assure stabilization of the flat surface with a complete cover of plants likely by the end of the second or third year. While the costs of constructing artificial flats preclude their widespread use, many natural sand or mud flats exist, particularly near harbors or channels that must be dredged periodically. These natural flats could easily be planted and stabilized with marsh plants using the techniques developed by Seneca and Environmental Concern. While what would be created is not the same marsh that took 3000 years to evolve, a productive beginning of the process would have been made.

The usual approach to dredging spoil disposal has been to dump the wastes on a diked tidal marsh, but since this is rarely allowed now that states are beginning to protect their wetlands, the spoils could be dumped in shallow water and planted promptly with reasonable assurance of stabilizing the spoils with marsh vegetation and at the same time increasing the potential productivity of the harbor or estuary. Of course, spoils heavily contaminated with hydrocarbons, heavy metals, or biocides pose special problems, as does the question of how the area to be covered with spoil compares in productivity with the marsh that will be established. These complications aside,

Fig. 16.17 • Hambleton Island in Chesapeake Bay before marsh restoration was begun was rapidly being eroded into a string of islets. (*Environmental Concern, Inc.*)

Fig. 16.18 • Six months after cord grass was planted, the young marsh was apparently well on its way to establishment. (*Environmental Concern, Inc.*)

tidal marsh synthesis or, in the broader sense, restoration, is a concept whose time has come.

Despite our growing awareness of the importance of intact natural systems of all kinds as sources of primary production, reservoirs of genetic information, buffers against the rising tide of pollution, and certainly not least their aesthetic and recreational appeal, we are running short on areas that can be saved. By the year 2000 all currently available land will either have been saved or developed. With rising demands for more recreational space from a larger public with far more leisure time, the additional land will have to come from areas currently regarded as wasteland whose environments will have to be restored, if not to their pristine glory, at least to aesthetically pleasing natural systems capable of reproducing themselves. Strip mined land, quarries, slag heaps, eroded farmland, true wasteland of all types will have to be pressed into service to satisfy the demand. While much work has been done in Britain and Germany, two countries that have already run out of land, few of these techniques have as yet been applied to analogous situations in the United States. We can be ready with the expertise and tools if the projects described and others like them are recognized as the hope of the future that they are and are fully supported with both public and private funds.

FURTHER READING

Blackmore, M. (ed.), 1968. *The Nature Conservancy handbook 1968*. Her Majesty's Stationery Office, London. Background on the origin of the Nature Conservancy and a descriptive listing of the Conservancy's holdings.

Bolling, A., 1971. "Tallgrass prairie park." *Nat'l Parks and Conser. Mag.,* March, pp. 6–10. A capsule history of the attempts to establish a prairieland national park.

Colbert, F. T., 1971. "The Lasiter ranch: applied range ecology." *Nat'l Parks and Conser. Mag.,* March, pp. 18–20. Suggests that application of ecological principles to a rancher's problems can be both practical and profitable.

Larimer, E. J., 1968. *An investigation of the possibilities for creating salt marsh in the estuaries of the Atlantic and Gulf Coasts.* 22nd Ann. Conf. S. E. Assoc. Game and Fish Comm., pp. 82–88.

Ovington, J. D., 1964. "The ecological basis of the management of woodland nature reserves in Great Britain." *J. Ecol.* **52** (Suppl.), pp. 29–37. Covers the types of environmental management and reconstruction being explored in England.

Schramm, P., (ed.), 1970. *Proceedings of a symposium on prairie and prairie restoration.* Knox College Biol. Field Station, Special Publ. No. 3.

Wells, T. C. E., 1969. "Botanical aspects of conservation management of chalk grasslands." *Biol. Conser.* **2**(1), pp. 36–44. A discussion of some of the problems of preserving an unstable habitat.

17

INDOCHINA: THE WAR AGAINST AN ENVIRONMENT

ON 22 APRIL 1915, French troops on the front line near Ypres saw
the vast cloud of greenish-yellow gas spring out of the ground and slowly
move down wind toward them, the vapour clinging to the earth, seeking
out every hole and hollow and filling the trenches and shell holes as it
came. First wonder, then fear; then, as the first fringes of the cloud en-
veloped them and left them choking and agonized in the fight for
breath—panic. Those who could move broke and ran, trying, generally in
vain, to outstrip the cloud which followed inexorably after them.[1]

The gas was chlorine, released from thousands of canisters by the Germans, who had patiently waited for just the right wind speed and direction to retaliate against earlier use of poison gas by both the
English and the French. The tactic could have been brilliantly successful: four miles of the Allied front collapsed into a swarm of panicked
troops fighting their way to the rear. The gas caused 15,000 casualties
that day, 5000 of them fatal. Had the German army fully realized the
impact of their innovation they could have plunged through the gaping
hole in the French lines and swept to the channel ports of Dunkirk
and Calais, perhaps writing a different conclusion to World War I.
But the Germans seemed as amazed as the French and sat gaping at
their handiwork. The Allies bitterly protested this uncivilized perver-
sion of honest killing, and then continued to use gas themselves.
Chlorine was soon replaced by mustard gas, then phosgene. By the
war's end, some twelve tear gases, fifteen choking agents, four blister
agents, three blood poisons, and four vomit gases had been released
into the European air (Figure 17.1).

Even in 1915 war was becoming efficient and impersonal; this

[1] Brown, F. J., 1968. *Chemical warfare, a study in restraints.* Princeton
University Press, Princeton.

Fig. 17.1 • A doughboy and his horse prepare for possible gas attack during World War I. (*U.S. Signal Corps*)

attitude had grown more and more marked in the years since then. The Revolutionary War expression "Don't fire until you see the whites of their eyes" must have had a rather ironic ring to the crew of a B-52 bomber dropping a load of bombs from over 30,000 feet on a target in Indochina during the Vietnam war. In dispassionate routine the missions were completed, and soon the crew was back aboard a carrier or at an airfield in Thailand, Vietnam, or Guam, having a beer, with about as much sense of the environmental impact of their actions as a gang of teenagers who have just tossed an empty six-pack out of the window of their speeding car.

In 1969, General William C. Westmoreland, then Army Chief of Staff, appraised the technological capability of modern war when he foresaw "battlefields on which we can destroy anything we can locate through instant communication and almost instant application of highly lethal firepower. . . ."[2] Since the firepower cannot always distinguish between the good guys and the bad, this approach has led to the creation of free fire zones where no one has any right to exist. This is total war.

Certainly the effects of war have never been limited to the human participants. Ancient records are filled with accounts of cut olive groves, burned crops, salt plowed into fields to ruin their productivity, and poisoned water supplies. In our recent past, the Southern states took decades to recover from such peripheral effects of the Civil War. But only in relatively recent times has available technology been applied systematically to intentional destruction of an environment.

[2] *The New York Times,* 15 October 1969.

The most thoroughly documented example of such environmental destruction is, of course, Indochina.

To discuss the details of America's involvement in Indochina would go far beyond the scope of this book; indeed, it may be years before historians and political scientists have a clear picture of the chain of events that led to the deployment of American troops in southeast Asia, despite a long-standing military reluctance to take on an Asian land war. Yet a quick sketch is necessary to interpret the significance of the environmental manipulations performed in the pursuit of this war.

HISTORY OF THE WAR

After more than a thousand years of domination by China (second century B.C. to tenth century A.D.), Indochina enjoyed 900 years of independence before coming under French influence in 1862. By the turn of the century all of present-day Vietnam, Laos, and Cambodia was under complete French control. After World War II the Vietnamese, moved by the same nationalistic impulses as the Indians, Africans, and Indonesians, took steps to achieve independence. These efforts were, of course, opposed by the French military forces. Since true independence appears to be impossible for smaller or weaker countries adjacent to a great power, control of Indochina by a Chinese Communist regime was feared. The French defeat at Dien Bien Phu in 1954 led to a partition of Indochina, and the United States entered Indochina as a military power in the early 1960s to support the anti-Communist puppet state recognized and then abandoned by the French.

The subsequent war was complicated by extensive guerrilla operations at night in difficult terrain, conducted by people who appeared during the day to be simple peasants working their crops. Inability to recognize the enemy causes problems: it rules out the typical war format of friendly territory on one side, enemy territory on the other, and a well-marked battlefront in between. In such a situation, two courses of action are possible: the population is screened to separate the good from the bad, then the good are protected and the bad destroyed or subjugated; or "they" are all destroyed, with the assumption that at least some are bad. Both techniques were used in the Vietnam war. Agrovilles and, later, strategic hamlets or prosperity villages were organized, separating loyal Vietnamese from the guerrilla National Liberation Front (NLF), both for their protection and to withhold supplies and manpower from the shadowy enemy. Free fire zones were also set up in strategic areas, on the assumption that anyone found there after the population had been officially evacuated was hostile and hence a suitable target. Both techniques have had an unusually severe effect on the environment.

THE ORIGINAL LANDSCAPE

For all the news datelines that sound so familiar after a decade of repetition—Saigon, Danang, Quang Tri, My Lai—Indochina is less well known to most Americans than the near side of the moon. A brief description will help. The coast of southern Asia is characterized by a series of rather mountainous peninsulas, a phenomenon one wag has referred to as "continental drip" (Figure 17.2). The easternmost of these, the Indochinese Peninsula, has a central spine of limestone and sandstone mountains, which drop sharply to the sea on the east and fall more gradually into the marshy deltas of the Song Koi or Red River on the north and the Mekong River on the south. The climate, characteristic of southeast Asia, is a tropical monsoon—relatively dry, warm winters and warm, rainy summers. Between April and October the rainfall averages 200 cm, with more in the highlands; temperature varies little in the lowlands, no more than 4°C throughout the year.

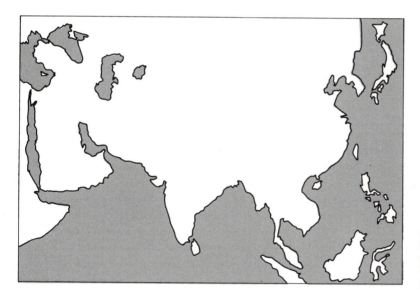

Fig. 17.2 • The coast of southern Asia is uniquely characterized by a series of mountainous peninsulas that project into the Indian Ocean.

Since Indochina has been inhabited by man for a very long time, no one is very sure what the original vegetation looked like; man's activities over the millennia, even before the trauma of the recent war, have greatly affected the landscape through paddy cultivation, rotating slash-and-burn or "rai" culture, lumbering, and other activities. But enough scraps of relatively undisturbed vegetation were left, at least until recently, to give some picture of what Indochina looked like forty years ago. This picture still holds to some extent, because although the war has disturbed the ecological relationships among species, it has not, as far as we know, wiped out any species altogether. About 50 to 60 percent of the area was forested. Some of the higher mountains had a dense forest of pine which, as the elevation decreased, became mixed with some oak, then with other deciduous species, making up a forest not very different in appearance from forest types seen in the southern United States. The lower elevations of the mountains, however, had a belt of rain forest made up of species characteristic of the wetter parts of southern Asia. The rest of Indochina was either under cultivation, or covered with a savannah of coarse grasses or with bamboo mixed with thorny shrubs. Extensive areas in the river deltas were covered with mangrove and nipa palm. Much of the savannah may have been forested in the past. Despite man's impact, the flora of Indochina, like that of all southeast Asia, was very rich. There were, for example, more than 2000 species of trees and shrubs alone in the Indochinese peninsula, compared with perhaps half that number in North America. The fauna was correspondingly rich, including not only a diversity of birds, but large mammals such as the elephant, tiger, wild boar, deer, rhino, and the rare Indochinese gibbon, as well as many more common primates, the Kouprey or Indochinese cow, and a subspecies of the gaur, another rather rare cow. The mangrove swamps and estuarine waters teamed with commercially valuable species of molluscs, crabs, and shrimp.

The soils that supported this diversity of species vary and include some areas of lateritic soils, particularly in the regions of highest rainfall. Lateritic soils, you will remember from Chapter 3, can be converted into laterite, hard red rock, if exposed to drying and high temperatures without the protection of vegetal cover.

This, then, was prewar Indochina. The war-induced manipulations of the last decade have changed this picture considerably.

THE IMPACT OF THE WAR

Often unable to concentrate their vast firepower on a well-concealed enemy, the United States forces instead attacked the concealing vegetation on a scale without precedent; first by the conventional means of bombing and some largely abortive attempts at burning; then by herbicides; and finally by bulldozing. The unseen enemy was attacked with high-altitude saturation bombing; harassing agents (primarily the tear gas CS[3]); and a variety of antipersonnel fragmentation bombs—the Rockeye missile, the Dragontooth and Gravel mines, bottle, butterfly, pineapple, and guava bombs, and of course napalm, white phosphorus, and thermit. The concealing vegetation that was destroyed was the natural environment of Indochina—pine

[3] The United States classifies some chemical weapons according to a letter code rather than a generic or common name.

forests, tropical rain forests, savannahs, mangrove swamps, and other miscellaneous types. The food-producing areas didn't escape, however; determined to reduce the food supply available to the NLF, the United States sprayed large tracts of cropland with herbicides and pockmarked much larger acreages with bomb craters. Although herbicides have caused the greatest furor in the press because they were used on an extensive scale for the first time in Indochina, the greatest impact on the Indochinese landscape has come from old-fashioned 500-pound bombs.

On a region the size of Texas the United States has dropped during the course of the war more than twice the tonnage of bombs dropped during World War II. While no one knows just how many craters have resulted from such saturation bombings, one estimate runs as high as 20 million in South Vietnam alone (Figure 17.3).

Craters

Fig. 17.3 • Manioc fields 20 miles southeast of Saigon "plowed" by bombs from B-52 bombers. The craters are 20 feet across. (*G. A. Orians*)

A standard 500-pound bomb generally produces a crater 20 to 50 feet wide, and from 5 to 20 feet deep, depending on type of fuse and ground conditions. Although the free fire zones contain the greatest proportion of the craters, they are also found in cultivated parts of the Mekong Delta, in the mountainous valleys near the demilitarized zone and Danang, in the coastal mangrove forests, and in the evergreen hardwood forests on the flats north of Saigon. In many of these areas a 20-foot-deep hole penetrates the water table and the craters hold water year round; in other places they hold water only during the rainy season. Heavily eutrophied, these pools may contain or be stocked with fish and may serve as reservoirs for irrigation. But they may also act as breeding ponds for mosquitoes carrying malaria and other diseases and the schistosome-carrying snail that can be so troublesome in tropical countries. At the very least the craters make working the fields very difficult because of unexploded munitions and the bits of shrapnel that cut the hooves of the buffalo used to pull the plows. Craters in forested country make it difficult to haul out logs, which must be cut into less than desirable lengths to negotiate the tortuous access roads.

Originally developed to control weeds in lawns and crops (see Chapter 13), herbicides seemed an ideal way to clear vegetation and thereby prevent ambushes along roadsides and the edges of waterways, infiltration of men and supplies across the demilitarized zone and the Laotian border into South Vietnam, and troop buildups in remote areas. Three major herbicides were used: Agent Orange, a mixture of 2,4-D and 2,4,5-T, which persists for several months; Agent White, a combination of 2,4-D and picloram, which remains active in the soil for up to several years; and Agent Blue, cacodylic acid, whose major ingredient is sodium dimethyl arsenate, another compound that accumulates in the soil. The first two herbicides were sprayed primarily on forest vegetation (Figure 17.4), the third primarily on food crops, particularly in the mountains. About 58 percent of the total amount of herbicides used was Orange, 31 percent was White, and 11 percent Blue. Starting in 1962, the herbicide program peaked in 1969; in 1967 it consumed virtually the entire United States' production of 2,4-D and 2,4,5-T.

Herbicides

Fig. 17.4 • Two C-123s in a tight turn spraying defoliants on a South Vietnamese hillside in 1968 during the recent war. (*U.S. Air Force*)

Although an eighth of the total land area of South Vietnam has been sprayed, the impact of herbicides on Vietnam's vegetation has been varied. Mangroves, for example, are extremely susceptible to herbicides (Figure 17.5); one spraying causes a complete kill with essentially no evidence of regeneration after at least six years. As a result, perhaps half of Vietnam's mangrove vegetation, 744 square miles, has been totally destroyed. Other hardwoods, however, are generally more resistant to herbicides. Usually, only 10 percent of

Fig. 17.5 • Mangrove is particularly sensitive to defoliants, often being killed with one spraying. Regeneration is very slow. (*Pfeiffer and Orians*)

the trees die with one spraying and the seedlings and saplings, which are the crucial stage of tree life, are not particularly affected. A second spraying within a year, however, heavily damages this understory and may kill 50 percent or more of the canopy trees. At least 4 percent of the forests of South Vietnam have been sprayed more than once (and another 15 percent once). Some estimates suggest that as much as 5 percent of the mature hardwoods has been destroyed. This wood, worth $500 million, would have supplied South Vietnam with timber for over thirty years.

The impact of defoliation goes beyond the simple economic loss of timber. Destruction of the normal vegetal cover can stimulate the growth of cogon grass and bamboo (Figure 17.6), weedy species of little value (if any) but of vigorous growth. Once well established, reproduction of trees is impeded, and elimination of these two species is difficult. Consequently the recovery of sprayed forests may be long delayed or even prevented in some places. Recovery of forests may also be held back by loss of nutrients. Most tropical forest

Fig. 17.6 • This rubber plantation in Long Khanh Province, South Vietnam was sprayed with herbicides in 1967. By 1970 cogon grass, a weedy species that is very difficult to control, had invaded and become well established. (*Arthur H. Westing*)

367

ecosystems contain most of their nutrients in their biomass rather than in the soil (geomass) (see Chapter 3). When the forest is destroyed, either by cutting and burning or by herbicides, the nutrients quickly leach away and are only very slowly replaced from long-term weathering of the subsoil.

The destruction of vegetation affects animal life too, particularly in the mangrove swamps, where the annual production of leaves is of critical importance to the detritus food web on which many estuarine organisms depend (see Chapter 7). Although reliable data are lacking on the populations of fish, molluscs, and crabs in the delta areas affected by spraying, work on similar vegetation in Florida suggests that animal populations, including commercially important species, would be adversely affected. Radical shifts in the kinds of birds might be anticipated too. Lack of cover would tend to reduce the abundance of all species, but populations of species that eat insects that in turn feed on leaves would be depressed, while species feeding on wood-eating insects would become more abundant. Since many widely scattered tropical tree species are pollinated by birds, bats, or insects, decreased populations of the pollinators might in turn affect rates of reproduction of the trees and reduce the food supply of fruit, nuts, and seeds which many species in a tropical forest need to survive. But perhaps the most devastating effect of defoliation in a tropical forest is the loss of cover on which so many species of both animals and plants depend.

Although Vietnam's commercial rubber plantations usually were not intentionally sprayed, numbers of accidental sprayings via wind drift have reduced latex production by as much as 30 percent, a loss level that can be sustained by large foreign-owned plantations, but sufficient to drive the small growers out of business. In 1960 South Vietnam had received 56 percent of its foreign exchange from rubber exports.

Despite its effects on plants and animals, the most disturbing consequence of the use of herbicides in Indochina has been the reports of increased rates of stillbirths and the birth of deformed babies in some of the most heavily sprayed provinces of South Vietnam. The herbicides were sprayed at ten or more times the normal dose rate used in the United States, and whole tankfuls were sometimes dumped if the spray plane ran into any sort of difficulty. This tragic effect on human fetuses stems from the presence of trace amounts of a teratogenic or fetus-deforming compound, dioxin, in the Agent Orange used. Although the level of dioxin in the Agent Orange used has been estimated at 10–15 ppm, since 93 million pounds of Agent Orange have been splashed about the Vietnamese countryside, roughly 1000 pounds of dioxin have found their way into the environment. Subsequently, dioxin has been found in a few parts per trillion (pptr) in South Vietnamese river fish, shellfish, and human milk. In some fish the dioxin concentration was over 500 pptr (half of a lab population of guinea pigs is killed by 500 pptr). Other evidence suggests that dioxin may form when Agent Orange-sprayed vegetal matter is burned.

In response to increasing public concern about the effect of herbicides on the people and environment of Vietnam, and the apparent conclusion by the Department of Defense that herbicides had only limited strategic usefulness, they were phased out of use in the later stages of the war, in 1970–1971. However, the environmental impact of this vast experiment will probably be felt for several decades.

Another reason for the decision to phase out herbicides in Vietnam was the realization that vegetation could be much more thoroughly removed by enormous 20-ton D-7 tractors with an 11-foot, 2½-ton plow blade, and occasionally by the even larger D-9 tractor. The program of bulldozing vegetation began in 1965 and was so successful that it was greatly accelerated in the next several years, until by 1971, 750,000 acres had been cleared. The massive tractor scrapes a 2000-yard strip along both sides of major highways, usually down to the subsoil. Forested areas suspected of harboring NLF have also been bulldozed—126,000 acres worth; this represents a loss of 20 million board feet of lumber worth $15 million. The scraped roadsides are highly susceptible to flash flooding and erode rapidly, silting nearby streams; they also remain barren or become choked with weeds that will make rehabilitation difficult. Where the topsoil has not been removed by scraping, cogon grass and bamboo quickly invade and inhibit natural regeneration of the bulldozed forest. The tractors do clear the bulldozed areas of landmines and some timber is salvaged for firewood and charcoal, but the benefits are insignificant compared with the environmental costs.

Bulldozing

Another clearing technique employed was the big bomb, the BLU-82/B, a descendant of the old War War II blockbuster. Even larger, the 15,000-pound BLU-82/B or "daisy cutter" contains a gelled aqueous slurry of ammonium nitrate and aluminum powder. Constructed to go off just above ground level, the bomb left no crater, but a 3-acre clearing suitable for an emergency helicopter landing field. The big bomb was also used to trigger landslides along the Ho Chi Minh Trail and to disperse enemy troop concentrations. Although not often used against personnel (as mentioned, one of the major problems of the war was that the NLF balked at presenting suitable targets for the vaunted American firepower), the shock wave could kill over a 3280-foot radius, i.e., an area of 120 acres.

The Vietnam war has been an extremely useful testing ground for various new devices to put an enemy out of commission, much as the Spanish Civil War in the 1930s provided the German Luftwaffe with a chance to evaluate its Stuka dive bombers before their use on Warsaw and Rotterdam. The Rockeye missile, for example, is capable of penetrating subterranean shelters before it shatters into

Antipersonnel weapons

369

thousands of fine slivers of hot steel, which travel at high speed and easily pierce the body. Since the Rockeyes are clustered in groups of 247 per canister, an average drop of Rockeye canisters can saturate an area of 20 acres with about 1000 of these deadly missiles.

Other antipersonnel items contain steel pellets or flechettes. A flechette is a steel nail with fins on one end to stabilize the trajectory and to make a bit of a mess as they enter the body and effectively shred various internal organs. Then, of course, there is an old standby, napalm, a jelly-like form of gasoline whose antipersonnel "value" was demonstrated in the Korean War. However, since napalm can be rubbed off, it has been supplemented with white phosphorus, which cannot be put out or rubbed off and so usually burns its way down to the bone. By combining napalm and phosphorus with a preparation called Thermit, which burns at a much higher temperature, a particularly hideous bomb has been produced.

The whole point of antipersonnel weapons is not to kill, but to cause irregular and difficult-to-cure wounds that tie up six to ten people plus facilities and supplies to provide care. The victim suffers agonizingly from his vicious wounds and a circle of expanding demoralization is meant to be set up in the remaining population. As the Pentagon would say, the weapons have a built-in "deterrent capability." To no one's surprise, about 80 percent of the casualties from antipersonnel weapons in the Vietnam war were civilians— insofar as there were civilians in the war. An American medical student in Vietnam described a typical fragmentation casualty as ". . . a lovely 28-year-old peasant woman who was lying on her back nursing a young child. The evening before, she had been sitting in her thatched hut when a piece of shrapnel tore through her back transecting the spinal cord. She was completely paralyzed below the nipple line. We could do nothing more for her than give antibiotics and find her a place to lie. A few mornings later she was dead, and was carried back to her hamlet by relatives."[4]

Despite such casualties, few of the flying fragments found their mark; the overwhelming majority flew into the nearest tree. If and when the tree is logged, the pieces of shrapnel from the fragmentation bombs, as well as from the 500-pound bombs used in saturation bombing patterns, raise hell with the saws at the sawmill, reducing the value of the damaged timber by 30 percent. This problem will, of course, last as long as the life of the youngest tree, a lifetime greatly shortened by the entrance of bacteria and fungi through the holes torn in the bark. Napalm, too, damages other targets besides its intended human ones; its sticky persistence has led to the burning of some of the susceptible pine forests in the highlands.

In 1963 weather modification had to be added to the growing

---

[4] Quoted by Pepper, W. F., 1966–1967. "The Children of Vietnam," *Ramparts Magazine* 5(7), pp. 54–55.

list of environmental modifications performed or attempted in Indochina. The seeding of clouds over Laos was reported to have been successful in causing rainfall over Vietnam, which interfered with enemy troop movements. Apparently a method has also been developed for generating an acid rain that interfered with the operation of North Vietnamese radar units controlling antiaircraft missiles.

Meteorological warfare

Weather patterns were also changed to provide cover for reconnaissance raids into North Vietnam and to increase the frequency of bombing missions in bad weather. While the environmental effects of this first round of experiments were probably minimal, the lid is off the cooky jar and a second or third round of experiments over Cambodia, Libya, or Cuba may not be so benign or controllable.

Vietnam, as is still true of most of southeast Asia, used to be a country of many widely scattered small villages with a few large cities. But since there was considerable military advantage to aggregating people into more easily defended and controlled settlements, the old pattern has rapidly changed. The widespread environmental disturbance has uprooted millions of people, who have flocked to the few urban areas. Saigon has changed from a beautiful provincial French city of 250,000 to an American slum megalopolis of 3 million with virtually no facilities to handle the influx of refugees. The tremendous impact of billions of American dollars has also had a staggering effect on the economy and social mores of the country. Streets thronged with motorized vehicles from bicycles to Buicks, burning the smoky mixture of gasoline and oil that is used as fuel, have so polluted the air that many of the city's trees are dead or dying (Figure 17.7). While it may seem picayune to quibble about urban pollution problems in Saigon while the rest of the country is literally and figuratively bleeding to death, the direct effects of the war will

Social changes

Fig. 17.7 • One of the less frequently discussed concomitants of war is the forced urbanization of cities filled with refugees from the ravaged countryside. Smog in downtown Saigon has destroyed many of the flowering shade trees for which Saigon was once famous. (*E. W. Pfeiffer*)

fade away gradually; the problems caused by the massive and forced urbanization can only grow.

THE WAR IN PERSPECTIVE

Criticism of the use of tear gas and herbicides in Vietnam was loud and shrill, matched only by the general furor touched off so long ago at Ypres over the use of chemical weapons there. Military apologists would have us believe that these are humane weapons. Citing World War I statistics, the apologists state that of the 27 percent American casualties due to gas, only 2 percent died, compared with 26 percent fatalities in non-gas casualties; and that only 4 percent of the gas casualties were permanently disabled, in contrast to 25 percent of those wounded by high explosives.

The outraged respond that high explosives are bad enough but that the use of tear gases or herbicides is a barbaric and immoral prostitution of human technology. Thus the decision to use such weapons ultimately is a matter of conscience. But a few points should be considered while searching for that resolution.

The crossbow so disturbed medieval moralists that a Lateran Council called by the pope formally outlawed this heinous weapon. Not so very much later the flower of French knighthood, the Chevalier Bayard, accepted crossbows but was appalled at the use of gunpowder. While he treated captured knights and archers with courtesy, bearers of muskets were summarily executed. Professional soldiers have often resisted new weapons as a corruption of their honor.

Some of the reaction against gases during and after World War I reflected the natural uncertainty about a new form of death. But much of it was the result of carefully cultivated propaganda generated by the Allies against the Central Powers to serve the war effort by encouraging anti-German sentiment. Thus anything new used by the Germans, particularly if it were successful, would be branded as inhuman. "By the time the Armistice was declared, gas propaganda had run the policy gamut—the illegal and inhuman act of a murderous aggressor in 1915; just and humane retaliation in 1916; blackout in 1917; and a triumph of Allied industry in 1918."[5] However, much to the chagrin of a chemical industry hoping to develop a market for gases and of a few military planners who had glimpsed a brave new world, the atrocity charges stuck in the public mind. Poison gas was outlawed internationally (although not by the United States) and efforts to expend public monies on further research were blocked for a time.

Despite the passage of fifty years, poison gas as an antipersonnel weapon remains stigmatized, while fragmentation bombs have been

[5] Brown, F. J., 1968. *Chemical warfare, a study in restraints.* Princeton University Press, Princeton.

accepted by many as the most useful weapons employed in the Vietnam conflict. Thus, there seems to be a strong thread of hypocrisy running through both sides of the controversy. If we accept the premise that killing people for whatever reason is immoral, then means are irrelevant. If we accept the inevitability of war it would seem, at least in the abstract, infinitely preferable to die in a few seconds from nerve gas than to bleed slowly to death. Why is it humane to drive people from tunnels with tear gas and into a saturation pattern of fragmentation bombs that shred them to pieces, while it is inhumane to kill them in their tunnels with a nerve gas?

The issue of humane killing now goes beyond the death of soldiers, however. What of the environment in which people must try to live when the ideological profundities have been settled? No matter how many people are killed by modern warfare, their number is insignificant when compared with the number of people yet to come who must live with the destruction and irreparable alteration of the land itself, land that must sustain life after the last corpse has been buried. The extent to which modern warfare brings about this environmental destruction is a deeper measure of its immorality than the death of combatants or even the increasing death rate for noncombatants.

How ironic in this age of rising ecological awareness and expectation that America will be remembered as the nation that preserved the whooping crane and turned Indochina into an environmental wasteland.

FURTHER READING

Boffey, P. M., 1971. "Herbicides in Vietnam: AAAS study finds widespread devastation." *Science* **171,** pp. 43–47. A brief but systematic review of the little we know about the effects of herbicides in Vietnam.

Branfman, F., 1972. "What kinds of weapons do we drop on Hanoi?" *The Washington Post,* 24 December. A short piece on antipersonnel weapons.

FitzGerald, F., 1972. *Fire in the lake: the Vietnamese and the Americans in Vietnam.* Little, Brown, Boston. A brilliant social and political analysis.

Kanegis, A., et al., 1970. *Weapons for counterinsurgency: chemical/biological, anti-personnel, incendiary.* American Friends Service Committee, Philadelphia. One of the few detailed descriptions available.

Littauer, R. and N. Uphoff (eds.), 1972. *Air war in Indochina.* Rev. ed. Beacon, Boston. An exhaustive analysis, the only one available.

Neilands, J. B., et al., 1972. *Harvest of death: chemical warfare in Vietnam and Cambodia.* Free Press, New York. Firsthand accounts of the impact of CS gas and herbicides.

Orians, G. H. and E. W. Pfeiffer, 1970. "Ecological effects of the war in Vietnam." *Science* **168,** pp. 544–554. A detailed report of the authors' trip to Vietnam in 1969.

Shapley, D., 1973. "Herbicides: AAAS study finds dioxin in Vietnamese fish." *Science* **180,** pp. 285–286. One of the first examples of hard data on the effects of herbicides to come out of the Vietnam war.

Weisberg, B. (ed.), 1970. *Ecocide in Indochina.* Canfield Press, San Francisco. A fascinating, often moving series of articles and papers dealing with many of the effects of war in Indochina.

Westing, A. H., 1971–1972. "Herbicides in war: current status and future doubt." *Biological Conservation* **4,** pp. 322–327. An eyewitness description.

Westing, A. H. and E. W. Pfeiffer, 1972. "Cratering of Indochina." *Scientific American* **226**(5), pp. 20–29. An analysis of the environmental impact of bomb and shell craters.

Whiteside, T., 1971. *The withering rain.* Dutton, New York. While the book is not limited to Vietnam, Whiteside marshals his facts carefully.

5

Man's
Urban
Environment

Anywhere, USA—and increasingly, Planet Earth. As populations grow, sprawl obliterates any meaningful relationship between man and his environment, and memory lanes become freeways. (*J. Paul Kirouac*)

18

THE HOUSE AS
AN ENVIRONMENT

And if after the hunt goes past
And the double-barrelled blast
(Like war and pestilence
and the loss of common sense),

If I can with confidence say
That still for another day,
Or even another year,
I will be there for you, my dear,

It will be because, though small
As measured again the All,
I have been so instinctively thorough
About my crevice and burrow.

ROBERT FROST, 1946

WOODCHUCKS have had thousands, perhaps hundreds of thousands of years to adapt to their environment by constructing burrows that provide protection from enemies and shelter from a long cold winter. Although man's original shelter was a ready-made tree or cave, migration into areas without convenient natural shelter together with culturation that provided new tools and materials, both required and allowed a variety of artificial structures to be constructed. For the thousands of years that man has occupied earth, these structures slowly evolved by trial and error into some starkly functional and often elegantly simple buildings. Once disdained by architects, this native architecture is being studied today as a superb marriage of form and function. A few examples from greatly contrasting environments will give some idea of the relationship between form, function, and environment.

DIVERSITY OF ENVIRONMENT, DIVERSITY OF HOUSE

One of the best adaptations to an especially hostile environment was the development of the igloo by the Eskimo. Dome-shaped and constructed of blocks of hard-packed snow, the igloo offered minimal resistance to the sweep of the wind during the winter when tents used in summer were unsuitable. Internally, the dome shape allows the largest volume possible with the least supportive structure and made the most effective use of point sources of heat, a simple oil lamp and the people themselves. The igloo was made by layering snow blocks in an arching spiral (Figure 18.1). By carefully packing the chinks with snow and by burning an oil lamp inside, the interior of the igloo was glazed into a shiny, smooth surface that not only sealed the igloo but reflected light and heat, much as aluminum foil might. When the interior was covered with furs and skins, the igloo was quite comfortable, at least compared with the conditions outside: at midnight, air temperature inside an igloo was 2°C (35°F) while outside air temperature was minus 34°C (34°F). During the day inside air temperature rose as high as 4°C (40°F) compared with minus 23°C (10°F) outside.

Fig. 18.1 • The igloo (Eskimo for "house"), once used as an overnight shelter for hunting parties, was a fine example of functional native architecture. Today, snowmobiles make it possible for hunters to return to their villages of more substantial structures each night. (*Courtesy of the American Museum of Natural History*)

The other environmental extreme, the desert, has also had a profound effect on the shelters of native peoples. The desert, unlike the arctic, poses problems of daily rather than seasonal extremes. The temperature ranges from 43°C to 15°C (110°F to 60°F) over a twenty-four-hour period, a range that puts an enormous strain on any cooling system. These extremes are modified, however, by the use of local materials, stone and clay (adobe), for houses with thick walls and roofs that insulate the interior of the building from the

Fig. 18.2 • These buildings at Taos Pueblo in New Mexico are built of soft clay bricks or adobe. Their thick walls and roofs effectively insulate the inhabitants from the sharp temperature extremes of the New Mexican desert. (*Courtesy of the American Museum of Natural History*)

blistering heat outside (Figure 18.2). Temperature in a typical adobe dwelling at 2 p.m. was 29°C (85°F) inside and 40°C (105°F) outside, while at night the temperature was 27°C (80°F) inside and 18°C (65°F) outside. During the day the thick walls and roof acted as insulators, heating slowly as they absorbed the sun's rays. At night the process was reversed; the desert air cooled quickly, but the warm walls and roof radiated heat throughout the night, protecting the occupants from the outside chill.

In the tropics, the problem is not temperature variation, either daily or seasonal, but the need for adequate shade, proper ventilation, and protection from heavy downpours. The native peoples achieve this by reducing or eliminating side walls and erecting a steeply pitched roof, thickly thatched. Thin, coarsely woven side walls take advantage of every breeze, and the thick thatch both insulates the shelter beneath from the heat of the sun almost overhead, and allows the daily rain shower to be shed quickly and almost silently (Figure 18.3). In contrast, the wooden, tin-roofed shacks that often replace the thatched structures are cramped, stuffy, and noisy.

Until fairly recently, areas with similar climates exhibited similar architecture no matter where the geographical location. A townhouse in Thailand (Figure 18.4) strongly resembled a house in New Orleans; a stone-adobe house in hot, dry Afghanistan (Figure 18.5) looked very much like a New Mexico pueblo; and a fishing village in Singapore Harbor (Figure 18.6) might be located along the Amazon River in Brazil.

The widespread migration of people into a New World with climates different from those they had known required adaptation to the local environment even when this meant a change in traditional

Fig. 18.3 • The thatch roof of a house in Thailand allows rain to roll off quickly and quietly while the loosely woven sidewalls allow circulation of air. (*Photo by Alice Mairs*)

Fig. 18.4 • This rather substantial townhouse in Thailand has a tile roof with a broad overhang, wide porch, and louvered shutters to shield from the sun and allow maximum circulation of air. (*Photo by Alice Mairs*)

Fig. 18.5 • Hot dry climates everywhere elicit the use of thick-walled structures that insulate well, as in these houses in Istalik, Afghanistan. (*Photo by Alice Mairs*)

Fig. 18.6 • An ideal marriage of form and function in a kelong, a structure from which men fish with nets, built on stilts in Singapore Harbor. (*Photo by Alice Mairs*)

building methods. For example, people from Plymouth, England, where winters are mild and summers cool, were obliged to adapt the type of dwelling with which they were familiar to the much colder winters and hotter summers of Plymouth, Massachusetts. In consequence, the rooms and windows became smaller, the ceilings lower, and the fireplaces massive (Figure 18.7). In contrast, the summers in Virginia and the deep South states necessitated large rooms with high ceilings and wide porches or verandas often sheltered by a portico or overhang—all of which allowed maximum insulation, shade, and circulation of air.

Fig. 18.7 • The New England winters are long and cold. Consequently this house, the famous House of Seven Gables in Salem, Massachusetts, has small windows and huge fireplaces in every room. (*Essex Institute*)

As people continued to move, following the frontier to Ohio, the plains states of Illinois and Iowa, then Oregon or California, their houses reflected the changing climate. Pioneers from New England recreated New England villages in northern Ohio, which has a climate similar to New England. Today some small Ohio towns strongly resemble parts of New England, even in their names. Farther west this flavor is lost as the farm and ranch houses built from the 1840s to the 1860s were adapted to the influence of local climate.

The basic reason for the wedding of form and function in native architecture lies in the sensitivity to and acceptance of the environment by most native cultures. But times are changing in most countries. The new and old coexist for a time, then the old way disappears. Unfortunately the new usually means a break with the past—the Eskimo gives up his igloo or tent for a wooden-frame house, the African his thatched hut for a tin-roofed shack (Figure 18.8). But much that is good may be discarded with the apparently outmoded

Fig. 18.8 • The traditional huts in this Bangladesh village are being replaced by the ubiquitous tin-roofed shack, a debatable improvement. (*CARE Photo*)

materials, techniques, and life styles. With so much of the world in a state of cultural transition between traditional tribal culture and a modern industrialized society, technology and ecologically sensitive architecture could help enormously to bridge the gap between the past and the future.

The Eskimo and his igloo demonstrate this point quite well. No one would want to confine an Eskimo to an igloo and deny him more advanced housing. But the igloo is basically a dome, an architectural form that has been ingeniously developed by American engineer-architect Buckminster Fuller. Using modern methods and materials the igloo could be transformed into a functional and comfortable house in sympathy with the local environment and preserving at the same time a traditional form that could help Eskimos bridge the gap between their cultural past and future. The same approach could be used in desert or tropical regions. Of course, architects have not been insensitive to these possibilities. Much has been learned from traditional dwellings and applied to regional design.

In the United States the gap between function and form widened more gradually. With the rise of mass communication—first national magazines and mail order catalogs, then movies, radio, and television—cultural homogenization began to erode local and regional differences in house types. The growing affluence of farmers and middle-class tradesmen in the nineteenth-century United States allowed a shift from a functional type of dwelling to a dwelling preoccupied with form. The simple farmhouse assumed Greek revival characteristics, and later, at the height of the Victorian era, townhouses became swathed in excessive wooden ornamentation (Figure 18.9).

Today you can drive around any town in the United States and

Fig. 18.9 • A redwood baron built this house in Eureka, California as a testament to the versatility of wood. Its Victorian exuberance is unmatched. (*Greater Eureka Chamber of Commerce*)

see bungalows built in the 1920s side-by-side with Tudor, Spanish, and French provincial styles from the 1930s, Cape Cod houses from the 1940s, and apartment houses incorporating a variety of styles.

The gap between form and function thus opened has been bridged to some extent by technology, which made possible the thousands of square miles of tract housing in the world, struck from a common mold and planted in the ground like some monster crop. There is, certainly, some regional differentiation—shake roofs in California, glass walls in Florida, small windows and full basements in Wisconsin. But one could easily switch houses between Fullerton, California and Oshkosh, Wisconsin and make them work in their new environment. What, then, is the price of making a house work in an environment to which it is ill-adapted—a thousand dollars a year in heating or cooling expenses? Before we can examine the cost, literal or figurative, of housing that is out of equilibrium with its environment, we must first look at what constitutes a suitable external environment for man's dwellings.

The Outer Environment

There have been two tendencies at work in the design of dwellings as packaged environments. One is toward the elimination of the connection between the inner and outer environments by building a windowless, possibly underground house, thereby solving most of the environmental problems that bedevil architects: maintenance of external surfaces; control of flies or mosquitoes in summer and drafts in winter; expense of heating and cooling; glare and shadows, with illumination of the exact intensity and brightness desired. With outside dust eliminated and inside dust filtered out by the obligatory air conditioning system, such a house would be extremely easy to keep clean. Occupancy would, of course, depend on a never-failing power source to heat, cool, ventilate, and illuminate; but auxiliary generators are available. Such an underground house would allow total landscaping of the lot above, particularly desirable where lots are small and expensive. When you pay several thousand dollars for a lot it seems somehow wasteful to cover most of it with your house.

The other approach runs in the opposite direction. The contact between inside and outside environments is broadened to the point where it is difficult to separate the two. This approach is achieved most successfully in mild climates where houses may literally enclose a piece of the outside environment in the form of a garden, such as was done in the atrium of a Roman villa. Large areas of glass wall can be used to remove the feeling of separation between inner and outer environments, or multiple stories and cantilevers can thrust the inner environment into the larger outer environment of treetops or sky.

This approach to environmental integration has been used in cold climates as well as warm climates, but greater dexterity of design is required in the former, since the inner and outer environments are so disparate that their integration must remain illusory. Only in a mild climate, such as that in Florida, along the Mediterranean, in Southern California, or the warm desert areas, can the two be physically integrated.

The setting

Since most dwellings are very much a part of the surface environment, the orientation on a lot is of critical importance to the distribution of natural light within the dwelling, the ease or difficulty of heat control, and maintenance of a dry cellar or garage. Far too often placement of streets on a tract by the developer reflects the most mathematically efficient way of subdividing property into the maximum number of lots. Houses are built on these lots either parallel to or at some standard angle to the street (Figure 18.10), giving the house buyer little choice about the exposure of his house to sunlight or to wind.

The house exists both in a seasonal environment in which the sun appears from different angles and for varying lengths of time throughout the year and in a daily environment of constantly changing patterns of light and shadow. On September 21 and March 21 in the Northern Hemisphere the sun rises at 6 A.M. and sets at 6 P.M.; on June 21 the day runs from 4:30 A.M. to 7:30 P.M. and on December 21 7:30 A.M. to 4:30 P.M. Fortunately one needs neither a mathematician nor a computer to predict the results of any particular dwelling orientation if one knows the aspect of the site and the shape of the building. There is a machine called a heliogram available to most architects that can reproduce the various seasonal changes in the lighting as the earth circles the sun. Using such a device a small

Fig. 18.10 • The houses in most subdivisions are aligned with geometric precision to milk the maximum number of units per acre from the land and to facilitate mass production construction techniques. Environmental and aesthetic values are not apparent until many years of shrub and tree growth mitigate the visual monotony. (*Freelance Photography Guild*)

model of a house can be orientated, taking into consideration light, temperature, and exposure factors.

Although an eastward-facing wall receives the same radiation as one facing west, the westward-facing wall is irradiated during the hottest part of the day, making it the warmest wall of the dwelling. When this is anticipated, some care can be taken to avoid placing a kitchen behind that wall or to devise shielding to check heat absorption. Because of the low angle of the sun in winter, the south wall is heated more in winter than in summer; on the other hand, the north wall receives most of its sunshine and the roof twice the radiation of any of the walls in the summertime.

But this constantly changing heat load on a house can be predicted, and steps can be taken to cope with it by the basic orientation of the house, design features, and landscaping. Since zoning often confines both the architect and builder by allowing only a certain orientation a defined number of feet back from the street, the architect often finds it impossible to arrange the rooms to accommodate both summer shade and winter sun. Well-designed plantings may help the situation. In brick or stone houses, the warm west wall can be shielded from the late afternoon sun by the strategic placement of trees or a deciduous vine such as Boston ivy. When the ivy sheds its leaves in the fall, the wall is favorably warmed by the winter sun. In the summer the ivy leaves form an insulating layer of air shielding the masonry from the heat of the sun.

The force of cold winter winds can be broken by planting hedges or belts of evergreens. Windbreaks have been built throughout much of the Midwest in the Dakotas, Iowa, Nebraska, and Kansas with excellent results; thirty-foot trees significantly reduce wind velocity 400 feet to the leeward. Dwelling design plus strategic placement of trees, shrubs, and vines can warm a house as much as 10°C (18°F) in winter and cool it by 9°C (16°F) in summer. This is equivalent to moving from New Jersey to upper New York State for the summer and down to Virginia for the winter. Although these expedients incorporate neither new materials nor techniques and indeed are only extensions of common sense, often they are not utilized. This is perhaps a result of ignorance but is more likely due to the wide availability of air conditioners and other products of technology.

Naturally, with modern heating systems and air conditioning, a standard tract house could probably be built atop Pike's Peak or in Death Valley and be made livable if not comfortable, but only at great expense. The whole point of adaptation of a dwelling to its setting by design, orientation, and planting is to gain maximum benefit from the natural heating and cooling afforded by the environment (Figure 18.11). But so far we have been complacently running roughshod over the natural landscape, using our technology to make every lot and every dwelling suit the builder's convenience.

Fig. 18.11 • Twenty years ago these subdivisions were as barren as the one in Fig. 18.10. But the growth of trees and shrubs is beginning to soften the harsh outlines of the streets and afford some measure of climate amelioration to the occupants of these houses. (*Thomas Airviews*)

The Internal Environment

Whatever its relationship to the outside environment, a dwelling functions as a shelter from the elements. Man is most comfortable at a temperature of about 22°C (72°F). This is explained by the rate of heat loss by the body. If it is too rapid we are cold, if too slow, we are too hot. At 22°C, apparently, the rate of heat loss is just about right. Since 22°C is a reasonably typical temperature in the tropics, we assume that man originated here and that he developed a need for substantial shelters when migration from tropical climates made necessary artificial means of recapturing this optimum zone of comfort. The diversity of houses that we have examined illustrates attempts to recreate this comfort level by whatever means were available.

Considering a shelter or house as an enclosed environment, what factors are important to our health and comfort? Certainly temperature, but also light, humidity, ventilation, and perhaps that intangible aesthetic that transforms a "house" into a "home."

Man can tolerate, with protection, the extreme natural range of minus 76° to 63°C (-105° to 145°F). Since the outside air temperature is only occasionally within man's ideal comfort range, 21° to 24°C (70° to 75°F), one prime environmental function of a house is to add or remove heat.

Temperature

Heat is intentionally supplied to houses by a great variety of sources: combustion of wood, charcoal, coal, oil, natural gas, or by electricity, directly or indirectly by heat pumps or heat exchangers. Since all of these sources are derived ultimately from the combustion process, air pollution is a direct consequence of heating houses. Of all the heat sources, electricity provides a much greater opportunity to control pollution. The emissions from a few power plant stacks are easier to control than pollution from millions of combustions from individual heat sources (Chapter 8). While it would be nice to have individual power plants, nuclear or solar, that would provide all our power needs, heat as well as electricity, such units are well in the future; the problem of pollution is not.

Artificial heat is not, of course, the only supply of heat in a house. As we saw in the previous section, conduction and radiation from roofs and walls heated by the sun are important sources of heat, which have been taken advantage of in regions with a fair amount of winter sunshine without severe cold. By manipulating the size, shape, and orientation of the windows, a well-designed house can be comfortably heated much of the winter using solar heat (see Chapter 10), which need be reinforced by conventional sources of heat only during long periods of overcast weather or extreme cold.

Other sources of heat, of minor significance in private dwellings but of some magnitude in a crowded store, apartment, or office building, are body heat and lighting appliance heat. A 100-watt bulb gives off about 500 Btus an hour and a human body about half again that. When thousands of people and lighting fixtures of much greater power than a single bulb are involved, the resulting heat can be of major significance. Several buildings have already been constructed to make use of this source of heat: a large post office in Houston, a sixty-story tower in Chicago, and other buildings in Pittsburgh and Portland, Oregon. The system is basically simple; air leaving the building and warmed by the activities inside is passed over a system of water-filled coils. Fresh air entering the building is warmed as it passes over the hot water coils. Some of the buildings have water tanks to store heat when the building is unoccupied. This heat recovery system works until the outside temperature falls to minus 12°C (10°F), then supplementary heating becomes necessary. This is a particularly obvious example of the economic and ecological good sense implicit in the use of what would otherwise be considered waste.

Heat removal is usually achieved by air conditioning, either central or room units. If both the roof and walls of a dwelling are well insulated, reasonable control over summer temperature can be achieved simply by opening doors and windows at night and closing them early in the morning. Since it is more important and difficult to prevent entrance of excess heat than to prevent its loss, the location and overall design of a house become important in controlling internal temperature.

Another obvious function of a dwelling is as a shelter from rain, but almost any type of structure will afford this protection. More

Humidity important is the control of moisture in the internal environment of a house. There are two ways to determine the moisture content of air: by measuring the vapor pressure or the relative humidity. Vapor pressure is a measure of the actual amount of moisture in the air at any given temperature; relative humidity is an expression of how much moisture the air is holding compared with how much it *could* hold at that temperature. A relative humidity of 80 percent at 27°C (80°F) means that the air is 80 percent saturated with moisture. This is not equivalent to a relative humidity of 80 percent at 10°C (50°F) because as the temperature of air falls its capacity to hold moisture decreases. For this reason 80 percent relative humidity at 32°C (90°F) may be unbearable, while 80 percent humidity at 15°C (60°F) may be quite comfortable, because warm air can hold so much more moisture than cold air.

In the winter we may have the apparent paradox of a house at 15 percent relative humidity and 24°C (75°F) losing moisture to an outside environment of 50 percent relative humidity at 13°C (20°F). Even at 15 percent relative humidity a house at 75°F contains more moisture than half-saturated air at 13°C (20°F), so moisture diffuses from a point of higher concentration inside to a point of lower concentration outside. Also, as the internal temperature of a house increases during the winter heating season, the relative humidity is bound to drop since the warmer air is capable of holding far more moisture than is generated by household activities alone.

The chief problem in dwellings over much of the United States in summer is not the heat but the humidity, to coin a cliché. Hence, any air conditioning system must dehumidify as well as cool the air passing through it.

In the winter, the problem is just the opposite. The small amount of vapor released from our breath, washing, and bathing tends to diffuse out through the walls and roof of the house, thereby reducing the relative humidity of a heated dwelling to as low as 10 percent—comparable with many desert regions of the world. This combination of high temperature and low humidity has a destructive effect upon furniture, causing wood to shrink and the glue to dry out. The effects on people are analogous; the nasal and throat membranes dry out, especially at night, creating conditions favorable to many respiratory viruses and bacteria. It is no coincidence that winter is the season we contract the most colds. Vaporizers can often alleviate the dryness of a small room, but it is more effective and convenient to do this centrally with a humidifier on a furnace. This can be easily accommodated in a hot-air type of heating system but is more difficult to install with hot water, steam, or electric heat.

Previously, illumination by candle or lamplight made the location, number, and size of windows or skylights to let in natural light of critical importance to household activities. Electric lighting assures plentiful artificial light, making the location of windows in a building much less important today than it once was. Windowless buildings can and are being built and, if anything, seem to be more effectively illuminated than windowed structures with their glare and uneven lighting. There is more involved, however, than good or bad illumination, for windows not only let in light but allow people to see out. There is a psychological satisfaction in being able to relate yourself to time and place by glancing out of a window and, if the view is at all pleasant, there is a mental respite from the pressure or boredom of routine.

Light

A controversy rages as to whether schools should have windows or not. There seems to be no good evidence that a lack of windows elicits student claustrophobia and a windowless school does afford more space. While one suspects that arguments for windows are based more on tradition and sentiment than on demonstrable undesirable effects, students already suffer from too great an isolation from the environment. Perhaps the constant awareness of the real environment viewed through a window might in a small, even insidious way, keep both student and teacher attuned to the relevance of life in a natural as well as a synthetic environment. Beyond the mere presence or absence of natural light, sunlight plays an important psychological role in determining mood. Sunlight pouring into a room on a cold winter day is relaxing and cheering and raises the spirits of those confined indoors.

Glass transmits 82 to 90 percent of the visible portion of sunlight but almost no ultraviolet rays, so it is impossible to get a winter tan by lying in sunlight under glass. However, special glass has been developed which will allow up to 60 percent of the ultraviolet to be transmitted. If it were simply a case of getting a tan, a sunlamp would do as well, but ultraviolet light, besides its tanning effect and importance to vitamin D synthesis, is an important germicide; most germs are killed by exposure for little over an hour to the ultraviolet in sunlight. Exposure of the interior of a house to the natural disinfecting power of unfiltered sunlight might be sufficiently important for normal sanitation and cleanliness to be considered in the design, orientation, and glass specifications of a modern house.

Windows also generate the greenhouse effect described in Chapter 8. The shortwave radiation of sunlight easily passes through the window glass, strikes an object or surface in the room and is converted into longwave radiation or heat that cannot pass back through the glass and so accumulates in the room. This effect is quite noticeable in a room with many windows admitting sunlight; for this reason houses with large areas of windows or glass walls must be carefully designed and situated so that this effect can be exploited in winter and carefully controlled in summer.

Ventilation

Beyond the perception of strong kitchen or bathroom odors or perhaps a vague notion of the air being "stuffy," most people are unaware of the need for ventilation in their homes. Starting with the usual mix of nitrogen, oxygen, carbon dioxide, water vapor, and dust, people give off twenty-three liters per person of carbon dioxide per day, plus variable amounts of water vapor, hydrogen sulfide, ammonia, propionic acid, and various organic compounds. Large quantities of grease aerosols are spread about by frying foods, carbon particles are added from burning foods and tobacco, and dust is constantly generated from bedding, rugs, and clothing. Occasionally malfunctioning furnaces or space heaters may release carbon monoxide in such quantities as to cause death by asphyxiation. More common is the combustion of cigarettes in poorly ventilated rooms where carbon monoxide buildup may lead to dizziness or impaired judgment.

Both burned grease and tobacco smoke may contain carcinogens. While there are no statistics on how many people have died of lung cancer by inhaling burned fat aerosols or someone else's tobacco smoke, there is no reason to believe that these materials are harmless simply because our relationship to them is passive. In general these man-made materials decrease our sensitivity to the stimuli that are integral to our functioning as alert and creative individuals.

Ventilation is, of course, particularly important in the winter when doors and windows of most northern homes are covered with storm windows to conserve heat. But it is necessary all year round where tall apartment houses abut each other. Most cities require air shafts in such buildings to assure interior apartments of at least some ventilation, but these shafts are rarely effective. Air conditioners often serve the multiple function in a city apartment of cooling and circulating the air, as well as removing some of the pollutants. The filter of an air conditioner removes only particulate matter, however, so the urbanite is still exposed to sulfur dioxide and smog, air conditioner or no.

A recent report of the Environmental Protection Agency comparing hydrocarbon levels inside buildings to levels in the street outside found that carbon monoxide levels exceeded federal health standards 23 to 47 percent of the time during the heating season. This is of particular concern to occupants of buildings situated in the air

Fig. 18.12 • Air rights construction is becoming increasingly popular in densely populated cities where land values are high. But it is difficult to control the entrance of air pollutants and noise from the traffic flowing below. These four apartments and transportation terminal rise over the approach to the George Washington Bridge in New York. (*The Port of New York Authority*)

space over highways, a practice that is becoming increasingly popular where construction sites are at a premium (Figure 18.12).

It is ironic that technology, with its potential for offering us a more comfortable personal environment, so often seems to be misused. Instead of providing innovative solutions to basic human needs, technology is often used in lieu of good design, and the basic environmental factors that define a house, within and without, are ignored. If a house is drafty, we push up the thermostat, if it is stuffy, we turn on the air conditioner, and if it is smelly, we squirt a perfumed aerosol about. Each of these problems could be solved through careful study of man's basic life needs, and by applying both the best in design and the most relevant technology to the satisfaction of those needs. But this rational approach will be indefinitely deferred if we continue to be satisfied with the mediocre in housing, which, by its very acceptance, denies the possibility of anything better.

There is much that the individual can do. With existing housing he can select a heating scheme which contributes the least amount of pollutants to the outside environment; he can insulate to control heat entrance and loss; he can humidify or dehumidify the air; and he can landscape to ameliorate temperature and wind velocity extremes. In

393

planning or selecting new housing he can be aware of the importance of choosing a site for temperature control of a house interior and for potential flooding or drainage problems. He can examine floor plans carefully to control summer and winter natural light values. Where large tracts are still in the planning and development stage, individuals can determine if local zoning regulations function merely to keep people or structures out, or if they inhibit the freedom necessary to properly orient a house on its lot. At that point they can determine what provisions are to be made for power supply (wires above or below ground), sewage disposal, water supply, potential air pollution, schools to accommodate more children. Special hearings are often required by planning commissions before a subdivision can be developed, and these hearings allow public opinion to be voiced on the desirability of the project and its impact on the outer environment of surrounding property owners.

In public housing in large cities the quality of the home environment can be improved and maintained by adoption of strictly enforced building codes requiring minimum standards in all dwelling units. Pressure can be brought to bear on public officials to consider

Fig. 18.13 • While this urban slum or favela in Brazil seems built into the air, the houses are actually clinging to the face of a very steep hill. (*Planned Parenthood—World Population*)

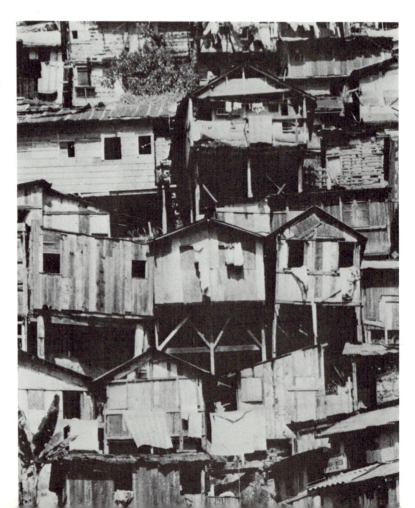

the outer and inner environments of public housing much more carefully than has been done in the past.

The key, then, to providing a comfortable and healthy home environment lies in recognizing our environmental needs, making sure they are included in our housing plans, then taking appropriate political action to assure that these plans are realized. When this is done, and we have the requisite knowledge and technology, no human need suffer the indignity of enduring a personal living space that would be regarded as unhealthy for a pet tropical fish, an orchid, or a computer (Figure 18.13).

FURTHER READING

Aronin, J. W., 1953. *Climate and architecture*. Reinhold, New York. Views the relations of houses to the climate of their site.

Fitch, J. M. and D. P. Branch, 1960. "Primitive architecture and climate." *Scientific American* **203**(6), p. 134. Informative discussion of representative types of housing adapted to cold, hot, and wet climates.

Geiger, R., 1965. *The climate near the ground*. Harvard University Press, Cambridge, Massachusetts. A fascinating text that looks at micro-climates and man's relationship to them.

Kira, A., 1966. *The bathroom, criteria for design*. Center for Housing and Env. Studies, Cornell University, Ithaca, New York. Critique of the most neglected room in any house.

Olgyay, V. G., 1951. "The temperate house." *Arch. Forum* **94**(3), pp. 174–194. Many tables, charts and graphs relating the house to the pattern of the changing seasons.

Ordish, G., 1960. *The living house*. J. B. Lippincott, Philadelphia, Pennsylvania. A charming study of the forms of life that have inhabited an old English cottage, reflecting changing styles.

19

THE URBAN-SUBURBAN ENVIRONMENT

REGARDLESS OF its suitability as an environment, the house functions for most people as an extension of their personal space, enhancing a sense of personal satisfaction and well-being. Unfortunately when houses are packed into dense concentrations or when apartments are stacked in huge impersonal towers, some personal space is sacrificed for common space; hallways, lobbies, streets, and parks become the living room of the urbanite.

Because of this concentration, the city has changed the natural environment in which it was situated, providing a new environment with unique demands. All environmental problems seem exacerbated by cities—air and water pollution, noise. Even the suburbs share an increasing burden of environmental problems. Yet well over half of the population of the United States lives in cities or towns that occupy only 1 percent of the total land area. While many of these urbanites would prefer to live in the suburbs and more than a few suburbanites would like to move to the country, a surprising percentage of people living in both suburbs and cities live there by choice and are quite happy to limit their rural experiences to occasional vacations. For, despite monumental problems, cities are exciting places to live, offering a quickened tempo for both cultural and intellectual opportunity. But these are subjective values. Viewed as the one man-made environment, urban areas are sharply differentiated from the countryside in many ways.

THE LOCATION AND FORM OF CITIES

Gertrude Stein once described a rather nondescript American city by saying, "When you get there, there's no there, there." Most cities that do have a sense of identity are strongly related to topographical forms. London and Paris are located on major rivers, Istanbul on a strategic strait, Rio de Janeiro and San Francisco on superb harbors. More recent cities have been shaped by railroads and major highway intersections, Denver, for example. But transportation routes themselves are strongly influenced by the dictates of topography (see Chapter 20).

Although the location of cities was originally related to the surface features of the countryside, once located, a city is subtly affected by the less obvious subsurface features. Venice was built on the marshy Po River delta in Italy both for protection from attack by land and for easy access to the sea, reflecting the extensive mercantile interests of the city. For centuries Venice prospered, but in recent times, the lowering water table beneath the city has begun to cause the slow sinkage of its buildings. As the piles on which the buildings rest are exposed to air, they decay and their erosion is quickened by the wakes of the motor boats, which have largely replaced the gondolas and rowboats. The city is now endangered.

Mexico City, seemingly so far removed from the plight of a city like Venice, has a similar problem. Built on a dry lake bottom at an altitude of 8000 feet, Mexico City has experienced a gradual lowering of the water table causing many buildings to sink; most spectacularly, the Palace of Fine Arts has sunk many feet below the street level. Removal of oil from an oil field under Long Beach, California, caused parts of that city to subside as much as twenty-nine feet, until water was pumped into the field to stabilize the ground level. While subsurface features can alter the future of a city once it is already there, they can also determine the shape a city may take.

Approaching New York City by car or plane, one can see the rather uneven distribution of tall buildings on Manhattan Island. There is one group of skyscrapers toward the southern end, the Wall Street area, and a second group in the midtown area from 34th to 60th Streets. In between, skyscrapers are conspicuously absent. While this is due partly to historic and economic factors, it is largely caused by the accessibility of stable bedrock (Figure 19.1). The taller the building, the firmer must be its footing in bedrock. Manhattan is fortunate to have a stable platform of hard rock under much

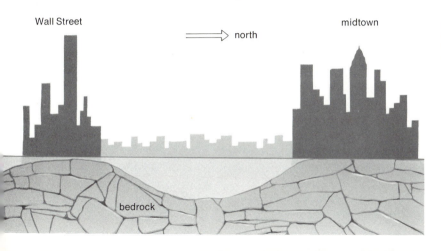

Wall Street

north

midtown

bedrock

Fig. 19.1 • The binodal development of Manhattan shows the relationship of tall buildings to bedrock.

of the city, but at different depths. To the north this formation comes to the surface in the midtown area, and is easily seen in Inwood, Morningside, and Central Parks. But as one goes south the bedrock begins to dip below ground. South of 30th Street it is deeply buried, until it lies hundreds of feet below the surface. Further south again, in the Wall Street area, it comes up within forty feet of the surface, and then, with the exception of the outcrop of Governor's Island in New York Harbor, it plunges far below the surface again. The expense of excavating beyond a depth of a hundred feet to reach bedrock has so far prevented the erection of very tall buildings in the section of Manhattan between 30th and Wall Streets.

Flying over the Los Angeles Basin you may have looked in vain for "downtown" Los Angeles. If you know exactly where to look you can make out a few multistory buildings in a sea of one- and two-floored structures (Figure 19.2). Los Angeles is located near the San Andreas fault on the earthquake-prone rim of the Pacific, and has so far avoided the risk of high-rise structures. Modern advances in engineering and construction techniques have partially overcome, if not the destructive potential of earthquakes, then the inhibitions of the builders. Therefore, this pattern is changing. In the Bunker Hill section of Los Angeles, a 100-story building flanked by sixty- and eighty-story companion towers is expected to be completed by the late 1970s.

Fig. 19.2 • Although recent construction of high-rise buildings is bringing Los Angeles into line with other large American cities, the sprawl is still quite conspicuous from the air. The freeways suggest a city caught in the coils of a monstrous serpent. (*Los Angeles Convention Bureau, Inc.*)

Fig. 19.3 • The verticality of Manhattan is apparent from directly overhead. The Hudson River is to the left, and the East River to the right. (*Thomas Airviews*)

There is another reason why Manhattan prickles with tall buildings (Figure 19.3), and Los Angeles still has so few. New York City has developed on a small island area. When Manhattan was developed in the early twentieth century, the value of land rose sharply and the available land was quickly bought up; the city could only expand upwards. Fortunately for the landowners this coincided with the development of the elevator. In contrast, in the Los Angeles Basin there was always open land just beyond any congested area, and this encouraged occupation of new land rather than more intensive development of the old.

THE MICROCLIMATE OF CITIES

As large cities grow in size, so does their impact on the environment. The overall climate of a city reflects the region where the city is located. But large cities can modify some of the climatological factors in their immediate vicinity, resulting in a relatively small-scale but important variation in climate, which is called a microclimate. When you think of the ways that a city differs from an equal area of forest or farmland, the reasons for the existence of microclimates become apparent.

Temperature

With increase in growth and three-dimensional complexity, the mean temperature in the city tends to rise, forming what has been

399

called an urban heat island. This results from the loss of evaporative cooling normally provided by vegetation and exposed soil, the gain of reradiated heat from pavement and building surfaces, and heat produced directly by factories and buildings. Walk down a city street on a hot summer's night and feel the hot breath of air conditioners from apartments or the heat shimmering up from the pavement. In wintertime, notice the heat rising from subway grills in the sidewalk, or the clouds of steam pouring out of manholes or storm sewer gratings. These heat sources have a cumulative effect. The rough profile of the city deflects the wind which might otherwise sweep some of this heat away.

The most striking temperature differences between the city and the country are seen in the daily high and low temperatures which tend to be attenuated in cities. This is emphasized when one listens to a city weather report. Expected lows are always lower in the suburban areas than the city center and the rural areas beyond are still lower. Conversely, although the city center gets quite warm on a summer day, the highs for the area may be in the suburbs or even in the country. The difference lies in daytime shading and nighttime reradiation.

Despite the radiative mass of the city, shading by tall buildings or narrow streets prevents sunlight from striking many potential radiation surfaces, thus lowering temperature maxima. On the other hand, the minimal shading in suburbs allows maximum heating of paved surfaces, roofs, and the walls of low buildings. Long after the country has become chilly on a summer's night and the suburbs comfortably cool, the city is still sweltering, perhaps only a few degrees lower at midnight than at sundown. By sunrise when the city is beginning to cool, the tall buildings are already being warmed. The mass of the city, then, acts as a buffer, damping the temperature extremes experienced in the suburbs or the countryside beyond.

While these microclimates do not produce dramatic changes in climate, over the years statistically valid changes have been noted. The city of Kyoto, Japan, experienced an increase in mean temperature of 1.8°C between 1880 and 1935, a period of rapid industrialization and growth. Now 1.8°C may seem insignificant, but remember, this is a mean. You should remember too that, as we saw in Chapter 7, an elevation of just 5°C in the mean temperature of the earth would be sufficient to melt all remaining ice caps, and to raise the ocean level by about 500 feet, flooding the world's coastal cities (Figure 7.1).

Precipitation

Particulate matter provides nuclei for the condensation of atmospheric moisture into precipitable drops or rainfall. The city's effects on rainfall are difficult to quantify, but there are some data which suggest that as cities grow in size the rainfall on the city proper increases. As the English city of Rochdale became industrialized and grew, its rainfall measured over ten-year periods increased: 1898–1907, 42.81 inches; 1908–1917, 45.83 inches; and 1918–1927, 48.65 inches. Interestingly enough, there was a statistically significant drop of 0.37 inches on Sundays over the time period that measurements were taken, when particulate matter in the atmosphere was lower than on the six working days of the week.

Particulate matter may affect surrounding towns as well as the urban area. Gary, Indiana, is covered much of the time by an enormous pall of smoke from its steel mills. As the cloud drifts eastward, it apparently triggers rain. LaPorte, Indiana, thirty miles southeast of the complex, recorded 31 percent more precipitation, 38 percent more thunderstorms, and 246 percent more hail days between 1951 and 1965 than comparable towns to the north and south.

Cities may also produce an orographic effect, that is, their domes of warmer air force the moisture-laden clouds upward into colder air which triggers rain further on. At times when the temperature of Buffalo, New York, is 10°C higher than its environs, precipitation plumes can be traced many miles inland—a dramatic demonstration of the climatological impact of cities.

Although rainfall seems to be greater over cities than their surroundings, at least over the long-term period, there is distinctly less snowfall in cities than the surrounding countryside. In temperate areas, especially along the coast, most heavy snow falls when the temperature is near the freezing mark. Often the city temperature is just high enough to change the snow into rain as it falls through the warm air enveloping the urban area. On such occasions the city may remain bare while the surrounding suburbs receive several inches of snow. When snow does fall on a city, the city-generated heat may cause the snow to melt as fast as it falls; even when it accumulates, it melts more rapidly than suburban snow.

Wind

While certain streets or areas of most cities can be quite windy because of local updrafts between buildings or the channel effect of towering buildings lining a narrow street, an equal number of places in a city are quite protected from wind. Because of the rough surfaces they present to the wind, cities tend to experience reduced *average* wind velocity in direct proportion to their size and density. Between 1909 and 1930 the mean wind velocity in Detroit, even when measured 259 feet above the street, decreased from 14.8 to 8.7 feet per second.

Solar Radiation

Regardless of its velocity, the air of cities, as we saw in Chapter 8, is especially rich in fly ash, dust, and particulate matter of all sorts. These materials have a climatological effect as well as an aesthetic and public health impact. Particulate matter can reduce the natural radiation received by a city by 15 percent, a reduction that is uneven across the radiation spectrum and tends to be greatest in the ultraviolet range. This portion of the solar spectrum is extremely important in the production of vitamin D in the skin (see Chapter 1).

URBAN SPACE

For anyone who regularly commutes between a city and its suburbs, the microclimate of the city is obvious enough. Even more obvious as one enters a city from its suburbs is the almost complete loss of contact with natural vegetation. With the exception of a few scattered parks and occasional tree-lined streets, the urban scene is too often one of stark, dirty buildings. Unfortunately, suburbs, rather than acting as green wedges pushing into cities to relieve their starkness, act instead as gray fingers extending from the city and extinguishing the interplay of man-made and natural features that soften and humanize man's environment. The problem is not purely one of a lack of open space, for parking lots and plazas can provide that, but of space softened by trees, greenery, water, and different textured surfaces that mitigate harshness and muffle city noise.

A lack of these amenities contributes to the poor mental health of many urbanites. A study made about twenty years ago by the Cornell Medical Center found that of a sample population of 1700 non-Puerto Rican, white adults between twenty and fifty-nine years of age, living in a midtown neighborhood of New York City, four out of five had symptoms of psychiatric disorders and one out of four was neurotic to the extent that his daily life was disrupted. This area was not considered a slum but a cross-section running from luxury apartments to working-class tenements. By contrast, another study in a rural Hutterite colony (a religious sect living in a tightly knit community) found only three out of 100 people with recognizable mental problems.

However, it must not be forgotten that while poverty, familial disorganization, and parental psychopathology are encouraged by urban and ghetto life, many of these social and psychological patterns were set up in the small towns and rural areas where numbers of present urbanites originated. The reason people have flocked to cities from the rural Deep South, the Ozarks, or Puerto Rico is not that cities, however falsely, offered so much, but that rural slums offered so very little. While urban slums are visible to millions, rural slums are usually hidden away in their remoteness and are known only to their inhabitants. For all the problems—jackhammers in the street; wailing police cars and fire trucks; subway trains that shriek, roar, and stall; grime and crime—the city provides a greater prospect for self-fulfillment and mobility than virtually any other environment.

Space assumes a far more important role in ameliorating the city environment than merely that of an aesthetic afterthought of the developer. To appreciate the importance of properly designed or sensibly retained urban space and to evaluate its role, we must first take a brief look at the relation between cities and their suburbs and the natural environments that once surrounded them.

Older cities realize more clearly than newer suburbs the value of open space. Many cities have tried to pace their growth with either internal park systems or peripheral green belts to insulate them from the surrounding suburbs. Cleveland and Washington, D.C. have attempted to preserve some open space by creating parks from the flood plains of many small streams that drain the region. These parks form necklace-like skeins running through the city (Figure 19.4), which help break up the monotony of heavily urbanized areas and provide open space for nearby residents.

Open space in older cities

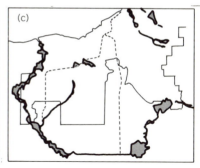

Fig. 19.4 • Many cities have wisely used their floodplains for parkland; three of the best examples are (a) Washington, (b) Baltimore, and (c) Cleveland.

Both London and Vienna have pioneered in the attempt to separate the city from its suburbs by a broad belt of undeveloped land, called a greenbelt, which encircles the city. Other cities have been able to set aside large blocks of land that today are irreplaceable: New York's Central and Prospect Parks, Philadelphia's Fairmount Park, and San Francisco's Golden Gate Park, for example. Many more opportunities exist to utilize open space which at present lies unused. Most waterfront cities on rivers, harbors, or lakes have turned away from their natural waterfront orientation and, by expressway development or redevelopment schemes, have isolated the public from the most obvious focus for their recreation. Landscape

403

architect Ian McHarg estimates that, of the twenty-two miles of Delaware River frontage available to Philadelphia, eight would more than suffice for port facilities; the remainder could be reclaimed for parkland, refocusing Philadelphia on the river. Much the same can be said of New York or Boston where miles of rotting piers testify to past use but present inactivity. In New York, after great pressure by citizen's groups, people were allowed to use a city-owned pier, the Morton Street Pier (Figure 19.5), giving them the only access to the Hudson River between 72nd Street and Battery Park, a distance of more than five miles, while dozens of other unused piers rotted into the Hudson. More recently the city government has attempted to rise to the challenge by recommending the use of five more strategically located old piers as unique recreational areas.

Fig. 19.5 • The Morton Street Pier provides the only public access to the Hudson River between Battery Park and Riverside Park at 72nd Street in New York; consequently it is well used.

One hopeful sign which indicates increasing awareness of the importance of recreational open space has been the development of vest-pocket parks and playgrounds in many large cities. Parks need not be large, with acres of grass and groves of trees. The simple juxtaposition of different textures of stone or small fountains or pools can be just as aesthetically rewarding (Figure 19.6). Ideally every block should have such a respite from the trauma of city life.

Measurements of sulfur dioxide on a transect across New York City's midtown district dramatically show Central Park's impact on diluting gaseous pollutants (Figure 19.7). An expanse of lawn or even a narrow strip is three to six times more effective in trapping dust than is a smooth flat surface of equal area. Trees are even more effective in helping to settle out particulate material. London's Hyde Park, though only one square mile in area, reduces the concentration of smoke passing through it by as much as 27 percent.

While ground surface areas are often unavailable for open space use, the rooftops of cities offer a vast reserve of "land" presently utilized only by nesting nighthawks. Although the cost of converting rooftops of present buildings into gardens would be prohibi-

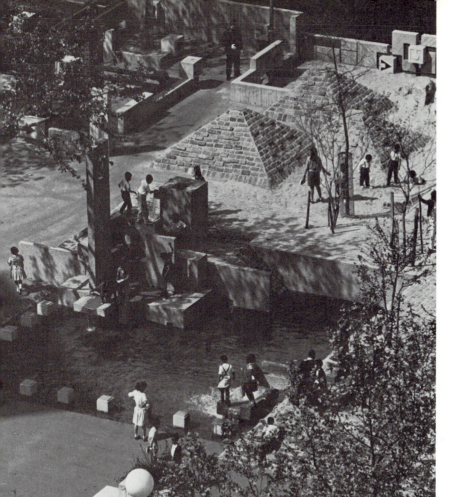

Fig. 19.6 • While there is little greenery in this playground, and no grass, the
imaginative use of textures,
forms, and water makes a far
more exciting playground for
children than the standard
grass and swings. The same
concept can be applied to
parklets scattered through a
city. (*Pomerance and Breines,
Architects*)

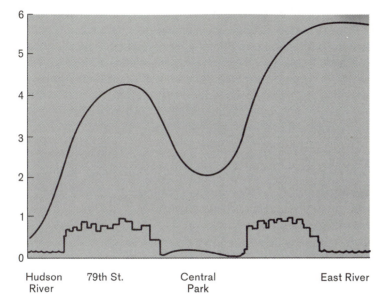

Fig. 19.7 • The dip in this
curve of SO_2 concentration in
the air over Manhattan graphically shows the effect that
open space can have in diluting the pollution being produced all around it. (*From
Shomon, J., 1971. Open land
for urban America. Johns
Hopkins Press, Baltimore, p.
30*)

tive in most instances, since the buildings were designed neither to support the additional load factor, nor to provide safe or convenient access, much could be done in planning future buildings to accommodate open space for public or semipublic use (Figure 19.8)—a garden for apartment residents or an open-air cafeteria or restaurant for office workers.

Fig. 19.8 • The rooftops of most buildings, which are usually sterile deserts of gravel or tar paper, can with foresight be utilized as open space as has been done with the Kaiser Center in Oakland, California. (*Kaiser Industries Corporation, photo by Bill Wasson*)

The conflict of natural and man-made environments is not restricted to larger cities and their suburbs. Growth in small towns, starting with the building up of a few main streets, is accompanied by much protest as lawns are paved and old trees removed. The smaller the town the more important the destruction of these symbols of past provinciality seems to be. Trees and unbroken concrete are incompatible. For a tree to remain healthy it must have space for its feeder roots to obtain moisture and nutrients and as lawns disappear and the pavement encroaches, the feeder root system is progressively curtailed until the limbs that exist in equilibrium with the feeder roots begin to die. Old trees are doomed under these conditions and young trees planted hopefully as replacements are severely limited in the size they can attain by the minimal space available.

Urban Renewal

While vest-pocket parks can be superb additions to the city, large-scale urban renewal projects often mar the city's fabric. The

American genius for euphemism was evident when the term "slum clearance" was replaced by "urban renewal." Slum clearance suggests removal of slums without implied solutions of what to put in their place or where to put the slum dwellers. Urban renewal, on the other hand, connotes a rekindling or recreation of the essence of urban living that has often been regarded as snuffed out in slum areas. In practice, urban renewal has often meant the inept replacement of an old slum with a new one or the displacement of low-income families into another slum so that the cleared land can make a profitable return, perhaps in the form of luxury apartments, high-rise office space, or cultural centers and sports facilities.

People in the worst slums interact and relate to one another in the only space they have, the streets. Such streets, thronged with people almost around the clock, are often far safer than elegant tree-lined streets in well-to-do suburbs.

When a slum is "renewed," buildings, streets, and whole blocks are swept away and frequently replaced by vertical towers separated by strips of green lawn. By day the renewed areas may be bright and cheerful but at night they may be deserted and dangerous. Sometimes even the building interiors are not safe; blind corridors, elevators, and stairwells provide ample room for antisocial activity.

Antisocial activity

Why the antisocial activity? In a street-oriented slum the activity and interest on the street seem to pervade the hallways and corridors of the tenements lining the street. If you were to wander in off the street you would find many doors open, children playing in the halls, men and women chatting from their doorways or leaning out of windows. The street and its extensions into the houses form a social outlet for the people, all the more important when more formal open spaces are unavailable.

In a typical renewal development, the commons of the street and halls is gone, lost in the impersonal bulk of buildings. Once the interaction and activity of the street have disappeared, the people turn inward. There is no reason to look out of the windows; there is no street activity, there is no fire escape to sit on, no front stoop, no sense of intimacy in the hallways. Wander into one of these buildings; doors are closed and halls are empty. Deprived of the opportunity to interact, people are withdrawn and resentful, often without knowing why. Antisocial acts become more frequent and less noticed. This phenomenon is by no means limited to low-income renewal housing. There have been repeated instances of violence in middle- and upper-class areas in New York, Chicago, and other large cities.

City planners sometimes seem unaware of the social importance of streets as environmental space, for although renewal projects provide open space, most of it is apparently unsuitable for social activity. More often a city planner or architect *does* understand the spatial needs of people (Figure 19.9) but is thwarted by archaic bureau-

Fig. 19.9 • This creative use of space in the form of an amphitheater and playground facilities does much to relieve the drabness of the public housing surrounding it (*Riis House Plaza, New York City*). (*Pomerance and Breines, Architects*)

crats and hard-nosed bankers who will not fund imaginative solutions, myopic labor leaders imprisoned by rules no longer needed to protect their men, and manufacturers with a vested interest in yesterday's technology. If mere space were the requisite, parking lots would be filled after business hours with happy people enjoying themselves.

Scale is also important, for the characteristic high-rise redevelopment project is constructed on an impersonal scale. Often much is made of its self-containment—entertainment areas, a supermarket, open space, and so on. But from the point of view of many tenants, the project is cut off from the city and remote from the quality of their old neighborhood and the rich life it supported.

This is not to say that slums are desirable. Far from it. Poorly heated in the winter, stifling in the summer, often overrun with cockroaches and rats, with trash in the hallways, the smell of urine on the stairs, peeling paint full of lead, cracking plaster, broken plumbing—a slum is a rotten place to try to live. But in their ardor to improve on this, city planners and model-happy architects often throw out the baby with the bathwater.

One logical approach to renewal of a slum neighborhood is to

Rehabilitation of existing housing

rehabilitate as much housing as possible. Many slum areas are made up of solidly constructed houses that, if renovated, could be made more comfortable than most new construction and at less cost. Some excellent examples of rehabilitation can be seen around the country, Wooster Square in New Haven, parts of the West Village and upper West Side in New York, Society Hill in Philadelphia, to name just a few. In most of these instances, however, the poor were simply shunted to another slum when the middle class discovered another "charming" neighborhood to "save." But the principle remains— *housing can be rehabilitated* (Figure 19.10).

Fig. 19.10 • Two townhouses built about the same time. The one on the left has been restored; the one on the right shows the wear and tear of conversion into tiny apartments. (*Photo by Dan Jacobs*)

A pilot experiment in New York pointed the way by cutting a hole in the roof of an otherwise sound but shabby tenement and lowering prefabricated bathrooms and kitchens into the replumbed and rewired building. The operation could be completed in a couple of days with minimum disruption of the occupants' lives by boarding them in a hotel until the operation was completed. Unfortunately, the experiment was abandoned when costs came to over $20,000 per family (enough to buy them a home in the suburbs). But with application of the vaunted American know-how, a mass production approach should bring costs down to the level where whole blocks could be upgraded, preserving the open-space value of the streets without unduly disrupting the social life of the community and maybe even improving on things by knocking out a house here and there for intensive-use parklets.

If, however, the quality of housing is low and beyond rehabilitation the most logical approach for a city planner to take is to find a city neighborhood that "works" like Boston's North End, the old French Quarter of New Orleans, parts of New York's Greenwich Village or Chelsea and find out *why* it functions smoothly, has a low crime rate, safe streets, and relatively happy people who interact peacefully with each other. What are the environmental factors at work there and how do they affect the people? Then the planner-architects should do their utmost to build in those features that seem to make people happy. Maybe the world's cities have had enough of idealistic models that work for no one and in some ways are more demoralizing than the slums they replace.

THE RISE OF SUBURBIA

Historically, cities have been compact and densely populated; yet because of their relatively small population, 10,000 to 100,000 until a few hundred years ago, they were able to coexist with the natural environment that surrounded them. Until the use of gunpowder became widespread in the thirteenth century and put an end to city walls, the medieval urbanite could walk through a gate into fields, forests or onto the shore, which were all in intimate contact with the city (Figure 19.11).

Fig. 19.11 • Mont St. Michel, both literally and figuratively an island, exemplifies the medieval walled city turned in upon itself but in intimate contact with its environment. (*French Government Tourist Office*)

Cities remained small because of the attrition of war and disease; the notorious Black Death or bubonic plague reduced the population of some European cities by 80 percent. But the self-contained medieval cities were most effectively breached by changing times; people spilled into the countryside and the cities grew beyond their ancient walls. The industrial revolution drew many people to new jobs, and the resulting factories and housing pushed even further into the surrounding countryside. New cities and towns also sprawled over vast areas where raw materials needed for industry were available, as in the Ruhr Valley in Germany, or Detroit and Chicago in the United States. The final impetus to growth of cities in both the United States and Europe was due to the striking advances in public health services and the advent of public transport systems, first the railroad, then the tram and trolley, finally the subway and automobile. This pattern of city growth is now being seen in other countries undergoing rapid industrialization, Japan and China, for example.

We think of cities in the United States today as being densely packed with people, but urban density has been falling for many years while the suburbs have grown. In 1910 New York City contained 64,000 people per square mile; by 1960 the density had fallen to 13,000 people per square mile. Rapid transit systems played an important role in this dispersion, as have changes in the standard of living. The average New York City household sheltered 4.5 people in 1910 but only 3.1 people in 1960. This means that one and a half times as much housing was required in 1960 as 1910 for a comparable number of people. These houses now surround the cities rather than constitute them. Of course, such statistics apply only to those who have come to enjoy both a higher standard of living and the ensuing greater mobility. Those who have neither remain packed in tenements.

Growth of a Megalopolis

The flight from the city began after World War I, continued through the 1920s and 1930s, and exploded after World War II. War-induced recovery from the depression together with the return to civilian life of millions of servicemen caused a tremendous upsurge in the rate of home ownership. Veterans' Administration and Federal Housing Administration loans and mortgages as well as federal tax deductions on mortgage interest made it, for the first time, cheaper to buy than rent; home ownership suddenly was within reach of greater numbers of people. Of course, some people moved from the country and small towns into suburbs, but the main impetus behind the growth of suburbs was the disenchantment and socioeconomic mobility of former urbanists.

By 1960, suburbs had coalesced into sprawling tracts, smothering the landscape and severing the cities' last link with open space (Figure 19.12). A city dweller no longer could take a bus or trolley to the end of the line and be in the country. To get to the country from a city's downtown became a major expedition, involving hours of driving in very heavy traffic.

By 1961 the large cities of the eastern seaboard of the United States were beginning to bleed together into one large increasingly urbanized mass that Jean Gottmann called a megalopolis. The largest megalopolis in the United States runs 500 miles from Portsmouth, Virginia, to Portsmouth, New Hampshire, and includes Washington, Baltimore, Philadelphia, New York and Boston; others are scattered over the rest of the United States and the world.

Will the United States soon be completely covered by merging megalopolises? By 1950 urbanization had occupied only 18 million of the 1904 million acres of the continental states. Even if urbaniza-

Fig. 19.12 • After World War II, the increased demand for housing resulted in an unprecedented sprawl of houses which today cover square mile after square mile of fertile land near large cities. (*Thomas Airviews*)

tion continues at a rate of 1 million acres a year, and if we assume a half-acre per house and a proportion of 40 percent houses to 60 percent services, streets, factories and other urban uses, by 2010 a population of 360 million would still occupy only 75 million acres of the land area of the United States. This might engulf 18 percent of current cropland, but with a doubling of productivity it might be possible to produce more food on less land, as is now done in some crowded European countries. Does this mean then that we can accept urbanization with the assurance that there will be plenty of land in the foreseeable future? It does not.

Loss of Farmlands

The best farmland is not randomly distributed: often it is associated with cities. Much fertile and valuable farmland has already disappeared beneath the suburbs, for example, in Nassau County in New York, Orange County in the Los Angeles Basin, and the Santa Clara Valley southeast of San Francisco. In the latter instance, 70 percent of the farmland in Santa Clara County was classified as prime agricultural land in 1949. But as the city of San Mateo, itself a suburb of San Francisco, grew, the rich valley in which it is situated became desirable to developers. The same characteristics that make land desirable to a farmer (flatness, depth of soil, accessibility, lack of large rocks or ledges), also make it attractive to the developer, who must keep in mind the expense of land preparation before construction is possible.

Many farmers in the Santa Clara Valley were unable to resist a price of $4000 an acre for their land, and soon bulldozers removed the orchards and subdivision housing began to spring up. It appeared that in a few years the entire valley would be completely developed and its superb soil buried forever. In an effort to save some of the valley a group of farmers persuaded the county to declare an exclusively agricultural zone to preserve the remaining farmland. This measure has held the line for a while, but farmers are constantly harassed by their new neighbors, who object to spraying operations and the noise of machine operations in early morning. As the value of this land increases to $15,000 an acre the process of urbanization may resume, making retention of the agricultural zone extremely difficult. In this way, 15 to 20 percent of California's best agricultural land has already been lost. On a national scale only 72 million of the total 465 million acres of arable land is considered prime, and one-half of this cropland has already been urbanized.

The problem, then, is not in the total number of acres either in open space or in urban use but the quality and distribution of the land that is being converted from one use to the other. The thousands of unused square miles that are still available in Wyoming or Nevada offer little for either the farmer or the urbanite. While it is not likely that the urban sprawl will soon be stopped, it can be regulated, with provision for more balanced land use.

Control of Land Use

The need for ecological control over land use has been demonstrated again and again. But assuring that an ecologist plays a significant part in land use planning does not necessarily lead to ecologically sound land usage, for there are strong local political and economic pressures and financial control always seems to be in someone else's hands. Only public opinion will sway those who control land development and persuade them that ecologically sound solutions are ultimately the best solutions.

Imagine, for example, a narrow rural valley with a stream flowing through it year round. The slopes are covered with trees and perhaps an occasional house built to take advantage of the rustic qualities of the environment. One day a bulldozer appears on the hillside above the valley and intensive development begins. During the construction period many tons of soil wash into the stream, followed by runoff from roofs and paved streets of the completed development. This rainwater no longer seeps into the ground and recharges the water table; instead it pours from storm sewers into the stream. Deprived of its source of recharging, the water table is lowered, requiring deeper and deeper wells for local water supply. The excess water in the stream causes flooding downstream where basements, previously dry, must be periodically drained or waterproofed. During periods of high water, septic tanks are flooded, the

Watershed problems

stream is polluted, and sewers become necessary to avoid public health problems. Because of previous silting the stream is dredged to handle the recurring flood waters, but this seems to make matters worse. Then the Corps of Engineers is asked to build an expensive dam upstream, which in turn permanently floods much valuable bottom land behind the dam. Wells are running dry now, so a municipal water system must be installed. What is left of the stream alternates between feast and famine, flooding one season, dry the next. Finally it is agreed to confine it to a culvert to avoid future problems. This chain of consequences is not at all unlikely, but is it inevitable? By no means.

If the area including the wooded valley had available a planning commission, or an ecologist who is trained to recognize potentially troublesome sites, the whole destructive process might have been avoided. Quite clearly there are some places where houses *should not* be built or at least built only with extreme care and foreknowledge of the dangers involved. The function of a planning commission is to remove the trial and error approach and to suggest on a rational basis which land is most suitable for development and which land is better left undisturbed. The developer sees a swamp as reclaimable land capable of being filled in and covered with houses: an environmentally informed planning commission might view that swamp as a sponge capable of absorbing 300,000 gallons of water per acre for every foot of rainfall added to it and wonder where all that water will go if the sponge is destroyed.

Floodplain destruction

The usual alternative to a floodplain as a sponge is the construction of flood control dams upstream. But such structures are effective only if the storm falls in the fixed catchment basin and the storage capacity of the reservoirs is adequate to contain the runoff. Even then as General Clark, former head of the Army Corps of Engineers, recently said, "We can't guarantee that no place will get flooded. People have to expect it every so often. Dams, levees and flood walls are not the cure-all against flood damage. We also need strong local laws to prevent encroachment on the flood plain of our streams. We do not seek, nor can we ever attain, 100 percent flood control."[1]

In June 1972, Hurricane Agnes produced severe flooding on the lower Susquehanna in Pennsylvania, which resulted in property damage of more than $3 billion. Considering that there are twenty-three dams and many miles of dikes north of Harrisburg, many people have asked why the flooding was so severe. The reason was the uneven distribution and extraordinary amount of the precipitation involved; as a result the least amount of rainfall fell in the vicinity of the largest dams. To contain another flood with the volume of Agnes in Pennsylvania it would take the equivalent of 522 new dams in the Susquehanna River Basin, 202 new dams in the Delaware River

[1] *The New York Times,* 27 May 1973."Flood experts call for curbs on development in low-lying areas." B. Drummand Ayres, Jr.

Basin, and 10 new dams in the Allegheny River Basin. This would require 129,000 acres of floodplain, much of which is already occupied by towns and cities, for the impoundments, and would cost more than the losses from Agnes.

A rational alternative is to return the floodplains to their natural function—the temporary storage of excess stream flow. This confines the destructive force of the flood to the main channel, gradually releases the excess water to the channel after the flood crest has passed, and deposits rich sediments on the floodplain soils. How can floodplains be returned to their original function? To do this all at once would be economically impossible. But suppose that after each flood, the most damaged roads, houses, factories were relocated on higher ground rather than irrationally being rebuilt on the same spot—in a few decades recurring flood losses could be dramatically reduced and the floodplain used for the recreational benefit of all.

Another problem that can be avoided by comprehensive planning is leapfrogging (Figure 19.12). As land close to the city rises in price the developer looks for cheaper land farther out on which he can still make a high profit. He leapfrogs the undeveloped but expensive land nearer the city and builds further out. This results in a chaotic mosaic of developed and undeveloped land that is extremely difficult and expensive to service with rapid transit, sewers, water, fire and police protection, and garbage collection. On a smaller scale, a farmer pressed for cash may slowly sell lots fronting along a road, thereby isolating his remaining land from the road, decreasing both its value and its accessibility to future development. Even more important, this practice puts great pressure on a rural road system incapable of handling the traffic thrust upon it.

A more traditional and piecemeal approach to the control of land use is zoning, which seeks to control the use of land by regulating the size of plots and the shape of buildings and their relationship to each other. Zoning is often used to prevent nuisance construction, such as backyard chicken coops, or to keep population density low and "undesirables" out by establishing minimum lot size. But such planning has often backfired. Instead of controlling the pace of development, it may accelerate it to the point where no open land is left.

Monotony and crowding from ineffective zoning practices are being overcome by the growing popularity of cluster or green-space housing. This plan allows houses to be built closer together in a tract, and frees the remainder of the land from development. The concept embodied in cluster zoning is quite old. Farm houses in parts of French-speaking Quebec, for example, are grouped in small villages allowing all of the surrounding land to be intensively farmed rather than tying up several acres per family with houses and outbuildings.

The open space saved from housing can be used for recreation

Leapfrogging

Suburban open space

415

and to preserve some of those values which people seek when they move from the city center to the suburbs in the first place—a semblance of a natural environment, free from parking lots, houses, and utility poles. For cluster housing to work, continuing provisions must be made to preserve the open land. Occasionally the whole concept is vitiated by later owners who decide to sell the commons to free themselves of maintenance costs.

Cluster zoning (Figure 19.13) is only one way to reserve open land in urban areas. Easements, which involve the purchase of development rights rather than the purchase of the land itself, can be used to preserve wooded or scenic land. Claiming more land than required for a building project and then placing the excess into an open space program is another approach. Gifts from large landholders or exchanges of land parcels are still other ways to obtain open space.

Fig. 19.13 • By clustering the housing and using forms sympathetic to the environment, much of the natural beauty of this seacoast at Sea Ranch, California is preserved. (*Aero Photographers*)

The question that often arises in suburban or rural areas is whether to develop land suitable for housing or for other uses. The town of Closter, New Jersey, when faced with a controversy over its attempt to acquire eighty acres for an open space program, made some calculations and drew the following conclusions: eighty acres would allow the construction of 160 houses, which would bring at least 200 new children into the local school system. Estimating that it would cost the town $720 per year to educate each child ($144,-000), $4000 to collect the additional garbage, $6000 for police protection, and $2000 for extra fire hydrants and miscellaneous services, the total expense to the community if this land were developed would be

$156,000. Calculated land tax revenues would amount to $100,000 a year. Thus the development of the eighty acres would have *cost* the town $56,000 a year. In the enthusiasm for growth of urban housing, so often encouraged by merchant's associations eager for new business, the ultimate costs of "growth and progress" are rarely spelled out. If they were, the desire for development might be tempered and effective plans made to control it (Figure 19.14).

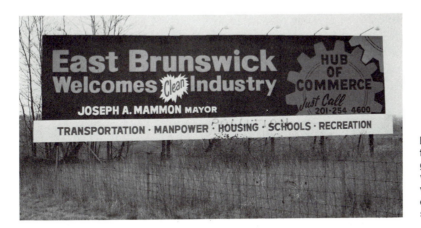

Fig. 19.14 • Attracting industry has traditionally been the goal of most small towns wishing to "grow up." The environmental and social costs of this growth are rarely considered.

NEW TOWNS

As cities have grown beyond manageable size and often have become remote from any possible natural environment, there has been an attempt to overcome the mistakes of old cities by starting new ones. The reasoning usually runs: a near-perfect environment would be within our grasp if we chose just the right site and carefully planned buildings and streets set in parklike grounds. But this planner's utopia continues to elude us.

Much has been said about the impossibility of successfully governing a city with over a million population. The ideal size submitted by planners varies from 30,000 to a couple of hundred thousand. Clearly a city of 10 million has exceeded manageable proportions. However, if the existing population of, let us say, France were evenly distributed into cities of a few hundred thousand people scattered throughout that country, some sixty new cities would have to be built, which would simply urbanize large areas of presently rural France.

The English pioneered the establishment of new towns in an effort to relieve pressure on London and other urban regions. Following the ideas of Ebenezer Howard, who developed the concept of the garden city, several new towns were planned in every detail and then established, but reactions from their inhabitants have been mixed.

One town constructed on the basis of twelve houses to the acre provided gardens front and back which the designer included to please working class families from big city slums who had never had more than a flower pot. But young couples resettled in these units felt isolated from old friends and family in the cities they left behind; young wives felt that the gardens insulated them from meaningful contact with neighbors and prevented the making of new friends to replace those left behind (apparently another example of the misunderstood role of space in the human environment); the husbands found there was no neighborhood pub to get away to and no recreation anywhere comparable to that available in the city. Then the need to commute daily to work took time and money, neither abundant to a working man. Many newcomers to these new towns, like the young couples mentioned, leave after a while, preferring the overcrowded cities. After a few years have passed, there is a mellowing. This can be seen in the English new town of Letchworth, yet there is a cloying sameness that one must expect from something conceived in its totality at one point in time. Diversity of building styles is, after all, a strong point of old towns. Letchworth is quite successful, however, compared with Cumbernauld near Glasgow, Scotland. This new town is built around a rambling building called the city center placed at the highest point of the city and apparently designed as a piece of conspicuous architectural sculpture. Unfortunately, environmental considerations seem not to have been well made and the center acts as a wind tunnel, presenting little environmental improvement over conventional cities that just happened. In addition, this fixed center provides the only focus, indeed relief, from the rather drab housing units which effectively insulate the city from the lovely Scottish countryside. Still, these considerations are trivial when Cumbernauld is compared with the Glasgow slums it was intended to relieve.

Another type of new town is an intentional suburb, a satellite town built without the presumption that it need stand alone. Two new Swedish satellite towns, Vällingby and Farsta, have shopping and cultural areas of their own but are connected by a high-speed rapid transit system (see Chapter 20) to downtown Stockholm, which offsets possible shortcomings of the new town (Figure 19.15). Tapiola, near Helsinki, Finland is another successful satellite town.

Some new towns, constructed on a grand scale, are the capital cities, Brasilia and Canberra. Brasilia has had a cold response from the public servants who are obliged to live there. After dark and on weekends the monuments, grand and imposing, are deserted, as are the recreational areas of many of the apartment blocks. The action is in the shanty town of temporary buildings constructed to house the workers who built and are still building Brasilia. Considering that when location is politically rather than environmentally determined and that civic activities are sharply limited, the boredom of a new capital is not too surprising. Indeed officials are often given a hardship allowance to redress their isolation from the natural center of a

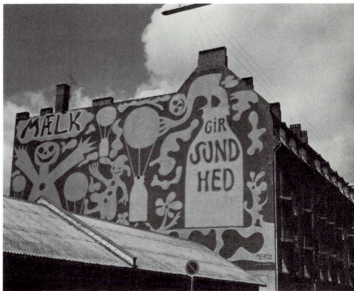

Fig. 19.15 • (a) Hässelby Gård, a suburb of Stockholm, successfully integrates mass transit, housing, and shopping facilities while accommodating pedestrian traffic and activities. (*Swedish Information Service*) (b) This cheerful sign in Copenhagen, painted in orange and blue, urges people to drink milk and brightens an otherwise drab urban scene.

country's culture.

We might be better off if housing and, collectively, cities could be flexible enough so that people could do their own thing, like children with building blocks. Somewhat short of this technological miracle but interesting, nevertheless, is modular housing which is being tried in many cities. By erecting a superstructure to which they are attached, modular units can be flexibly and creatively arranged to fit the whole structure into any existing neighborhood (Figure 19.16). But it makes as little sense for a planner to try to design a new town without real knowledge of its intended occupants as it would for an ecologist to try to recreate a salt marsh without any knowledge of the animals that he might expect to live there. However, Howard was quite probably on the right track, with his notion of city units separated by green belts, for the largest and most densely populated city inevitably breaks up into blocks and neighborhoods, as thoroughly isolated from the city as a whole as suburbs miles away. If one could add the amenity of comfortable housing and convenient open space for a spectrum of recreational outlets, Howard's garden city might be considered to be a form of neighborhood. Perhaps the successful neighborhood can provide a prototype for the long sought utopia of planners.

419

Fig. 19.16 • The superstructure of this modular construction can be clearly seen in the vertical pillars. Modular units are then fitted into the three-dimensional grid, giving unusual flexibility. (*Warner, Burns, Toan and Lunde, Architects*)

The Future of the City

Cities and their suburbs are likely to continue to sprawl across the United States. All people seem to have an insatiable need for some bit of territory over which they have absolute control. It thus becomes the responsibility of architects and planners to build this privacy and individual control into urban areas. But the chaotic sprawling suburbs with which we are familiar need not be chaotic or sprawling. Preservation of some open space through parkland, agricultural reserves, scenic easements, and cluster zoning can provide breathing space and some sense of relationship between man-made and natural forms. Long-range planning is imperative on a regional basis before any amelioration of the grimmer aspects of urban life can be hoped for. With some idea of which areas should grow, when, and how, transport problems can be anticipated and included in long-range plans.

The present environmental problems of urban areas are not unsolvable, however. Indeed there is every reason to expect that there will be successful cities in man's future if we take the long overdue step of infusing a few billion dollars into a full-scale study of the total environment of cities. We need to know how transportation systems can be designed to meet the needs of *people;* what makes one neighborhood exciting to live in and another boring; how we can make the best use of space in our urban designs; what environmental factors are involved in chronic mental problems and how they can be controlled; what human needs are not met in present housing, both old and new?

These are questions that can be approached logically and answered rationally; the solutions can then be considered in any future plans for redesigning old cities or building new ones. There will soon be too many of us to continue the cruel idiocy of trial-and-error experimentation with people's lives, especially at a time when more people are moving from the country to urban areas than ever before. By the year 2000, it is estimated that over three-fourths of the American people will live in towns and cities with populations of over 5000. If a start is made now toward solving the age-old environmental problems of cities there is a chance that these cities may not only be habitable but enjoyable.

FURTHER READING

Adams, F. S., 1973. "Hurricane Agnes: flooding vs. dams in Pennsylvania." *Science and Public Affairs* **29**(4), pp. 30–34.

Detwyler, T. R., M. G. Marcus, et al., 1971. *Urbanization and environment, the physical* geography of the city. Duxbury Press, Belmont, California. A good review in some detail of the reciprocal effects of city and environment.

Eldredge, H. W. (ed.), 1967. *Taming megalopolis.* Vols. I and II. Praeger, New York. An encyclopedic collection of papers dealing with a variety of urban problems.

Gottmann, J., 1964. *Megalopolis; the urbanized northeastern seaboard of the United States.* M.I.T. Press, Cambridge, Massachusetts. Gottmann's book, though a tome, is a classic work, the first to recognize the implications of regional cities.

Hall, E. T., 1966. *The hidden dimension.* Doubleday, Garden City, New York. Incisive examination of man's private and public use of space.

Herber, L., 1965. *Crisis in our cities.* Prentice-Hall, Englewood Cliffs, New Jersey. Popular but literate account of urban problems.

Jacobs, J., 1965. *The death and life of great American cities.* Random House, New York. A common-sense point of view that cuts through much of the persiflage surrounding city planning.

Mumford, L., 1961. *The city in history.* Harcourt, Brace & World, New York. Leisurely and thorough, Mumford blends scholarship and style.

Rudofsky, B., 1969. *Streets for people; a primer for Americans.* Doubleday, New York. Well-illustrated book pointing out the use of streets as urban space.

Srole, L. et al., 1962. *Mental health in the metropolis, the midtown Manhattan study.* Vol. I. McGraw-Hill, New York. A pioneering, thought-provoking study of the effects of urban life on mental health.

Srole, L., 1972. "Urbanization and mental health: some reformulations." *American Scientist* **60**, pp. 576–583. Some further thoughts balancing the somewhat negative thrust of the midtown Manhattan study.

Tunnard, C. and B. Pushkarev, 1963. *Man-made America: chaos or control?* Yale University Press, New Haven, Connecticut. If you read any book on this list, read this one. Beautifully thought out and illustrated, it gives penetrating insight into man's attack upon the environment in the guise of urbanization.

U.S. Department of Health, Education, and Welfare, 1969. *The climate of cities: a survey of recent literature.*

U.S. Department of Housing and Urban Development, 1970. *New communities, a bibliography.*

U.S. National Commission on Urban Problems, 1969. *Building the American city: report.* Praeger, New York. Much information about ghetto areas, zoning problems, urban renewal.

20

TRANSPORT: LIFELINE AND NOOSE

CITIES HAVE BECOME so vast that any vestige of self-sufficiency they may once have claimed has long since disappeared. In order to function they rely exclusively on complex transport systems that move their managers, workers, raw materials, goods, and services. But these transport systems are also responsible for many of the urban ills which threaten the system they support. The dual nature of transport as a lifeline and a noose is a problem rooted in the past.

ENVIRONMENT AND THE EVOLUTION OF TRANSPORTATION

Although each of the successive modes of transport has had its impact upon the environment, the environment in turn has had a definite effect in shaping and directing both the type of transport and its lifespan. To see this in its historical perspective let us examine the interaction of transportation and environment in the United States since 1800.

One of the most important early innovations in the United States was the development of a sturdy, commodious wagon by German settlers in the Conestoga Valley near Lancaster, Pennsylvania. Freely adapted from the crude wagon of their homeland, the Conestoga wagon quickly came into wide use in the early nineteenth century to transport freight between the rapidly growing cities of the eastern seaboard. Settlement followed the easy wagon routes, north along the Connecticut and Hudson Rivers, southwest following the Great Valley, and west along the James and Rappahannock Rivers into western Virginia and Maryland. This developing country was not trackless however; buffalo, elk, and other large game had long followed the easiest grades over mountain passes and along river valleys. These trails were used by the Indians and finally by the white men

Wagon roads

with their wagons. Other trails were cut and developed into wagon roads by army campaigns—Braddock's and Forbes' roads to the forks of the Ohio and later Daniel Boone's wilderness road that led over the Cumberland Gap into the frontier country of Kentucky beyond.

As the wilderness trails extended westward, the people remaining behind needed more than single-file tracks or crude wagon roads; all-weather roads able to handle heavy freight traffic all year round were imperative. Using a road building technique developed by a Scotsman by the name of McAdam, a paved road was laid between Philadelphia and Lancaster in the late 1700s, soon followed by the National Post Road leading west from Washington roughly parallel to the present Interstate 70. Macadam pavement in the early nineteenth century did not mean asphalt or concrete, of course, just a graded layering of crushed stone. But anything that replaced mud, dust, and frozen ruts was a welcome improvement. Because the federal government was loath to subsidize this construction, such improved roads were opened as toll roads. In fact, around 1800 such a speculative fever took hold that nearly 6000 miles of toll roads were authorized in New York State alone. But the resulting high rates for freight movement spurred the development of another transportational innovation of the early nineteenth century—canals.

Following with great interest the enthusiastic development of canals throughout Europe, especially in France and England, speculators were ready to compete with toll roads in the growing business of transportation. Politicians were also eager to open new areas to settlement, an attitude due in part to a sharing of the general enthusiasm for growth and progress, but also from a desire to profit handsomely from speculation in land made more valuable by increased access.

In 1825 the Erie Canal, which finally connected the Great Lakes to New York City, was constructed for $7.5 million. This was the first of many steps that ultimately made New York City the largest and most important city in the Western Hemisphere. In nine years tolls had paid off both construction cost and interest for the canal; travel time between New York and Buffalo was reduced from twenty to eight days, and freight rates cut from $100 to $10 per ton. Small wonder that the succeeding two decades saw an unprecedented wave of canal construction; over 1700 miles of canal were constructed in Pennsylvania alone, while Baltimore eagerly invested $11 million in the B & O Canal.

In 1811, Robert Fulton, inventor of the steamboat, formed the Mississippi Steamboat Company to carry shipping between Pittsburgh and New Orleans. Soon steamboats were on the Ohio, Mississippi, and Missouri Rivers and shipping grew rapidly, greatly increasing the importance of New Orleans and the new towns of Memphis, St. Louis, Nebraska City, Council Bluffs, Leavenworth, and St. Joseph.

Post roads

Canals and riverboats

But by 1860 the heyday of the steamboat and the canals as a major form of transportation in the United States had come and was quickly to pass, doomed by the rapidly expanding railroads.

The same year the Erie Canal opened, 1825, Colonel John Stevens had built a steam locomotive. Within four years the Baltimore and Ohio Railroad connecting the Atlantic Seaboard and the Ohio Valley opened for business (Figure 20.1). The options of post road, canal, riverboat, or rail that were successively open to the East were, for many years, closed to the trans-Mississippi West. Until the railroads could span this almost unsettled vastness, animal power was the only practical form of tranportation. The Conestoga wagons because of their high sides, size, and canvas-covered tops were still the first choice of settlers pushing west from the river towns of Independence, St. Joseph, and Leavenworth; later, similar wagons made by Murphy, Carson, and the Studebaker brothers joined in opening the West to trade and settlement. The Oregon, Bozeman, and Santa Fe Trails and the Central Overland Road, developed in part by Army surveyors to establish and service outlying forts, were the most heavily used.

The railroad

Fig. 20.1 • One of the first passenger trains in service in America. The locomotive was built by a watchmaker in York, Pennsylvania in 1832, the handmade coaches in Philadelphia. The locomotive ran for sixty years and is still capable of running under its own steam. (*The Warder Collection, New York*)

Soon, however, the railroads had pushed west to the Missouri River, and having eclipsed all but the Erie Canal in the East, began to compete with the steamboat trade on the western rivers. New cities such as Chicago, St. Paul, and Omaha, with rail connections to the East and serving as jump-off points for travel farther west, rapidly began to outshine the older river towns of Nebraska City, St. Joseph, and Leavenworth. Steamboats were slow and, because they were obliged to follow the rivers, indirect. Like the earlier toll roads in the East, their rates were high enough to be easily undercut by the railroads.

The impact of the railroad on the West was immediate and long-lasting. The destruction of the vast buffalo herds was quickly accomplished from the base provided by the railroads, which made possible and profitable the transportation of hides, meat, and bones to the East. Railroads had not only the power of life and death over any town along their prospective routes in the more settled East, but they could and did found towns en route further west, using land granted by the government. By 1900 the railroads reigned supreme, but the next actor was waiting impatiently in the wings. Henry Adams once said that his generation of Americans was mortgaged to the railroad; ours is mortgaged to the automobile.

HIGHWAYS AND THE LANDSCAPE

As the railroads took over trade once carried by wagons and stages, less attention than ever was paid to roads, but around the turn of the century the rise of bicycling as a sport of fad proportions again focused attention on the pitiful state of road maintenance and construction. The point was heroically made by the epic journey of Thomas Stevens, who bicycled from San Francisco to the East Coast in 1884, 3700 miles in three months. Today high school boys make coast to coast trips on their bicycles every summer but they have the benefit of roads which Mr. Stevens lacked. The League of American Wheelmen lobbied actively for construction of roads, ostensibly for bicycling, but they were cognizant too of the importance to the local farm economy of farm-to-market roads connecting outlying farms to railroads. By the time the automobile made its debut, the public had been aroused to the need for improved roads, and with the added stimulus of the car, good roads were soon forthcoming.

The Growth of Highways

Initially towns were connected center to center to allow maximum accessibility. As long as there were few cars, there were no problems: traffic and parking were easily accommodated by two-lane interurban highways. These town-to-town roads were gradually expanded into a network. The influence of local topography and land usage on their pattern is obvious to the road map reader. Even today you can easily identify a section of the country by the pattern of its highways. For example, as you can see from a road map, the roads in central Pennsylvania tend to arc southwest to northeast following the curving ridges and valleys of the Appalachian Mountains (Figure 20.2a); in Kansas, although the land is flat and roads could theoretically go in any direction, the land is all farmed, and since no one wanted a road to cross his field, the roads follow section lines with many right-angled turns, giving the appearance of a giant checkerboard (Figure 20.2b); in West Virginia the roads twist and turn as they move across a plateau heavily dissected by streams and rivers (Figure 20.2c).

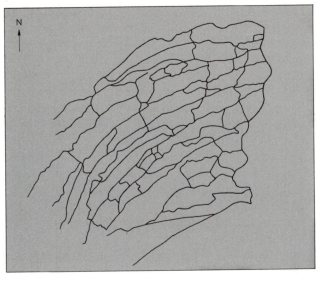

Fig. 20.2 • (a) Roads extend over the landscape of least resistance, so their pattern
on a map gives a graphic indication of the
landforms of the area they traverse: in Pennsylvania, the roads follow the valleys between
the ridges that curve from the southwest to
the northeast.

Fig. 20.2 • (b) In Kansas, where the land
is relatively flat but intensively farmed, the
roads follow section lines to avoid bisecting
farmland, and so the pattern is a checkerboard.

Fig. 20.2 • (c) In West Virginia, the land-
form is a deeply dissected plateau. Roads
must descend steep slopes to cross rivers
and narrow valleys and then climb back up
to the plateau surface again, or else follow
the winding routes of the watercourses. This
results in a pattern of squiggly lines and explains why it takes so long to drive across
West Virginia, a relatively small state compared to Kansas. Of course, interstate highways with their many bridges and minimal
grades eliminate both the problem and the
pattern.

As the towns and cities continued to grow, often along new or improved highways, and the congestion of their sprawl was added to the increasing number of cars, it became more difficult to get from one town to another. As more and more people became mobile, both through their ability to own automobiles and their widening spheres of interest, they were no longer satisfied to drive from town A to town E, through B, C, and D, but wished to go from town A directly to town E. As truck transport developed and began to give the same competition to the railroads as they in turn had given to the rivers and canals, still more pressure developed for through traffic routes that bypassed the congested towns and cities. Their construction was begun in the 1930s in the more densely populated areas and continued after World War II.

But population growth and commercial development began to attack these bypasses almost as soon as the concrete was dry; clearly what was needed was a road that separated opposing traffic, controlled access, and whose intersections were separated. Divided lanes were available long before the automobile, Commonwealth Avenue in Boston, for example; but the Bronx River Parkway in New York, the Merritt Parkway in Connecticut, and the Pennsylvania Turnpike in Pennsylvania were the first major roads in the United States to combine all three characteristics and become expressways. About the same time, Germany began her system of autobahns, planned by teams of engineers and architects who were sensitive enough to environmental considerations to include nesting places for birds in their bridges. When World War II broke out road building came to a halt for six years. But afterwards, there was a rash of toll expressway or turnpike building, mostly in the northeastern United States. Then in 1956 the federal government began an enormous program of interstate expressways designed to connect virtually all cities with 50,000 population or more (Figure 20.3). Over 41,000 miles of these limited access divided highways were to be provided. The original cost has since grown to $60 billion, and the completion date of 1972 has been extended to at least 1976. As a result of the interstate system, we have gone the full cycle in many cities, from roads built from city center to city center, to bypasses, and back to downtown expressways.

Modern expressways are planned with great care and the roadways fitted to the landscape according to very specific rules. Unfortunately, until quite recently, these rules had nothing to do with the particular aesthetic values of the countryside over which the expressways were rolling. The route was, and sadly still is in many places, determined by the expense of a particular route modified by its projected use. If this directed that an expressway should go through an historic site, a grove of redwoods, a state park, a marsh, or pond, all were sacrificed to the supposed greater good of the transportation corridor that replaced them.

Highway design

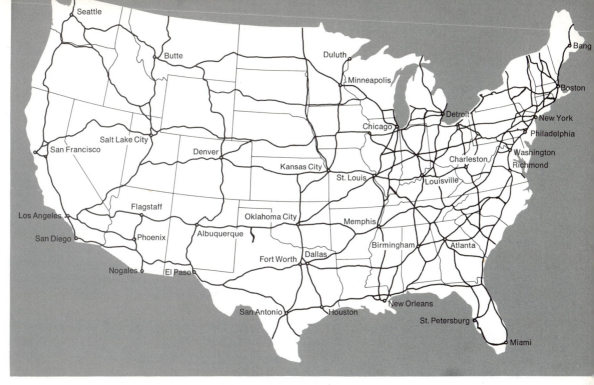

Fig. 20.3 • The interstate highway system connects all major cities with over 41,000 miles of high-speed expressways.

Expressways and standard two-lane roads are blueprinted with as much care as a building, for every drain, curve, grade, ramp, overpass or underpass must be carefully specified. This can result in a well-engineered route with no sharp curves or steep grades—the choice between an attractive road in harmony with the landscape and one that seems to have been carelessly laid down lies with the designing engineer. A properly designed road must be articulated horizontally, vertically, and laterally with the countryside it traverses.

Only rarely can roads be set arrow-straight onto a landscape; even when they can, as on a coastal plain or flat plateau, this is usually undesirable because of the danger of monotony to the driver. Most roads are designed on the basis of tangents (straightaways) and curves, much like the tracks of model trains, which come in straight and curved pieces. Engineers, by fitting these tangents to the curves, carry the road along the proper route, climbing or avoiding hills, approaching towns or rivers. When this is done with long tangents and short curves the road looks awkward and the curves appear too suddenly (Figure 20.4); the driver is alternately lulled, then forced to react. One alternative is to shorten the tangents and lengthen the curves; this improves the driver's reaction but it is as difficult to fit in short pieces of tangents as it is short curves.

Perhaps the best alternative, at least in hilly country, is a continuous curvilinear configuration in which the road constantly curves

429

Fig. 20.4 • By fitting long straightaways and short curves together a road appears to lurch drunkenly across the landscape.

Fig. 20.5 • Continuous curvilinear roads present a much more graceful fitting of a highway into the landscape. (*Thomas Airviews*)

gently to make the best fit with the landscape (Figure 20.5). Such roads are no more difficult to build, although they may be slightly more expensive to design. But a road with long tangents and short curves may ultimately cost more because of a greater accident rate and because of the potentially greater expense of fitting the road to the landscape by cutting and filling—a process that can be exceedingly expensive. Aesthetically, in the proper topographical setting, a continuous curvilinear design creates a far more attractive harmony between man-made and natural environments than any alternative.

Interstate expressways, by law, and most other federal and state routes, by practice, have grades limited to 5 percent, that is, a five-foot rise in 100 running feet. This vertical fitting of road to landscape can also be achieved by tangents, curves, and curvilinear design. Often a road using the long tangent and short curve approach in its horizontal alignment does the same thing in the vertical plane. The result is a jumble of awkward transitions: downhill runs that seem to bend sharply at the bottom, uphill segments that break into nowhere at the crest of a hill (Figure 20.6). With a curvilinear or at least a long curve and short tangent design, these awkward transitions can be avoided; once more the gain is both in aesthetic appearance and safety.

Fig. 20.6 • Highways must be carefully aligned in the vertical as well as the horizontal plane. The transition at the bottom of this hill is too abrupt.

The first expressways followed natural contours, and curved in harmony with the landscape, but these roads, although their windings and curves might have been pleasing on foot or in a Model T, were exceedingly dangerous when packed with fast cars driven by people in a hurry. Testing the power of earthmoving machines, road engineers replaced these parkways with beeline routes cutting through hills, rolling on filled beds over valleys with no attempt to repair, much less avoid, the damage to the landscape. Steep cuts were not only ugly scars that defied revegetation (Figure 20.7) but, prone to erosion, they showered the roadway with rock and mud with every rain, and required constant maintenance. Today enough care is taken in the construction of state and even local roads to avoid excessive cutting and filling, even if the road is longer as a result. Necessary cuts are beveled and planted, improving their appearance and stabilizing them against erosion.

Fig. 20.7 • Highway cuts should be beveled and sloped to allow plantings to stabilize the cut. This poorly designed cut will erode and drop soil and stones on the highway indefinitely.

Roadside planting

Most interstate expressways in suburban areas, at intersections, and occasionally along lengthy segments have plantings, well-intentioned and expensive, but sometimes very poorly executed. The whole point of expressway landscaping is to fit the road into the landscape with as little disturbance as possible to the natural environment and at the same time, by ensuring good visibility, to allow maximum speed and safety.

Two areas in which many roadside plantings fail are scale and appropriateness. While it is a fine gesture for a local garden club to help beautify our expressways by planting pansies in the median strip, the effect is utterly lost at sixty-five miles per hour. Many landscaped plantings also fall into this category. Trees or shrubs are strung out in lines or isolated groups that even when mature will have a scattered, chaotic appearance; mass planting on a bold scale can overcome this problem. Ironically this has been done most impressively in the semidesert of Southern California, where nothing would grow along the freeway without extensive irrigation. But mass planting of lush ivies, bayberry, oleander, iceplant, and other ornamentals makes some of the Los Angeles and San Francisco freeways among the most beautifully landscaped in the world (Figure 20.8).

This brings up the other planting problem—appropriateness. If the planting is in a dry region where native shrubs are few or lacking in proper scale but where irrigation is available, exotic ornamentals might very well be appropriate. In a well-watered region with an abundance of native trees and shrubs, it makes little sense aesthetically, ecologically, or economically, to plant expensive exotics. The Long Island Expressway, for example, runs for much of its length through an oak-pine scrub forest. Instead of planting species native to this local forest, scarlet oak, bear oak, and pitch pine (Figure

20.9a), to coordinate the highway with its environment, the "beauti-fiers" have planted little clumps of Austrian pine and tulip trees (Figure 20.9b), which effectively divorce the road from its natural environment. So planted, the road could as well be in Maine or Georgia.

Fig. 20.8 • Despite the dry climate, or perhaps because of it, California freeways are among the most beautifully landscaped in the world. Note the effective use of mass planting. (*California Division of Highways*)

Fig. 20.9 • (a) These beautiful pitch pines overlook the Long Island Expressway right-of-way. (Photo by Dan Jacobs) (b) The right-of-way planting, however, ignores those species well adapted to the dry sandy soil of central Long Island and features instead this lonely tulip tree, which is more at home in a rich, well-watered forest. (*Photo by Dan Jacobs*)

The great Brazilian landscape designer, Burle-Marx, trained in Europe like most of his contemporaries, returned to Brazil and rebelled against the slavish use of European ornamentals in Brazilian parks and gardens, whose landscapes ignored the incredible richness of the native Brazilian flora. Overcoming provincial prejudice, Burle-Marx made use of native materials with brilliant success. The same principle can be followed on American highways.

Another approach would be to allow natural revegetation to take place wherever possible. This would sharply reduce maintenance since plantings of ornamentals often require much tending. An example of this natural approach to highway planting can be seen on another Long Island highway, the William Floyd Parkway. Rather than removing the oak-pine woods evenly along the right-of-way, the designer made the grassy shoulder of variable width, sometimes stretching to the edge of the right-of-way, sometimes allowing trees to approach the road, giving a flexible, scalloped appearance that ties the road beautifully to the wooded land through which it passes.

An unplanned result of planting grass as a stabilizer along highway margins is that the lush pasture formed attracts deer, especially in heavily wooded country. The wild deer are easily frightened by headlights, and the casualties both of deer and injured motorists are high. Construction of fences has been helpful but is quite costly. An ecological solution to the problem would be to find an aesthetically pleasing roadside stabilizer or a planting that would be unpalatable or unattractive to large, grazing animals.

The Environmental Impact of Highways

Because the development of highways in the United States has been gradual and roads have a historic lineage going back to cow tracks, game trails, or Indian paths, we tend to lose sight of the incredible impact road construction can have in undeveloped regions. A potentially dismal example is provided by the construction of the trans-amazon highway. The 2000-mile highway is being constructed through one of the last and largest undeveloped forests in the world. Running 200 miles south of, but parallel to, the Amazon River, it will connect Recife and João Pessoa in Brazil with the borders of Peru and Bolivia. Another route, this one north-south, is being constructed to connect Santarém on the Amazon with Cuiabá (Figure 20.10). Motivated by the Brazilian government's long-standing and reasonable desire to develop the heart of Brazil, but precipitated by the plight of the impoverished population of drought stricken northeastern Brazil, the road, much of it two-lane and unpaved, will be complete in a few years and will allow resettlement of people from the Northeast to Amazonia. Apparently, a magnificent opportunity to plan an orderly and ecologically sound development of the Amazon Basin is being lost in a boom syndrome reminiscent of the frontier days of

complete ———
under const. ∞∞∞∞∞
planned – – – –

Santarém

João Pessoa

Recife

Cuiabá

Rio

Sao Paulo

TRANSPORT: LIFELINE AND NOOSE

Fig. 20.10 • The trans-Amazonian highway is but the opening wedge in the carving up and destruction of the Amazon rain forest as a natural system. The road will act as a stimulator of development in Amazonia; indeed, it is being constructed almost solely for that purpose, like the transcontinental railroads in the Great Plains of the United States. (© *1972 by The New York Times Company. Reprinted by permission*)

the American West: hurried construction of a transportation route "opening" the frontier; wild speculation in lands bordering it; vague promises of riches for the lucky settlers; and potential environmental disaster likely for all. At present rates of clearance, the entire Amazon rain forest will be cut in thirty-five years and ill-fated attempts made to farm the lateritic soils (see Chapter 3). It needn't be this way, of course, and there is still time for the Brazilians to pull the fat from the fire. But considering the time it took this country to begin to appreciate the impact of our highways and other manipulations on the environment, a change in attitude is unlikely, at least until the resource has been exploited and destroyed. Despite our accomplishments in environmental restoration (see Chapter 16), the Amazonian rain forest, once destroyed, will not be restored in our lifetime.

EXPRESSWAYS IN THE CITY

Despite the formidability of building a road through the Amazon rain forest, building roads in rural areas is relatively uncomplicated when compared with the problems that arise from the planning and construction of expressways through urban areas. No one, rich or poor, wants to sacrifice his property, house, or business to highway construction. Road designers know that city parks are the least vigorously defended parcels of land in most cities; consequently interstate expressways are invariably routed through them. Probably the most absurd example of the lengths to which this practice is carried is the North Expressway in San Antonio, Texas:

435

For more than five years a bitter controversy has continued over the location of an expressway route through San Antonio's famed Brackenridge Park and related open spaces. This park . . . includes not only undeveloped land and a major flood control facility and parkland but recreation and sports areas, picnic grounds, zoo, college campuses, a renowned landscaped Sunken Garden and outdoor theatre, a city school stadium, a municipal golf course, and stretches of the natural water course.

The proposed expressway curves and winds through this open space system, crossing an Audubon bird sanctuary and Olmos Creek, a tributary waterway in its natural state. It moves along a picnic ground and recreation area obliterating a Girl Scout camp and nature trail. It plunges across the Olmos Flood Basin and rises to enormous heights to go over the Olmos Dam. It severs the campus of the College of the Incarnate Word and stretches across the lands of the San Antonio Zoo. It blocks off the half-built public school gymnasium, slides along the rim of the Sunken Garden, hovers over the edge of the outdoor theatre—squeezing between that and the public school stadium—blocking a major entrance. It slashes through residential areas and slices along the golf course and a wooded portion of the natural watercourse of the San Antonio River.

How many irreplaceable trees of magnificent size and age, how much spoil of adjacent area, how much auxiliary space for interchanges, drainage, and other highway structures will be needed have not, as yet, been calculated fully.[1]

After park upon park had been compromised, if not ruined, people began to protest, first on their own, and then with the help of their lawyers. This led to a Supreme Court decision in 1971 ordering a review of a segment of I-40 through Overton Park in Memphis, Tennessee to determine whether approval to route the highway through the park was "arbitrary, capricious, an abuse of discretion, or otherwise not in accordance with law." Routine violation of parklands is now somewhat less likely.

The Highway in the Ghetto

Another promising corridor for the routing of urban expressways has been slum districts. The poor, it was considered, could always find another slum or be relocated in some form of public housing. The land left over from expressway construction could be redeveloped with luxury apartment or high-rise office buildings and the whole package called urban renewal. But the poor are beginning to find their tongues too and as a result urban expressway construction in city ghettos is also being blocked. The interstate highway system may be nearly compete in 1976, its new target date, but in rural areas, not in the cities.

[1] Zisman, S. B., 1967. "Open spaces in urban growth." *Taming megalopolis* Vol. I, Eldredge, H. W. (ed.), Praeger, New York.

Citizen protest over leveling of parks and slums has brought about some rethinking of the role of highways in the urban environment. Expressway construction is a big business backed by a formidable lobby of construction firms, labor unions, steel and concrete producers, car manufacturers, gasoline and tire dealers, and organizations of car owners. The more roads that are built, the more workers are employed, the more profits are made. If cities are congested, the reasoning runs, we need more expressways to handle the traffic. When these are filled we add more lanes, then double decks, then parallel roads. Ignoring the limited capacity of city streets, more expressways bring more congestion leading logically to the situation described by Victor Gruen, who calculated the effects if everyone coming into Manhattan today by various forms of mass transportation was required to use a car:

If 1,000,000 transit passengers were to drive in to work or to shop, they would occupy approximately 750,000 automobiles. If, having reached Manhattan, they were satisfied merely to stand still, bumper to bumper, they would cover 15,000,000 square feet of road surface. To make space for them within the main business area of Manhattan, we not only would have to eliminate all sidewalks, but would have to demolish every last structure, and then double-deck a part of Manhattan Island now covered by office buildings, hotels, theatres, stores, etc. Inasmuch as standing still, bumper to bumper, would be a highly undesirable and fruitless activity. and as people obviously will want to move around, we will have to provide three times that space; that means we will have to build six layers of transportation area covering the entire business core. If we also desire space for taxicabs, trucks, service vehicles; if we consider that there might be occasional accidents and stalled cars; and that it will be necessary to build some ramps, stairs, elevators and escalators, nine levels will be required. On top of the ninth level, we could then start from scratch and construct new buildings to house those facilities which we had to demolish in order to make space for motorization.[2]

Cars and Cities

Gradually the limited capacity of cities to absorb automobiles is being recognized and beltways are being constructed to divert cars away from the city's center rather than force them through the city's business district. Relieved of the burden of through traffic, a city can then have the flexibility to innovate. Some cities have separated cars from people, constructing underground plazas and shopping malls. Ironically it is always the people who are put underground, leaving the vehicular traffic to enjoy the light and air of the surface. Another possibility followed by Fresno, California has been to ban all vehicular traffic from certain main shopping streets, converting them

[2] Gruen, V., 1964. *The heart of our cities; the urban crisis: diagnosis and cure.* Simon and Schuster, New York.

into pedestrian malls with sidewalk cafés, parks, paved areas, planters—an exciting inducement to shop in the adjoining stores (Figure 20.11). Adequate peripheral parking is, of course, essential to the success of any such plan.

Any alternative to expressways through cities must involve use of mass transit. Buses, subways, and suburban trains have long been belittled by the highway lobby for obvious reasons. One result has been the neglect of mass transit, which has discouraged its use to the point of reinforcing the opinion that mass transport is outmoded and inadequate, and that the automobile is the only mass transit of the future. Gruen's calculation should disabuse anyone of that fantasy.

Fig. 20.11 • Once typically filled with traffic and noise, this street in the business district of Fresno, California has been transformed into a pedestrian mall. (*Fresno County and City Chamber of Commerce*)

The key to the issue of more highways versus more mass transit seems to lie in the Highway Trust Fund. Set up by Congress and financed by a special gasoline tax, this fund grows by $5 billion a year and has been devoted solely to constructing more highways— which attract more cars that pay more gasoline taxes to build more highways. While many apparently feel that this large user's fund should be spent only on highways, others feel that at the very least cities should have the option of using their share to refurbish, rebuild, and construct mass transit systems in lieu of more highways. Considering the billions that have been poured into highway construction in the last twenty years, compared to the miniscule funds used to support mass transit, it should hardly come as a surprise to learn that the use of mass transit systems has drastically declined; in 1929 the average American used public transport 115 times per year; in 1958, fifty-four times. The reason is simple—mass transportation has not kept pace with the automobile in its technology, convenience, or comfort. When an effort is made to provide convenient, clean, and comfortable *rapid* transit, including buses, trains, and subway cars, many commuters are happy to leave their cars at home.

The problem, characteristically, is deeper than simple merchandising, however: environmental considerations and their psychological ramifications enter the picture. Very often the only privacy a typical white-collar worker has is the daily commuting trip in his car. The road may be jammed bumper to bumper but the personal space of his car is *his*. It represents a glimmer of independence in an otherwise thoroughly routine life. For this reason there will always be those who drive their cars as close to their work area as possible. But even they can be accommodated at least in part by some form of mass transport. Large parking lots or garages on the edges of a city would allow suburbanites to drive part way, since mass transport is often ineffective in low density suburbs, and take a train, bus, or subway to the central business district. This approach is now being tried in Hartford, Connecticut. Comparing capacity in persons per lane per hour, cars on a city street can carry 1400 people, buses 7500. On an expressway cars can handle 3500 people, buses 15,000, rapid transit trains 60,000.

Man and machine

ORVs AND AMENITY

Off-road vehicles (ORVs) are a logical culmination of America's fascination with gadgets, greatly increased leisure time, and swelling affluence. While the motorcycle is hardly new, snowmobiles, trail bikes, and all-terrain vehicles have been introduced in the last ten years and air-cushion devices hover offstage (Figure 20.12). These innovations came upon us suddenly, catching lawmakers unawares with the environmental and social problems caused by thoughtless and often careless use of these vehicles.

Perhaps the most obvious problem is noise. Like the pleasure a small boy gets from taping a piece of cardboard to the wheel of his tricycle to make a noise, a considerable part of the pleasure in operating many types of ORVs is the sense of power communicated by noise. A snowmobile's whine sounds like a chain saw and can be heard for miles. Considering that ORVs as a class can invade any environment, their widespread use allows little respite for people who prefer natural sounds in natural environments. The industry is quite capable of decreasing the noise level of ORV engines by at least 50 percent; perhaps it would act if it were convinced that the public would buy a quiet ORV. Unfortunately restrictive legislation

Fig. 20.12 • While snowmobiles can be fun to use for some, they can be dangerous and far noisier than necessary. (*Wide World Photos*)

439

will probably be necessary to force industry to act.

Erosion and destruction of vegetation are the next most important effects of ORVs. In some parts of California almost 60 percent of the vegetation has been destroyed by motocycle use. Mountain meadows, grasslands, and deserts are particularly attractive to ORV users and are particularly sensitive to their impact. Snowmobiles are especially hard on tree seedlings and golf greens. Littering is, of course, extended by ORV users far beyond the edges of major highways and settled areas.

Game harassment, while practiced by only a few, is likely to become a critical problem. Animals in the dead of winter or carrying young are likely to be weak, and harassment can cause death or abortion of young. Streams and lakes with limited accessibility during summer can be quickly and methodically fished out in midwinter, and nesting birds can be encouraged to abandon their nests if frequently disturbed. Finally, destruction of vitally needed predators—foxes, coyotes, wolves, and mountain lions—particularly in remote areas, is becoming a problem.

Vandalism, whether cutting fences or breaking and entering of isolated cabins, is still another problem. From the point of view of the ORV operator, safety is of some concern. Anyone young or infirm can get very far from help in an ORV. Reckless driving resulted in 82 fatalities in the 1969–70 snowmobile season alone.

But perhaps the most difficult problem is how to meld the conflicting interests of a cross-country skier and a snowmobiler, or a backpacker and a trail biker.

There are several ways to lessen these problems: legislation setting up rules and regulations that can be enforced by required registration fees for all ORVs; vehicle use zoning for ecologically sensitive areas, seasonal prohibition of use where necessary, and channeling of use into established routes or trails; federal legislation requiring uniform standards for safety equipment, permissible noise, and pollution control devices. There is no reason why ORV enthusiasts and nonmobile outdoorsmen cannot both satisfy their inner needs, but quite obviously this cannot happen without regulations, standards, and their strict enforcement.

MAN TAKES TO THE AIR

For many cities the epitome of growth and progress is the size of their airport, the number of planes leaving and taking off, the number of people who funnel through their transportation centers to other parts of the nation or world. These activities *do* generate money, which accrues to the city, but the problems generated by airport complexes are seldom faced realistically or defined economically. Air and noise pollution have already been discussed in Chapter

8. The problems that concern us here are the congestion caused by overuse of ground facilities and the huge space demands of expanding international airports.

No city in the world has more congested airports than New York. Kennedy, LaGuardia, and Newark airports, all bursting at the seams, must handle such dense air traffic that holding patterns often have dozens of jets endlessly circling the metropolitan area waiting for the opportunity to land. Something is clearly wrong when you can fly from Miami to New York in two hours, then be obliged to circle the airport for one hour waiting to land.

There is no reason why more people cannot fly directly to Europe or South America from any number of cities throughout the United States. Many cities have large airports which are capable of handling much of the traffic that passes through the ever-narrowing funnel of New York's Kennedy Airport. Why, if you live in Altoona, shouldn't you be able to fly from nearby Pittsburgh to London or Rome? Couldn't customs facilities be set up at these airports as well as at Kennedy, Boston, and Washington? If this feeder traffic were to be diverted, present overcrowded facilities would be adequate until more basic technological changes could be worked out.

But the standard answer to overcrowding at airports is to expand them or build new ones; consequently the search goes on for a fourth huge jetport for the New York area. No matter where a fourth jetport is located in the New York metropolitan area, even if a pre-existing military base is used, the environment of miles of surrounding countryside will degenerate through the resulting development, noise, and air pollution. But the sky can accommodate just so many planes in holding patterns before the whole system breaks down into chaos.

Since each new generation of jet planes is larger, requiring longer runways, and ever-larger terminals to accommodate increasing loads of passengers, economic alternatives to the present system must be found. But if huge airports are to remain the standard in air transport, efficient mass transit systems capable of moving goods and passengers quickly to and from the airport, regardless of its distance from the city, become imperative. At the present, only two American cities, Cleveland and Boston, have this kind of link with their airports. The time for developing solutions to the problems of congestion and space is growing short. While New York casts about for a site for a huge new jetport, other cities are beginning to eye natural environments that cannot be replaced; Chicago, Lake Michigan; San Francisco, its bay; and Miami, the Everglades. With truly *rapid* transit systems regional airports could be constructed at some distance from congested cities, pooling the resources of a region and at the same time sparing those natural environs of large cities that do still exist.

FUTURE FORMS OF TRANSPORT

In the future there is promise of more than refurbished trains or air-conditioned buses, for there is no reason why technology cannot be applied to the problem of moving people rapidly *on* the surface of the earth as well as above or beyond it. Popular science magazines are filled with fascinating schemes for magnetogasdynamic propulsion, gravity tubes, and various suspension systems—air cushion vehicles, magnetic suspension systems—all of which are based on perfectly plausible physical principles capable of working if technology is assisted with investment dollars. More immediately useful are highspeed trains capable of moving people from one end of megalopolis to the other at 150 miles an hour. The Tokaido Express which runs between Tokyo and Osaka in Japan is probably the best example of the application of present technology to a standard form of transport (Figure 20.13). The Express has also supplied some interesting statistics; traffic is growing at the rate of 26 percent per year; at the same time air traffic between Tokyo and Osaka has fallen from 22 to 6 percent of the travel load. This certainly is a boost for supporters of highspeed rail connections between nodes in the megalopolises of the United States. Such connections, it would seem, would also ease congestion at airports and perhaps reduce the pressure for more and larger facilities.

Fig. 20.13 • The Tokaido line connecting Tokyo and Osaka carries commuters at speeds up to 125 mph, making the 320-mile trip in three hours and ten minutes. (*Japanese National Railways*)

But to achieve even 150 miles an hour in a train requires complete rebuilding of the roadbed. With some modifications of the existing track the metroliner service between Washington and Boston has become not only possible but profitable, weaning commuters away from their cars. Speeds rivaling those in Japan must await further roadbed improvements, though. Surely some such system, metroliner or monorail, can provide an interim form of mass transit suitable for our sprawling megalopoeises until the new technology becomes available. But even interim systems require vast sums of money.

One thing is certain, however: there is no *single* answer to the transport problem in urban areas. To preserve the fabric of cities and restore some degree of amenity to their environment the ultimate solution will probably involve mass transport systems designed to move people rather than vehicles within the city and from selected collecting points beyond the downtown area. It will require beltways that allow traffic to bypass the limited capacity grid of city streets. It must offer some system of pedestrian and vehicular separation, either by converting districts or neighborhoods into pedestrian islands, limiting the type of traffic or the timing of its access on other streets, or putting vehicular traffic underground, giving pedestrians their proper place in the sun.

RIGHTS-OF-WAY

The movement of goods and people requires corridors or rights-of-way which traditionally have been devoted to a single use. No one would think of taking a casual stroll down an expressway or having a picnic on a railroad track. But there are other rights-of-way whose uses are not so clear-cut. The development of huge regional power plants has led to miles of power transmission lines. Pipelines, originally built to transport oil, now move natural gas, chemicals, and even coal, crushed and mixed with water.

The Alaska Pipeline

In 1968 huge deposits of oil and natural gas were discovered in Triassic sediments some one to two miles below the tundra of Alaska's North Slope. Roughly 10 billion barrels of oil await tapping and transport to an increasingly oil-hungry market (see Chapter 10). Since neither road, rail, nor convenient sea routes are available, a pipeline has been urged by the consortium of oil companies that is exploiting the field. The most direct route seems to run 125 miles over the North Slope, 90 miles over the Brooks Range, rising to 4800 feet at Dietrich Pass, then 350 miles over the interior plateau and across the Yukon River, 75 miles over the Alaska Range, crossing the Denali Fault, 100 miles across the Copper River basin, 40 miles over the Changach Mountains, and finally 40 miles to Valdez, where tankers will carry the crude to refineries in the Pacific Northwest, California, or to export markets. A staggering engineering feat considering the topography alone, the construction is complicated by permafrost (permanently frozen subsoil) over a portion of the route. Since the oil will have to be heated to at least $57°C$ to keep it from forming thick, waxy deposits on the inside of the pipe, there is the problem of melting the permafrost and rupturing the pipe. At least 83 million cubic yards of gravel will have to be excavated for the route, and of course roads constructed to allow the pipe to be transported and laid. Despite precautions in the crossing of known faults and rivers with highly variable flows, the prospects for future pipe rupture and contamination of a virgin wilderness with crude oil are great. Construction alone would be highly destructive.

But what of the gas? Since Alaska will not allow it to be wasted by burning and natural gas is in growing demand, a pipeline will have to be built to carry gas as well as oil to markets. But there are no facilities for gas liquefication in Valdez, nor are there special tankers available to carry liquefied gas to its markets; so the gas pipeline has been routed tentatively through Canada to the midwestern United States. Why not combine the two pipelines in a single corridor? This would avoid most of the mountainous country and major fault systems, and would eliminate the need for a fleet of tankers hauling oil and potentially polluting the northeast Pacific. But the combined Canadian corridor would be twice as long as the Alaskan, and in traversing more permafrost terrain and more wide rivers, massive oil spills must be anticipated. The issue is not easy to resolve. Clearly the oil will have to be exploited sooner or later, for we cannot assume a significant early decrease in energy demands, and pipelines seem to be the most logical way to transport the oil and gas. Environmentally, the issue seems to boil down to how much we can minimize the impact of both pipeline construction and potential oil spills.

For a lucid exploration of the economic background of the TAP see Cicchetti (1973). One proposed solution involves a new technique called the cold dispersion pipeline system. In this method, devised by S. S. Marsden at Stanford University, the oil would be suspended in a brine solution and chilled rather than heated. Such a pipeline would reduce the tendency to thaw the permafrost around the pipeline, would require refrigeration units only in permafrost areas, could carry natural gas as well as oil, and could be built on a narrow right-of-way. Since a few years' investment of intensive research into such new methods of pipeline construction and maintenance should certainly reduce the environmental costs of the pipelines, perhaps significantly, it would be tragic for the public to be misled into thinking that we must have the oil now, whatever the environmental cost.

Management of Rights-of-Way

Regardless of the commodity being transported, rights-of-way must be managed so that the service they perform is not impeded by the environment. The basic reason for highway plantings, you will recall, is to stabilize the right-of-way by preventing rock falls or mud slides. Powerline and pipeline rights-of-way, which are often roadless, are usually covered with vegetation of some sort that must be managed to prevent damage to the facility. When land in well-watered areas is cleared, the disturbed ground is soon covered by weeds, then grasses, shrubs, and finally trees in succession. The goal of right-of-way management is to maintain vegetation at some particular stage in this succession, usually the grass or shrub stage.

According to Frank Egler, a distinguished plant ecologist who has studied plant succession and its manipulation for many years, the direction of the succession is greatly influenced by the types of plants already present at the time of abandonment. If these can be manipulated, the end result can be both predicted and controlled.

The standard approach of power companies, pipeline operators, and highway departments has been to spray their rights-of-way with herbicides to kill trees and shrubs selectively, leaving only grass (Figure 20.14). This works well up to a point; shrubs and trees *are* temporarily eliminated. However, the sprayed trees and shrubs turn brown and become an unsightly fire hazard. Because the spray does not kill the roots of the unwanted woody plants they continue to sprout and spraying must be repeated every year or two. Even if these existing woody plants could be eliminated from the unnatural grasslands, seedlings would constantly be a problem.

Fig. 20.14 • When power lines follow line-of-sight routes, they can be very destructive of aesthetic values.

445

The blanket spraying operation presently favored by chemical companies and by many right-of-way operators has proven to be a mistake. Egler has demonstrated in trial plots and field observations that shrubs encouraged by the selective killing of competing trees form a much more stable ground cover than grass, and one which requires far less maintenance. In addition, shrubs provide a greater variety of browse and food for wildlife than grass. While highway rights-of-way are no place for grazing animals, other types of rights-of-way offer no such conflict of interest. Managed according to Egler's technique, a powerline or pipeline right-of-way would have a strip of low perennial plants—goldenrods and bracken ferns—which can keep out tree seedlings with a little selective spraying. Beyond this there might be another strip of low shrubs—viburnum, shrub dogwoods, blueberries, and low junipers—and beyond that on the forest edge there might be alders, shrub willows, dogwood, and redbud. The particular species would, of course, vary from place to place as with highway shoulders and medians, but the idea of using a variety of plants from the natural flora would be the guiding principle everywhere.

Instead of the expense of spraying large amounts of herbicides over the entire roadside, shrubs could be maintained indefinitely by one man walking the right-of-way once every few years with a knapsack sprayer. Once the shrubs are established, maintenance may even be unnecessary, for seedlings of potentially competitive trees would be shaded out. One section of a powerline right-of-way in New York State covered with a shrub called sweet fern has had no maintenance for almost forty years. Since Egler estimates that close to 10 million acres in the eastern United States alone are being used as rights-of-way, any program that would increase their use by wildlife or even people would certainly be worth looking into in a progressively park-hungry world. The reason for the delay in applying ecological principles to rights-of-way management ultimately lies in ignorance and apathy among the utility companies who, encouraged by chemical companies, seem to view every shrub as a potential tree.

Powerline Aesthetics

Power lines, unlike pipelines or roadsides, have one major aesthetic liability: their ugly towers, resembling overgrown erector sets may completely dominate the countryside they traverse (Figure 20.15). In 1967, 7 million acres were taken up by powerline rights-of-way, and this figure is likely to triple in the near future unless alternatives are found.

The obvious solution would be to put the lines underground, like the utility lead-in wires of a house. But high-voltage lines are expensive to bury using present technology. If direct current transmission, which has much lower line loss, were used, it would have

Fig. 20.15 • Power-line pylons can be designed to be less obtrusive. These two pylons share a common right-of-way in Delaware.

to be inverted back to alternating current for distribution, requiring expensive equipment. These problems are being researched and by the year 2000 gas dielectric, resistive cryogenic, and superconducting systems now under development will allow most power transmission lines to be placed underground.

Until that time the Hudson River Valley Commission, charged with preserving scenic and aesthetic values in the Hudson River valley, recommends the following guidelines for preserving scenery that might be marred by power lines:

1. Avoid, where possible, prominent scenic features.
2. Follow lower slopes or valleys between hills.
3. Do not cross the contours of hills at right angles and avoid the steep slopes that expose the power line to view.
4. Jog alignments when crossing major roads to reduce the view of the line from the road.
5. Bend the lines to follow the edges of different types of land use rather than plowing through the complex.
6. Route lines along existing pipeline and railroad rights-of-way wherever possible.

The unsightly stringing of power lines across rivers can often be avoided by using existing bridges and either running the wires underneath in conduits or stringing them alongside on outriggers. Finally, the towers now fulfilling engineering requirements but ignoring aesthetic ones could also be designed to harmonize better with the countryside.

If power needs continue to double every ten years, we shall have still more power lines. If they are constructed with some of these considerations in mind with regard both to the appearance of their

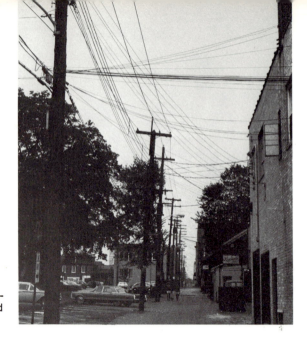

Fig. 20.16 • Utility wires in towns, even where confined to alleys, add a note of busy confusion and are very subject to ice storm damage.

hardware and to the environment through which they pass, then many of the lines to be built in the next thirty years can at least be made as inconspicuous as possible, until technology can give them a decent burial.

Out of sight, out of mind

Although power lines, because of the high voltage which they carry, resist underground placement for the present, the same is not true of the 220-volt lines that enter our houses or the lines strung up and down our streets (Figure 20.16). Fortunately large cities were driven at an early date to place their wires underground (Figure 20.17), simply because of the vast numbers of lines. Many new communities around the country are being constructed using underground

Fig. 20.17 • Large cities were forced to eliminate overhead wires many years ago, as this 1881 cartoon suggests.

utility lines from the start, as a selling point. The cost is either being shared by contractor and utility or is being passed on to the buyer in the purchase price. One factor that prevents older neighborhoods from switching from overhead to underground lines is that the cost would have to be borne by the householder directly, a burden that many would view as unnecessary, particularly when they are also faced with the replacement of septic tanks by expensive community sewage systems, in a nationwide effort to combat water pollution.

Unfortunately, we have grown accustomed to seeing utility poles running along streets and highways, much in the same way as we have grown up with the sound of airplanes in the sky overhead. It is probable that if both were to disappear, most people would never know the difference, particularly when the landscape is ugly to begin with (Figure 20.18).

Fig. 20.18 • (a) In older housing developments utility wires were strung along on poles as suited the convenience of the developer or utility. (b) Today most new subdivisions put their utility lines underground, which allows the planting of trees that soon soften the aspect of the ugliest neighborhood. (*Photos by AT&T*)

We have long grown accustomed to thinking that the ugliness and destruction that result from transportation are inevitable, the price we must pay for progress. This is part of the old struggle-against-nature syndrome, which seeks to overwhelm the environment by the brute force of technology and then to make that technology a convenient scapegoat for the environmental chaos that has resulted. Whenever the interaction of environmental factors is understood, technology can smooth the way and lead to a resolution of potential man-environment conflicts. Each of the transportation modes discussed previously can be fitted into the natural environment with minimum adverse effects if we can develop an ecological sensitivity that allows the environment some basic rights. If these rights are protected and if technology gives us the ample means of doing this, we can coexist with nature despite all our roads, airports, and transmission lines. But if we continue to ignore the modest demands that the environment makes upon our use of it, then all the technology that we can devise will not suffice to put right the endless problems that will be generated.

FURTHER READING

Baldwin, M. F., 1970. *The off-road vehicle and environmental quality.* The Conservation Foundation, Washington, D.C. Excellent review.

Cicchetti, C. J., 1973. "The wrong route." *Env.* **15**(5), pp. 4–13. Concise economic review of the problem with an evaluation of some alternatives.

Gruen, V., 1967. *The heart of our cities.* Simon & Schuster, New York. Absorbing account of the potential still left "downtown," by one of America's foremost city planners.

Howlett, B. and F. J. Elniger, 1968. *Powerlines and scenic values in the Hudson River Valley.* The Hudson River Valley Commission of New York, Tarrytown, New York. This well-illustrated little booklet contains much sound advice on the placing of powerlines.

Hummerstone, R. G., 1972. "Cutting a road through Brazil's 'Green Hell.'" *The New York Times Magazine,* 5 March. Some case history background on a development reminiscent of the laying of transcontinental railroads in the United States about 100 years ago.

Orski, K., 1973. "Fast trains." *Env.* **15**(3), pp. 6–10. An up-to-date look at the possibilities of rail transport over the next twenty years.

Owen, W., 1966. *The metropolitan transportation problem.* Rev. ed. The Brookings Institution, Washington, D.C. Critique of the automobile as the solution to all transportation problems.

Rapuano, M. et al., 1968. *The freeway in the city*. U.S. Department of Transportation Report, Sup. Doc., Washington, D.C. With planning, highways need not destroy cities.

Robinson, J., 1971. *Highways and our environment*. McGraw-Hill, New York. A handsomely illustrated and thorough survey of the theme.

Siddall, W. R., 1967. *Transportation geography, a bibliography*. Rev. ed. Kansas State University Library, Manhattan, Kansas. Fine bibliography on all phases of transportation.

21

SOLID WASTES: MIDDENHEAP INTO MOUNTAIN

As IF our transportation-induced problems were not enough, we must also cope with the staggering quantity of solid waste materials generated by an increasingly affluent society that habitually attempts to throw away the unthrowawayable. Man's first communal living areas were probably caves, well sheltered from the weather and potential enemies, man or beast. Solid wastes—excrement, bones, old tools, charcoal, and ashes—were quite literally thrown away. When the resulting piles or middenheaps got a bit lumpy underfoot, or became too offensive, it was a simple matter to cover up the mess and start all over. After a few hundred years a small cave would fill up, forcing the occupants to find new quarters, but some of the larger caves accommodated thousands of years of continuous occupation and hundreds of feet of wastes.

The cities, though they came much later in man's history, were hardly more subtle in their handling of solid wastes. Discarded materials were freely scattered just outside the city wall or around the dwellings. When destroyed by fire or war, ancient cities were often rebuilt upon their solid wastes and rubble. Over hundreds or thousands of years, the once-level sites of cities slowly became mounds, often abandoned and covered with sand or soil, but occasionally still inhabited by a few hundred people perched above the surrounding countryside on their huge waste heap (Figure 21.1).

An astonishingly complete historical record of daily life in caves and cities can be pieced together by careful archeological digging. But exploration of past cultures by examination of their waste is by no means limited to prehistoric times. Williamsburg, Virginia was restored after very careful archeological scrutiny uncovered the nature of building materials, techniques, and tools, deduced from the

Fig. 21.1 • This mound in Iraq represents 3500 years of human occupation and originally stood seventy feet above the surrounding plain. (*The University Museum, University of Pennsylvania*)

scraps and wastes associated with the construction of every building. Excavations in the older parts of eastern seaboard cities often bring to light old bricks, pieces of bottles, and crockery that tell much about the past. New York's Central Park, once a shanty town of squatters' huts, has pieces of glass and crockery scattered throughout, some dating from the 1840s and 1850s.

TODAY'S WASTE

In many ways, however, the solid wastes with which we are faced today are quite different from those of even the recent past. There is, for example, a much greater variety of more stable wastes. A simple glass bottle can at least be broken to take up less space, and a steel can will rust away in a few years; but a plastic bottle or an aluminum beer can lasts indefinitely. Even more serious is the rate at which solid wastes are accumulating. The average person throws away five pounds of wastes per day, or 1800 pounds a year, 121 pounds more per person in 1966 than in 1958; and by 1976 it will be 136 pounds more than in 1966. Of the 73.5 million tons of waste expected to be produced in 1976, only 7 million tons will result from growth in population. The next 15 million tons will represent the wasteful consequences of the packaging revolution.

We have come a long way from a plain brown bag or newspaper-wrapped fish; today and increasingly in the future, items will be packaged "for the convenience of the consumer." Take for example the plastic bubble packs hung on hooks by the thousands in all stores. The merchandise, a few nails or an electric socket, is

453

plainly in view, but the package usually serves as advertising space for the manufacturer; it is less handy to carry, is difficult to open, and its contents surely cost the consumer more than the same item bought unpackaged. The so-called convenience foods with their aluminum foil covers and pans, pressurized cans of just about every liquid—all add immeasurably to the trash load to be disposed of and to the cost of the material sold. Those tiny picnic-sized containers of salt, for example, inflate the price of the product to eight times the price of salt bought by the pound.

The full impact of the packaging revolution, which began after World War II, can be appreciated only when the cost to the public is calculated. In 1960 it was estimated that the average American family spent $500 a year just for packaging. In 1966 the disposal of wastes cost close to $375 million a year; 75 percent of this was spent for picking up and transporting household wastes to their disposal site. If you include the costs of air and water pollution from present disposal practices and the loss of the space that present landfill techniques are rapidly usurping, the final cost of packaging wastes is even greater than it appears.

The Kinds of Wastes

Of course, discarded packages are not the sole ingredient of household wastes, any more than household wastes represent all wastes produced by contemporary society. Before we look at disposal techniques we should have some idea of which wastes are causing the biggest problems and why. Since industrial wastes have already been considered in Chapter 8, our prime focus here will be consumer, rather than producer, generated wastes.

What is the nature of household solid wastes? One breakdown has been given by an American Chemical Society report (Table 21.1). Notice that the largest category, miscellaneous paper, accurately reflects a part of the packaging revolution.

Table 21.1
Breakdown of Household Wastes in the United States*

WASTE	WEIGHT PERCENT	WASTE	WEIGHT PERCENT
Cardboard	7	Leather, rubber,	
Newspaper	14	molded plastic	2
Misc. paper	25	Garbage	12
Plastic film	2	Grass clippings, dirt	10
Wood	7	Textiles	3
Glass, ceramics, stone	10	Metal	8

* Subcommittee on Environmental Improvement, 1969. "Cleaning our environment, the chemical basis for action." American Chemical Society, Washington, D.C.

At one time waste paper, particularly newspaper, was carefully saved by most households to be sold to the rag man for small but welcome change. Later it was given away to neighborhood charity collection drives, and now it is thrown away with the trash because no one will cart it away. What has happened? The demand for waste paper by the paper industry has certainly declined: in 1945, 35 percent of the waste paper generated was reused; by 1966 this had fallen to 21 percent, and by 1980 less than 18 percent is likely to be reused.

Paper

The slackening demand has probably been due to the increased efficiency and lower cost of harvesting virgin pulp, but also the small margin of profit in handling waste paper, resistance to change on the part of large paper users who might use recycled paper, and the financial interests that large newspapers have in virgin pulping operations. These factors have been matched by a sharp increase in the cost of handling the collection, transport, and processing of waste paper. The packaging revolution has contributed mightily to the pileup of waste paper because of the huge increase in volume and the contamination of paper with aluminum and plastic coats. Furthermore, special inks and treatments, because of their variety, can be extremely expensive and difficult to separate from the useful paper pulp. Some success has been achieved in efficient and inexpensive de-inking of waste newsprint, but fancy paper wrappings and containers frequently are not worth the trouble. To make matters worse, special coatings, however attractive to the consumer, retard the natural decomposition of waste paper in landfill dumps and lead to delays in reuse of the land.

The tin can, which has been a mainstay of the food industry for over 100 years, is made not out of tin but steel, covered with as thin a layer of tin as the manufacturer can economically apply. Tin does not react with the contents of the can and has a pleasing shiny surface which, unlike the underlying steel, does not quickly rust. Although tin cans have been around for many years, their life in the natural environment is relatively short. The thin tin coat weathers away in the open, exposing the underlying steel to fairly rapid rusting. In humid climates most tin cans will rust away in a few decades— sometimes much less time (Figure 21.2a). Campers have long made it a habit to throw their tin cans into the campfire before burying them because this helps remove the tinned layer and exposes the cans to rusting more rapidly.

Cans

Ironically, tin coating, which allows rustable steel to be used and so predisposes the can to environmental decay, made the cans virtually worthless as iron scrap because of the contamination with tin, which is a problem, especially with the newer basic oxygen steel-making furnace. For this reason cans are thrown away rather than salvaged.

However, there is one growing use for scrap cans. With ever stricter controls over smelting effluents and with the decreasing qual-

Fig. 21.2 • (a) As objection-
able as they are, tinned steel
cans *do* rust in a few years
and disappear. (b) Aluminum
cans, however, persist indefinitely.

ity of copper ores in the United States, the copper industry is shifting from smelting the copper by roasting the ore to dissolving the copper from its ore. This is done with sulfuric acid passed over shredded tin cans. The iron in the cans, which is more chemically active than the copper, replaces copper in solution. The copper is precipitated out on the cans, collected, and refined. The remaining iron is still waste, of course, so the process can hardly be called recycling. But at least some use has been made of something presently regarded as useless. Over 100,000 tons of cans per year are used in this process, mostly in the Southwest. However, this absorbs only 2 to 3 percent of the available waste can supply, and at best in the future will only take care of a sixth of the cans discarded each year.

But the tin can now seems to be on the way out because chrome and resin-coated steel has become cheaper than tinned steel and is more acceptable as scrap to the steel mills. Aluminum has also made great strides in the container market. As with the new steel cans, aluminum cans can be reused as scrap but they are so readily discarded that their collection is unlikely. Once thrown away piecemeal, their retrieval expense is as much as thirty cents a can. While the new steel cans will corrode if the tough chromium or resin coats are broken, the aluminum can is practically indestructible when dumped into the environment (Figure 21.2b). At present 25 percent of the aluminum produced in the United States each year winds up as trash.

Whatever the change in metal cans, glass as a packaging material will endure. Since 75 percent of glass is silica sand, the most abundant substance in the earth's crust, cost of raw material is low and likely to remain so. At the present rate of bottle production, sand reserves are calculated to last around 3 billion years. However, some glass companies have expressed concern that, should bottle production increase significantly in the future, the sand supply might become dangerously depleted a little over a billion years from now.

Glass bottles are reasonably destructible. Despite the danger from broken glass, breaking at least reduces the bulk of a glass bottle or jar. But volume is not so much a problem as number. Once, most glass bottles were stoutly built with the intention of their being used many times. It was later realized, however, that bottle return is a nuisance to the consumer, the middleman, and the bottler, who must at times clean some incredibly dirty bottles.

A few years ago some merchandising genius discovered that for every returnable bottle, used twenty times on the average, twenty nonreturnable bottles could be sold; this discovery resulted in the "disposable" bottle. In 1959, 2.7 percent of the bottles produced were nonreturnables; in 1967 their slice of the bottle market had grown to 13.5 percent. Yet the disposable bottle is just beginning to be used; in 1966, 65 billion bottles were filled but only 28 billion bottles were produced. In the next few years far greater quantities of

Bottles

bottles will be added to the disposal problem. But if the total energy requirements for returnable and nonreturnable bottles are examined, disposable bottles are not nearly as cheap as has been supposed (Table 21.2).

Responding to the problem of roadside litter (66 percent of roadside trash is bottles and cans), Oregon passed a law in 1972 prohibiting the use of nonreturnable cans and bottles in the state. Bottles now carry a two- or five-cent deposit, and pull-tab cans were outlawed altogether. As a result of this legislation, roadside litter has declined by 81 percent in Oregon. Despite opposition by manufacturers and brewers, similar legislation is being considered in many other states and by Congress as well.

Table 21.2
*Energy Expended (in BTUs) for One Gallon of Soft Drink
in 16-ounce Returnable or Throwaway Bottles**

OPERATION	RETURNABLE (8 fills)	THROWAWAY
Raw material acquisition	990	5,195
Transportation of raw materials	124	650
Manufacture of container	7,738	40,624
Manufacture of cap (crown)	1,935	1,935
Transportation to bottler	361	1,895
Bottling	6,100	6,100
Transportation to retailer	1,880	1,235
Retailer and consumer	——	——
Waste collection	89	468
Separation, sorting, return for processing, 30 percent recycle	1,102	5,782
TOTAL ENERGY EXPENDED IN BTUS PER GALLON		
Recycled	19,970	62,035
Not recycled	19,220	58,100

* From Hannon, Bruce M., 1972. "Bottles, cans, energy." *Environment* 14(2), pp. 16–17.
Copyright © 1972 Scientists' Institute for Public Information.

Fortunately the glass industry is not oblivious to the problem it is generating. Several kinds of self-destructing bottles are being tested by industry-sponsored research. One is a bottle that can withstand normal droppage but when struck hard fractures, like auto safety glass, into harmless granules which are much more easily disposed of than the more bulky bottle. Broken glass of any sort is a good landfill material because it is dense, less subject to settling, and is insoluble. A second approach to disposable bottle-making is the production of a water-soluble glass, covered inside and out with a waterproof layer to last out the shelf life. When the layer is broken the glass would quickly dissolve. Perhaps we shall end up flushing soda bottles down the sink; if so, the problem will not really have been solved. Instead of getting rid of a pound of solid glass bottle, we would ask a sewage plant to get rid of a pound of synthetic material dissolved in the sewage effluent.

One enterprising use of waste glass has been in the production of glassphalt, a paving material that uses ground glass as the aggregate. This has the advantage of being available in large quantities in urban areas where other forms of aggregates are scarce and expensive, and being amenable to cold-weather paving when asphalt is difficult to work.

Throwing away tin cans or bottles is one thing; throwing away a junked car is quite another (Figure 21.3). At one time the problem was unknown, for as recently as the 1950s one could get $40 or $50 for an old car. Wreckers could make their profit by stripping off useful parts and valuable scrap, because a typical car contains sixty-six pounds of aluminum, thirty-seven pounds of copper, thirty-five pounds of zinc, and twenty pounds of lead, and then he could squash the remainder into a small bale worth about $40. The steel industry at the time used a 50:50 mixture of scrap steel and iron ore to make new steel and consumed over 40 million tons of scrap in the process. But conversion in the steel industry to the use of the more efficient oxygen furnace changed the bright picture for scrap steel. Not only did the new furnace use only one-half the amount of scrap steel as the old one, but the scrap had to be much cleaner, with a much lower content of metals other than iron, especially copper.

Automobiles

Fig. 21.3 • Abandoned cars are bulky and difficult to store before being recycled.

459

By the early 1960s the new urban phenomenon of abandoned cars was causing consternation. The decrease in the value of a ton of scrap to around $10 plus costs of $12 a ton to bale and another $10 a ton to transport made a dying auto worth $5 at most. Rather than pay $25 to haul a defunct car to a junkyard, many owners elected to leave them where they stopped running. In most cities all valuables are stripped in a few days and the car is often burned to a rusty wreck shortly afterwards (Figure 21.4). In 1965, 190,000 cars were abandoned in this fashion, close to 50,000 in New York City alone.

Fig. 21.4 • Stripped and burned, this car slowly rusts and accumulates trash that is only sporadically removed by the city. (*Photo by Dan Jacobs*)

The slack market for scrap steel together with the rising labor cost of separating nonferrous from ferrous elements has caused a pileup of junked automobiles that can no longer be ignored; the junked autos seem to be everywhere, not just in junkyards, but on roadsides, in farmyards, city lots, and back yards.

Technology is finally rising to the challenge, and several impressive car-eating machines, capable of dealing with the backlog, have been developed by large scrap firms. One such machine can chew 1400 cars a day into small pieces which can be efficiently sorted into steel and nonsteel piles by magnets. This scrapyard separation produces a grade of scrap that is much more salable than the customary bales of mixed metals.

Regardless of the handling techniques, two problems remain: the cost of transporting junked cars to modern regional processing yards, and air pollution from burning the cars, which has been thought necessary to eliminate plastic, rubber, and upholstery materials. The first problem might be handled by small hydraulic

Fig. 21.5 • A novel approach to the disintegration of automobiles is to crush them with this 24-ton juggernaut. (*Wide World Photos*)

crushers that can flatten a standard sized car to a thickness of one foot (Figure 21.5) and thus make it possible to ship large numbers of cars by flatbed truck or rail car to regional collection centers. One alternative has been the development of portable car eaters which work their way around a region, junkyard by junkyard, eating as they go. Air pollution from burning out cars has been eliminated by reducing cars to small pieces of scrap in a powerful hammermill, leaving the nonmetals to be recycled, used as landfill, or incinerated. This approach would tend to concentrate this phase of the business in regional centers that, because of their business volume, could afford the necessary antipollution equipment.

Usually by the time a car is scrapped the tires have been removed for salvage. But most cars in their lifetime run through at least three or four sets of tires and these too create disposal problems: altogether, 100 million tires are discarded every year (Figure 21.6). Because many states have minimum tread requirements enforced by yearly inspection, tires are often thrown away with a worn tread even though over 80 percent of the rubber in the tire remains. Some are recapped, but more and more people who have experienced poorly recapped tires prefer new tires. In contrast to the 10 percent of tires recapped in the United States, 80 percent of all tires are recapped in less affluent foreign countries. In addition, reclaimed rubber processing for various purposes has fallen from 30 percent in 1945 to 10 percent or less today in the United States.

Most old tires end up in landfill, for which they are poorly suited. Occasionally they are burned as a means of protecting citrus orchards from frost or for burning green timber in right-of-way clear-

Tires

461

Fig. 21.6 • Most discarded tires end up as just another item in the general debris that is slowly accumulating in all environments.

ance. But burning rubber in the open air causes about the thickest, blackest smoke of any combustible and is outlawed in many places.

One promising solution to the pileup of discarded tires is being developed by Firestone Rubber and the United States Bureau of Mines. Shredded tires are placed in a closed vessel and heated without air to 500°C. At this temperature the rubber yields 140 gallons of liquid oils per ton of scrap that are similar to and as useful as those from coal. Also 1500 cubic feet of gas with a heating value equivalent to natural gas is produced. The residue solids have a very high carbon content that might be useful as a basis for other products.

YESTERDAY'S DISPOSAL TECHNIQUES

Despite some progress in the handling of a few of the major waste categories, techniques for disposal of the overwhelming mass of solid wastes have hardly been improved since Roman times. The town dump is still the site that receives all the solid wastes, usually dumped into a gully, canyon, or quarry pit. Such open dumps are infested with rats, mosquitoes, and flies. They often catch fire and smolder, producing foul-smelling smoke, and the trash tends to be blown about by the wind.

Landfill

One step beyond the open dump is the so-called sanitary landfill (Figure 21.7). Of the 12,000 landfill sites around the country, less than 6 percent qualify for the term sanitary; the rest are little more than open dumps. To carry out sanitation efficiently, it is necessary for a bulldozer to compact each truckload as it is dumped, then for a dragline operator to quickly cover it with earth, and then for another bulldozer to compress and smooth the site to make room for the accommodation of another load of waste.

The advantage of a properly maintained sanitary landfill is the prompt burial of garbage and trash before it has a chance to catch fire, be blown away, or attract vermin. A good site can accommodate a great deal of trash and can be situated on old strip mines, sand and gravel pits, or abandoned quarries which are presently wasted land found in almost every state. When the site is filled to capacity, a two- to three-foot layer of earth is laid over the last trash layer, and a park or recreational area can be created.

Fig. 21.7 • In a properly run sanitary landfill, trash is covered with earth every day by a bulldozer. This reduces breeding sites for vermin and eliminates danger of fires.

Since only 6 percent of the landfill sites qualify as sanitary, obviously the ideal conditions described are not often achieved. Too often trash is not covered immediately but once a day or every other day. Much paper and light trash gets blown about and the site with its surrounding properties frequently becomes an eyesore. Most sites that are selected do not need the improvement of sanitary landfill; salt marshes, for example, which are favorite landfill sites along the coasts, are valuable ecologically and aesthetically, and yet they are in danger of complete elimination, simply because they are handy and "empty" (see Chapter 7). Further, the anaerobic decay of the buried trash causes low and erratic settling, thereby making construction of buildings unwise for many years after the site has been filled. When a landfill site is unsuitable, the water table may be contaminated with obnoxious seepage from the buried trash, endangering nearby wells.

However, sanitary landfill can be a useful disposal technique, especially for small towns, if well-sited on true wasteland, and scrupulously tended to—conditions much more easily promised than realized.

Incineration

Another ancient disposal technique for getting rid of waste is to burn it. In its primitive form, incineration swaps one problem, solid waste disposal, for two, air pollution and solid waste disposal. When anything burns, particularly a heterogeneous mass such as trash, smoke is always produced in great quantity; after the combustibles are consumed the noncombustible glass and metals remain to be carted away, usually to a landfill site or out to sea.

Their poor design and operation make most incinerators, whether backyard, apartment house, or municipal, simply roofed burning dumps. When the new air pollution standards were set up, cities were forced to close down incinerators, because such equipment could not pass their own standards. Apartment house and office building incinerators, although handy for occupants and welcomed by city sanitation workers, pour large quantities of soot, fly ash, and smoke into the atmosphere.

Another pollutant being generated in increasing quantities by incineration is exceedingly corrosive hydrogen chloride which is released when polyvinyl chloride (PVC), a component of plastic, is burned. Only 36,000 tons of PVC were burned in 1966, but over 1 million tons were produced in the diverse forms of garden hose, rainwear, shoe soles, floor coverings, and so on. Since 50 percent of the PVC is chlorine, the potential combustion of this plastic in the future makes emission control of incinerator stacks imperative.

Incinerators can, however, when properly designed and operated, cope with large amounts of combustible trash. Many European cities have been able to combine disposal of trash with generation of heat, and have put to use the thermal energy that in United States incinerators goes up the chimney as waste. One reason that American cities have lagged behind Europe in incinerator technology has been the availability of cheap fuel in the United States compared with its shortage in Europe. A power company in New York City can get quality coal for $5 a ton to generate heat for electricity; in Munich, coal may cost $20 a ton, which inspires more creative efforts to use alternate sources of heat. But as air pollution regulations become more strict and fuel becomes scarce, more and more attention will be given to designs which both eliminate smoke release and use the heat produced to subsidize the cost of antipollution devices. One useful proposal is to build small incinerators capable of handling the solid wastes of about 150,000 people, locating these strategically so that transportation costs could be cut. The steam could be used to generate electricity, easing peak period power loads or perhaps in a desalinization scheme if the city were located in a dry climate.

The 20 million tons of fly ash that are potentially precipitable from incinerator chimneys makes extremely fine bricks and concrete, but because building supply firms seem unaware of this raw material, and incinerator operators seem to have little ability to push a resource, most fly ash either goes up the chimney or, if precipitated out, is dumped as fill along with the noncombustible solids.

One tin can or piece of copper wire is worth very little; but the 125 million tons of incinerator ash produced each year contain 9 million tons of metal worth $400 million! The metal in some ash piles has been valued at $12 a ton, making it a more valuable source of metal than many ore deposits. But this metal continues to be dumped as trash since its recovery is presently uneconomic; perhaps in a few decades the trash heaps of cities will be eagerly mined for their valuable supplies of metals.

Composting

Another old method used widely in Europe but much less frequently in the United States is composting. Compost is the rich, dark remains of organic garbage which has been aerobically decayed, losing its odor in the process, and thus making a valuable soil conditioner of materials previously buried in landfill projects.

Despite their obvious advantage in making an otherwise wasted resource useful, composting plants have some drawbacks. Often poorly designed, like incinerators, they produce odors which make them unpopular neighbors. To be properly run they need reasonably pure organic garbage, which means that such garbage must be isolated in the homes of the district served and picked up independently of paper and solid trash. Tests show that less than 10 percent of households involved in such a program seem able to keep their garbage uncontaminated. This means that the garbage must be sorted before composting, or screened afterward, which adds considerably to the processing expense. In addition, most people have a misconception of the nature of compost. Like vermiculite and perlite it is a good soil conditioner; it loosens the texture of heavy soil and increases the absorptive capacity of sandy soil. It is, however, rather low in nutrients and cannot be considered as a fertilizer. Often composting plants have difficulty in locating a steady market for their product even if it is dried and bagged, which has led to abandonment of the process by many cities.

Responsibilities of Solid Waste Disposal

When we look at the performance of industry in squeezing the last bit of iron from low-grade taconite deposits or the last hydrocarbons from crude oil, we are impressed with the level of technological skill coupled with economy of operation. Until recently this expertise ended with the marketing of the product. Of course some firms have always stood by their products with warranties, or double-your-money-back guarantees, but in no instance did the concern of the manufacturer extend beyond the reasonable life span of his product. Today, however, population growth together with the packaging revolution has begun to overwhelm our antiquated system of waste collection and disposal. Many manufacturers and industrial organizations are coming to the conclusion that they bear some responsibility for the design of their products beyond the point of successful marketing. But a great part of any changing attitude is due to the realization that if manufacturers do not volunteer changes that will help the nation to cope with the rising tide of waste, they will soon be *required* to make changes. The detergent industry, which switched in 1965 from nonbiodegradable to biodegradable surfactants, did so because of the threat of state laws outlawing the older type of detergent. It is not a coincidence that the glass industry has begun to explore the possibilities of self-destructing bottles at a time when dozens of state legislatures have considered banning, limiting, or taxing no-deposit-no-return bottles. But industry is no more to blame than a federal government which was willing to spend $300 million a year on chemical and biological warfare research, whose need is questionable, while it was spending only $15 million on the study of solid waste disposal systems, which we need desperately.

An Archaic Disposal System

Although the nascent efforts of industry to make more disposable disposables are commendable, the waste handling system is badly flawed at both ends: initial collection and ultimate disposal. As we saw earlier, most of the expense of waste disposal lies in the cumbersome, noisy, and ineffective method of carting by trucks. Many years ago human wastes were deposited in outhouses, which were cleaned out periodically by a man with a horse and wagon. But today in cities, human wastes are deposited in receptacles connected by sewers to central disposal plants. Can you imagine emptying your chamber pot at the curb on the way to work in the morning? Yet as we pause to watch the space stunts on TV, our beer and soft drink bottles go into a trash can that is emptied into a truck and hauled away to some dump or landfill.

NEW APPROACHES

Approached from this point of view, our whole system of waste collection is badly out of date. Buying a fleet of shiny new trucks which make less noise and hold more trash is no more an improvement than replacing a horse and wagon with a truck. Plainly what is needed is a solid waste pipeline analogous to the sewage waste disposal system now accepted as a rather basic element of urban life. One pilot system worked out in Sundeberg, a suburb of Stockholm, Sweden, uses a vacuum sealed pipeline that collects the solid wastes from 5000 apartments. Since the pipe openings are twenty to twenty-four inches wide, most household wastes are easily accommodated. The air stream in the pipe pushes materials thrown into the system along at fifty feet per second to an incinerator over a mile and a half away. The incinerator then supplies heat to the apartments in winter. The ability of the system to move heavy materials was graphically demonstrated recently when an automobile battery was found in the incinerator. The great advantage of the Swedish system is the elimination of presizing; virtually everything can go down a two-foot-wide pipe. The drawback is the difficulty of moving solids by vacuum tube over distances much greater than a mile or two.

A research team at the University of Pennsylvania has overcome this distance difficulty by using a water slurry system to carry wastes. Using a much smaller pipe than in the Swedish system, wastes are pulped and shredded before being transported. Solid wastes would be fed into a combination pulper and shredder, which would chew paper, cans, bottles, bones, and whatever, into small fragments which could then be flushed through an independent system to a plant where the waste could either be incinerated or dried and economically shipped by rail to outlying landfill sites.

More attention is being given to the reclamation of fibers from such slurries. Paper fibers a few thousanths of an inch in diameter are often one-sixteenth to one-eighth of an inch long. Because of these unique qualities these fibers can profitably be reclaimed from other contaminants by fine screens. As soon as the technology can be worked out, fiber recovery will have much promise in the reduction of solid waste volume.

Recycling Wastes

But even when the transportation problem is solved we are still faced with the problem of what to do with the wastes, for all we have really done is to collect them and to reduce their volume somewhat. The key to the whole problem of solid, liquid, or gaseous waste disposal is not disposal at all, but recycling. We still have the old caveman mentality about wastes; we think that all we have to do when the local environment fills up with wastes is to throw them a little farther. So we build our smokestacks higher, carry our nerve gas canisters farther out to sea, and truck our garbage to the next town. Now the signs are everywhere that our total environment is beginning to fill up. By the end of this century, a slim twenty-five years from now, there may be no more landfill sites and the sea may be too polluted to receive any more trash. Then what can we do?

The soundest approach is to develop systems capable of handling the entire waste stream. One partial recovery system being developed by the Bureau of Mines first incinerates the wastes, reducing the volume significantly. The residues are then screened to remove ash and dirt. Then ferrous metals are removed by magnets and the remainder of the residue is pulverized by a hammermill and screened to remove glass. Aluminum is separated from the other metals by suspension in a liquid flotation apparatus. The glass isolate can be sorted by a high intensity magnetic field, which responds to the ferric impurities in glass, or by a photocell coupled to small air jets. Systems such as this have much promise in handling a continuous flow of the diverse waste materials generated in urban centers.

But the central difficulty with recycling has always been that wastes by definition are worthless, certainly not worth spending any money on reclaiming. This is in part the result of our current level of affluence and the apparent inexhaustibility of sources of raw materials. Ultimately, of course, certain raw materials will become scarce and today's waste will be tomorrow's resource. Can the natural environment survive that often postponed date? The evidence now apparent to us all indicates that it cannot. We cannot wait for the valueless to become valuable, as in time it must; we must deliberately *place* value upon the valueless, by whatever means, regardless of the tyranny upon our social system, bearing in mind that the environment has until now borne the brunt of our disregard. Virtue is never easily inculcated, and recycling materials in our urban environments will not come easily either.

Numbers of concerned citizens have struggled valiantly to spur the recycling of wastes by laboriously collecting bottles, paper, and aluminum cans; then sorting, baling, or crushing the materials, and attempting to find firms that will take them. As commendable as these efforts are, they are doomed to failure because the effort is applied to the wrong end of the problem. Government must be made aware of the role it must play in making the whole concept of recycling economically viable. Once this is done, entrepreneurs will quickly spring up to turn a buck by recycling materials. Also, business must adopt a new design ethic that will include not only the economics of manufacture and aesthetics of the finished product, but the ease with which the product can be recycled. Automobiles could be much more easily recycled if this were considered in the original design.

One approach government might follow could be a special tax on virgin materials, which would heighten interest in raw materials currently regarded as waste. This would both conserve the virgin materials and reduce the waste burden. Another approach might be to set up tax incentives for industries that find new ways to use wastes associated with either the manufacture or ultimate disposition of their products. Still another would be to eliminate the inequity in transport costs between scrap and virgin materials. Recycling is hardly new. Even today 60 percent of the lead and 40 percent of the copper in use is from scrap. As population levels off, as inevitably it must, 90 to 100 percent of most currently used materials will have to be recycled, just as living organisms have recycled the water, oxygen, and carbon dioxide for the past billion years. In following suit with our man-made refuse, we will finally be returning to that steady-state environmental balance known to all other organisms on our planet.

FURTHER READING

Anonymous, 1972. "Degradable plastic lids are now commercial." *Chem. & Eng. News* **50**(25), p. 32. Notes the intriguing possibilities opened by biodegradable plastics.

Blake, P., 1964. *God's own junkyard.* Holt, Rinehart and Winston, New York. Fine collection of photographs illustrating our messy environment.

Carr, D. E., 1969. "Only the giant car-eater can save us." *The New York Times Magazine,* 4 May. Good review of the waste car problem.

Darnay, A. J., 1969. "Throwaway packages—a mixed blessing." *Env. Sci. & Tech.* **3**(4), pp. 328–333. The packaging revolution comes home to roost.

Dickson, E. M., 1972. "Taking it apart." *Env.* **14**(6), pp. 36–41. A case for designing products for their ultimate recycling.

Grinstead, R. R., 1972. "Machinery for trash mining." *Env.* **14**(4), pp. 34–42. Discusses some of the current solid waste disposal/recycling techniques.

Hannon, B. M., 1972. "Bottles—cans—energy." *Env.* **14**(2), pp. 11–21. An economic analysis of recycling containers versus disposing of them.

Lewicke, C. K., 1972. "Glass recycling makes strides." *Env. Sci. & Tech.* **6**(12), pp. 988–990. Some interesting statistics on the sources of waste glass.

Packard, V., 1960. *The waste makers.* McKay, New York. Planned obsolescence and the economics of throw-away.

Subcommittee on Environmental Improvement, 1969. *Cleaning our environment, the chemical basis for action.* Amer. Chem. Soc., Washington, D.C. Sound suggestions by a panel of chemists.

Zandi, I. and J. A. Hayden, 1969. "Are pipelines the answer to the waste collection dilemma?" *Env. Sci. & Tech.* **3**(9), pp. 812–819. Technology to the rescue.

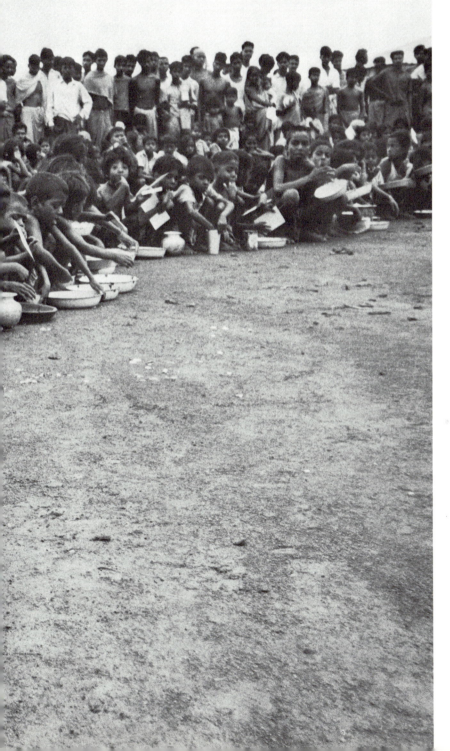

6

The People Problem

Children lining up for their food ration in Bangladesh during the recent India-Pakistan war. The scene could well become more widespread and not limited to wartime. (*CARE Photo*)

22

SUPPORTING EARTH'S
POPULATION

IN 1798 THOMAS MALTHUS published *An Essay on the Principle of
Population* . . . in which he wrote:

The power of population is indefinitely greater than the power in the
earth to produce subsistence for man.

Population, when unchecked, increases in a geometrical ratio. Subsistence
increases only in an arithmetical ratio. A slight acquaintance with numbers
will show the immensity of the first power in comparison with the second.

This implies a strong and constantly operating check on population from
the difficulty of subsistence. This difficulty must fall some where and must
necessarily be severely felt by a large portion of mankind.[1]

Unfortunately for the currency of Malthus' theory, disaster did not
overtake England; population did increase, but so did subsistence—
spectacularly so. Yet Malthus was correct in his basic analysis of the
problem.

HUMAN POPULATION GROWTH

Unlike animals, which breed until limited by predation, starvation, or disease, man has the unique ability to limit both his death
rate and his birth rate. Even so, it took from the beginning of man's
evolution until 1850 for his population to reach a billion.

Early man, in the hunting and gathering stage, required about
two square miles per capita to support himself. At this cultural stage,
the accessibility of food stabilized man's population, until the agricultural revolution removed that limitation. But the earth's resources
are not uniformly distributed, and while agriculture flourished in some
areas, other areas were nonproductive. This led to population in-

---

[1] Hardin, G. 1969. *Population, evolution, and birth control.* W. H. Freeman, San Francisco.

creases in countries with adequate resources, but limitation by starvation in others. By the eighteenth century, when Malthus made his observations, land and food shortages in many European countries had led to some desperate social conditions.

But by several quite fortuitous developments Europe was saved and Malthus' theory was temporarily discredited. First, the discovery of North and South America, then Australia and New Zealand, had provided outlets for surplus population. These new empires generated capital that permitted the industrial revolution to sweep Europe more quickly than it might otherwise have done. In addition, these colonies provided a market for goods produced by the mother country. Then efficient agriculture, stimulated by the industrial revolution, allowed not only a population increase but a slow rise in prosperity. Advances in medicine lowered the death rate, but the birth rate fell also, for the newly industrialized population desired to limit its families in order to feed and educate them better.

Convinced of the desirability of constant population growth, governments had always encouraged people to have large families to run the machinery in factories, and when necessary to man the guns that served the national interest. While it was certainly useful to governments and industry to have a large labor pool, it was not long before one of the laborers, the Englishman Francis Place, came to the conclusion that it was in the best interest of laboring men to limit their numbers and thereby increase their value and power. From this humble background arose the concept of family planning, which will be discussed in Chapter 23. While the acceptance of family planning increased gradually and steadily through the first half of the twentieth century, the catalyst to its rapid expansion came with the surge in population growth after World War II. It became obvious then that concentrated study and action were necessary to control a world population beginning to strain world resources seriously.

If we review the ebb and flow of the population of western European countries we can see a pattern of demographic transition from one stage to another that is being repeated throughout the world. The first stage was a population with high but fluctuating birth and death rates. No matter how many children were born, few survived the hazards of childhood or disease to become reproducing adults, so the population was either stable or slow in growing. In the second stage, because of improved medicine and health care, the death rate began to fall much more rapidly than the birth rate. The decrease was not regular; in fact, in some modern instances, the drop was precipitous. This was seen when malaria was controlled in Ceylon after World War II; the death rate fell from over twenty per thousand in 1946 to ten per thousand in 1954, although the birth rate remained high. In the third stage, both birth and death rates were low and the population was again slow-growing or stable.

England passed through just such a demographic transition be-

Pattern of demographic transition

tween 1650 and 1900; at the beginning of this period, both birth and death rates hovered around thirty-five per thousand per annum. By 1750 the death rate had fallen to twenty-one per thousand while the birth rate remained at thirty-four per thousand. By 1880 the birth rate had fallen to sixteen per thousand, nearly matching the death rate of twelve per thousand.

How can we explain this demographic transition? When people are poor in an agrarian context and the death rate is high, children are an economic asset in the struggle to wrest a living from the soil and parents are dependent upon them for security in their old age. Since many children die before reaching a productive age, many must be born to close the gap. When the death rate falls rapidly as it has in most countries in recent years, rural women, long used to almost annual pregnancy throughout their childbearing years, quickly find themselves with larger families than they had anticipated. Thus, the burgeoning birth rate and annual growth rate in many underdeveloped countries do not necessarily represent a *desire* to have large families; rather, more children are surviving than previously.

As farms are mechanized, fewer hands are needed, so rural people flock into the cities for greater opportunity than farm life provides. Now numerous children become an economic liability, expensive to feed, clothe, and educate; hence fewer are desired and fewer are born. Since farm mechanization and movement to cities reflect increased prosperity and industrialization, the process of development tends to reduce the birth rate. Indeed, communist dogma states that overpopulation is an illusion of conservative capitalism, and that, if development is rigorously pursued in underdeveloped countries, the birth rate will quickly fall in the rising tide of worker prosperity. This proposition is naïve, however, for the industrial development that took fifty to a hundred years in western countries is not likely to be significantly accelerated in underdeveloped countries. In addition, these countries already have far larger population growth rates than European countries had at a comparable stage and may well be swamped by the population increment over this time span, before they become sufficiently industrialized.

Even when the birth rate has been significantly reduced by war, disease, or starvation—the "natural" limits—it is still subject to unpredictable fluctuations and even to fashion. By the mid-1930s Germany, France, and England became concerned at the close approach of birth to death rates, and governments, fearing underpopulation for the first time since the Great Plague of the mid-fourteenth century, started to subsidize large families directly and indirectly. At that time one or two children were considered fashionable, and anyone with four or five was often hard pressed to support them and subject to a subtle disapproval by friends and relatives. After World War II the return of social stability led to a jump in the birth rates of most western countries, and families of three or four children again became acceptable. But after a few years the birth rate of most developed countries began to fall again. The United States was the exception for quite a while. As late as 1957 its birth rate was around twenty-five per thousand. With a death rate of nine per thousand, the annual rate of growth[2] was 1.8 percent, greater than Italy, Japan, or India at that time. Had the growth rate continued at this level for only 150 years, the population of the United States would have climbed from 150 million to over three billion. But fortunately in 1958 the birth rate began to fall (no one is quite sure why), dropping to 17.4 per thousand in 1968.

In late 1972 the fertility level in the United States dropped to the replacement level for the first time in history. However, it would have to stay at this level for seventy years to result in zero population growth. This is because the number of births depends on both the fertility rates and the number of women in their childbearing years. Since a large number of females have already been born, the population base will continue to grow until this group of females stops having children.

Not until all countries achieve and sustain this low fertility level can world population be stabilized. Unfortunately, at present two-thirds of the world is in stages one and two of the demographic transition.

While the overall growth rate of the world is about 2 percent, there is a great disparity between the developed and relatively stable nations of Europe and the rapidly growing underdeveloped countries of South America, Africa, and Asia (see Table 22.1). At just 2 per-

Table 22.1
Some Population Statistics of Selected Countries and Regions
*(mid-1969)**

AREA	POPULATION (HUNDREDS OF MILLIONS)	RATE OF GROWTH (percent)	DOUBLING TIME (years)
Europe	456	0.8	88
U.S.A.	203	1.0	70
Russia	241	1.0	70
Asia	1990	2.0	35
Africa	344	2.4	28
Latin America	276	2.9	24

* Data from the United Nations Population Reference Bureau.

cent growth, 180,000 babies are born every day, adding 65 million to the world's population every year. Small wonder the doubling time of world population is only thirty-five years and continuing to decrease (Table 22.2). Projections of the present growth rate beyond a few years are extremely difficult because of the many factors, almost unmeasurable let alone predictable, that affect the number of children

[2] Annual rate of growth = (birth rate − death rate)/10

Table 22.2
*Years Required to Double the Population**

YEAR	POPULATION (BILLIONS)	NUMBER OF YEARS TO DOUBLE
1	0.25	1650
1650	0.50	200
1850	1.1	80
1930	2.0	45
1975	4.0	35
2010	8.0†	?

* Dorn, Harold F., 1968. "World population growth: an international dilemma." *Environments of Man.* Addison-Wesley, Reading, Massachusetts.
† Projection of United Nations estimates.

born in any given year. Yet, one question arises from this projection; considering the thin line between feast and famine for many of today's 3 billion people, how many people can the earth support? While natural resources in the form of timber, ores, oil, and such are important considerations in determining the supportive capacity of earth, demands for these commodities could be sharply reduced by intensive recycling of these materials, or even lowering our standard of living. This is not possible with food. Therefore, a broad discussion of natural resources has been intentionally omitted to allow fuller treatment of the food problem. (Energy resources, however, are discussed in Chapter 10.)

FOOD RESOURCES AND PEOPLE

As a general rule one acre of fertile land per capita is required for a nation to be self-sufficient in food production. France, Sweden, and Denmark, with about one acre per capita, all achieve this self-sufficiency. India and Pakistan, on the other hand, lie very close to the boundary between self-sufficiency and dependency with 0.8 acres per capita. China with 0.5 acres is a chronic food importer, while England with 0.4 acres per capita despite high yields and intensive farming must import significant amounts. In contrast the United States has 2.9, Australia 4.0, and Canada 6.5 acres per capita, enabling all three countries to export large quantities of food. Despite this margin of food production at least two-thirds of the world's people suffer from undernourishment or malnutrition.

There is a distinct difference between undernourishment and malnutrition. Undernourishment is quantitative—an insufficient intake of calories leading to simple hunger; malnutrition is qualitative—the number of calories may quash hunger pains, but there is not enough protein in the diet to avoid chronic debilities or secondary infections and diseases. Children up to the age of six have an especially great need for protein. When this is not available a protein deficiency syndrome, kwashiorkor, develops, which is characterized by digestive tract disturbances, skin sores particularly on the legs, swelling of some parts of the body, wasting of others, and a reddening of the hair (Figure 22.1). A few years ago during the Nigeria-Biafra civil war magazines were filled with the haunting pictures of Biafran children suffering and dying from malnutrition. But older children and adults suffer from malnutrition too, and the susceptibility to illness, lethargy, and apathy it induces place a steady drain on the human resources of developing countries.

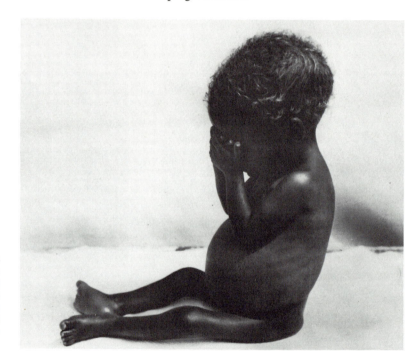

Fig. 22.1 • Kwashiorkor is the African name for a disease of young children whose food contains too little protein. The spindly legs and swollen belly are characteristic symptoms of this condition. (*FAO Photograph*)

However, protein supplements must be given with some care lest the cure be worse than the problem. In northeast Brazil some children suffering from malnutrition who were given a dried skim milk supplement developed blindness as a result. The reason? Without vitamin A, which is necessary in the production of lubricating fluids in the eye, the cornea becomes ulcerated, then softened and finally opaque. Since vitamin A is fat-soluble, skim milk is low in this vitamin; so when the milk protein stimulated growth without providing the necessary amount of vitamin A to support that growth, the children's bodies mobilized their remaining supplies of vitamin A and the eye disorder followed. Another even more bizarre instance was seen in Colombia, where a population was found with a 40 percent infestation of *Entamoeba histolytica,* a protozoan that generally burrows into the intestinal wall, causing a serious condition called amoebiasis. However, despite the high level of *Entamoeba* infestation, the incidence of amoebiasis was negligible. The answer to this puzzle was found in the high-starch diet of the people. Because of the low protein intake, production of starch-digesting enzymes was reduced, allowing a much higher level of starch to persist in the intestine. The protozoans were found to be feeding on this starch rather than attacking the intestinal wall. If this population had been given protein supplements without concurrent efforts to control the *Entamoeba* infestation, the incidence of amoebiasis would probably have soared, causing more problems than the lack of protein.

Whatever the effects of inadequate diet, only part of the blame can be laid to economic inequities. Other more peripheral problems prevent full realization of a nation's agricultural productivity. Religious attitudes, such as those that protect cows and monkeys in India, have been very difficult to resolve. These animals consume huge quantities of food that could contribute directly to agricultural production. Nationalism, understandably important to nations that have only recently emerged from the shadow of colonialism, has led to irrational attitudes and goals that have absorbed much scarce capital that underdeveloped countries could put to much better purposes.

Faced with the problem of widespread hunger in the world today, what must be done, using presently available technology, is to increase productivity of underdeveloped countries until they either become self-sufficient in food production or sufficiently industrialized so that they have enough foreign exchange to purchase food on the world market.

Present Technology

Only North America has maintained an average annual 2 percent growth rate in agricultural production over a long enough period (thirty years) to assume continuance. Today food production is increasing at 2.3 percent a year in the United States, but at less than half that rate in Latin America and Africa. In the Middle East, rates in many instances are substantially below the rates of population increase. There are several possibilities for improvement.

Of the 700,000 species of plants in the world nearly 80,000 are edible, but only 3000 are normally considered crop plants. Three hundred species are in abundant use, but only twelve species or genera provide 90 percent of all edible crops. The leading food plants in order of their importance are rice, wheat, corn (two-thirds of the annual crop is eaten by livestock), potatoes, soybeans, sorghum, barley, rye, millet, cassava, sweet potato, coconut, bananas, and plantains. Obviously an incredible number of potential food plants growing in all conceivable environments is unexploited. More will be

Plant breeding

said about developing new crop plants, but in the short term much can and is being done to improve the major crop plants already in worldwide use.

For example, the International Rice Research Institute at Los Baños in the Philippines developed a variety of rice over a five-year period that has increased production of rice in one instance in the Philippines from 710 to 10,000 pounds per acre where intensively cultivated (Figure 22.2). This strain, IR-8, has been introduced to other Asian countries as well and has increased average rice yields from two to four times. In the few years since its development IR-8 has been planted on 20 million acres in Turkey, India, Pakistan, and the Philippines. With the introduction of IR-8, Pakistan became self-sufficient in rice production for the first time in many years.

Fig. 22.2 • Although IR-8 rice yields heavily, it does so only when fertilized. The rice on the right has been fertilized; that on the left is unfertilized. (*FAO Photograph*)

In a parallel development Mexico gathered together a "germ bank" of over a thousand different varieties of corn, and was able to develop several high-yielding new varieties adapted to high elevations, low soil moisture, and other special environments. This improvement in corn yield, together with irrigation where feasible, greater use of fertilizer, and better transport facilities, has enabled Mexico to raise its agricultural productivity by 4 percent per year in recent years. Unfortunately, this gain has been mitigated somewhat by a population growth rate of over 3 percent per year.

These developments have been particularly important because, until recently, most of the agricultural expertise was derived from temperate zone climates, soils, and crops, which had limited application to crop cultivation in the tropics.

Fig. 22.3 • *Triticale,* a new
grain resulting from a cross
between wheat (*Triticum*) and
rye (*Secale*) is widely adaptable to different climates. The
grain on the left was grown in
Mexico, the grain in the center in Manitoba, and the grain
on the right is ordinary wheat.
(*University of Manitoba, Department of Plant Science*)

Another improvement in a basic cereal crop has been accomplished by crossing wheat and rye to form a new species called
Triticale (Figure 22.3). Perfected in Canada, this new grain is highly
resistant to lodging (the tendency of grain plants to be blown flat by
wind, making them unharvestable), has a higher protein content than
most other cereals, is a good livestock feed, has a greater yield than
either of its parents, and combines vigor with wide adaptability to
climate.

Other recent plant breeding developments include a wheat with
an increased yield of 30 percent, a leguminous (a member of the pea
family) livestock feed well-adapted to the tropics, and a new variety
of corn, opaque-2, with a much augmented supply of a very valuable
amino acid.

But this green revolution has not taken place without problems
—biological, economic, and political. The "miracle grains" are susceptible to disease and insects. What resistance exists is limited to a
few genes, making the grains quite vulnerable to a new strain of
pathogen, as happened to the potato in Ireland in the 1840s or corn
in 1970 in the United States. High yields are obtained only with
intensive use of biocides, fertilizers, and irrigation—all requiring
enormous amounts of capital to produce and supply, and exacting an
increasingly apparent environmental toll. Paul Ehrlich estimates that
for India to fertilize at the intensive level used by the Dutch, almost
half the fertilizer produced in the world would be required. Economically, the green revolution can only take place if heavily subsidized by a government. This causes increased production but increased internal prices as well. On the international market the grains
are usually sold at a loss because the world price is less than the
government price paid to stimulate farmers to grow the new grains.
Finally, there is a philosophical problem. The theory of Malthus

quoted at the beginning of this chapter is sometimes called the dismal theorem. In 1956 the economist Kenneth Boulding described an "utterly dismal theorem"—any technological improvement ultimately increases the sum of human misery by allowing a larger proportion of people to live at the same level of misery and starvation. An example to ponder: in the Ireland of 1600, 2 million people lived in wretched poverty. With the introduction of the potato (a green revolution?) in the 1700s Ireland's population rose to 8 million by 1835. As a result of the potato blight in the 1840s, 2 million people starved to death, 2 million emigrated, and 4 million remained in wretched poverty!

Intensive agriculture

New high-yield plant varieties cannot even be grown, however, without improvements in their cultivation. The land must be properly prepared by tractors capable of breaking the soil more quickly and effectively than water buffalo or oxen; it must be fertilized with the right formulation applied in the proper concentration at the best time; and the crop must be quickly and efficiently harvested, processed, and stored. Finally, if the increased yield is to be passed on to the people, distribution routes and new transport techniques and systems must be developed; otherwise isolated people may be reduced to starvation simply because there is no effective way to get food to them.

Presenting peasants with improved seed, then, is clearly only the beginning. The orchestration of the necessary accompanying technology requires time and usually outside help and money. Once begun, however, it can generate its own continued momentum. In the meantime, people continue to go hungry. While the diet of undeveloped countries is roughly comparable with developed countries in the caloric intake per day, there is a critical lack of protein in the diet of the former.

Protein supplement

At least three protein concentrates have been developed: incaparina, laubina, and fish protein concentrate (FPC). Incaparina is a mixture of cottonseed flour, whole corn cooked and ground, yeast, calcium carbonate, and vitamin A. The resulting powder contains 25 percent protein and can be mixed with water and drunk like dried milk or it can be added as a supplement to food. Laubina is made of wheat, chick peas, bone ash, sucrose, and vitamins A and D. Like incaparina, it is compounded of materials whose taste is known and recognized by the people it is intended to serve, and this goes a long way to ensure its acceptance, for people are exceedingly conservative about the food they eat. Rice-eating people would often rather starve than eat wheat or barley, which are unknown to them. Americans are no exception. When was the last time *you* ate roasted grasshoppers, lizard, or rattlesnake meat? Many confirmed beef eaters will not touch even the liver, much less heart, kidneys, tripe, or brain.

Fish protein concentrate has the advantage of being produced from just about any kind of fish, especially trash fish not normally eaten. When the fish is ground up and the oil and water removed, the end result is a fine, free-flowing powder. Odorless, tasteless, and containing 80 percent protein, FPC can be produced for about 25 cents a pound. As a powder it can be added to a great variety of baked and cooked foods to supplement the diet with protein. Just about any country with a fishing industry, fresh or salt water, should be able to generate this superb source of animal protein cheaply enough to raise protein consumption to at least minimal levels, buying time until the slower techniques of increasing agriculture production are implemented.

Another underexploited source of protein is wild game. One of the symbols of progress in an underdeveloped country is the establishment of a livestock industry to supply protein for internal consumption or for export income. Where no suitable stock exists, as was true in New Zealand, domesticated grazing animals may be profitably introduced. But New Zealand is an exception; most other grasslands of the world have numbers of native grazers well-adapted to the environment and the available forage (Figure 22.4). Until recently the value of wild game as a protein source, equal to or surpassing any stock that might be introduced, was ignored.

Then some pioneering work in east Africa by Pearsall and Dasmann disclosed that wild game is more efficient at grazing and even more productive of protein than introduced stock. The difference lies in the food habits, water requirements, and mobility of game. Cows are quite fussy, eating only certain grasses and leaving the rest. If the number of animals is increased, overgrazing takes place, leaving

Game ranching

Fig. 22.4 • Gnus and zebras are only two of the many grazers and browsers that efficiently harvest a broad variety of herbs and grasses in east Africa. (*Courtesy of the American Museum of Natural History*)

stock and range in very poor condition. Since cows are not very mobile they tend to overgraze and they require sizable water supplies.

Game, on the other hand, is beautifully adapted to the many niches in the environment: elephants eat the bark and roots of trees; giraffes feed on the higher branches of trees; eland, a species of large African antelope, feed on the lower branches, and so on. Because of this diversity, a much broader range of vegetation is utilized by game than domesticated cattle, and thereby becomes potentially available to man. Wild animals are much more mobile than stock, and when the browse becomes thin, they are capable of traveling miles in search of better range; therefore, overgrazing in the natural environment is rare. Finally, most of the wild grazers in East Africa have low water requirements; for example, the gemsbok and oryx take no water at all and the zebra can go from one to three days without water. One of the major drawbacks to the introduction of stock in many parts of Africa is their susceptibility to disease caused by the tsetse fly. Wild animals that evolved with the fly are immune.

Recognizing these advantages, Dasmann made a cost analysis study comparing stock and native grazers on a large ranch in Rhodesia. He found that while a cow was supported by one acre in England and was ready for market in a year, in Rhodesia the comparable figures were thirty acres and four years; while native grazers supplied only a pound more of meat per acre per year, there was a major difference in production cost: the profit margin on the beef was $1416 per year while the profit on the game amounted to $8960 per year. This differential was due to the greater cost of maintaining species introduced into an alien environment. Harvesting of large native game that is abundant and in no danger of extinction is clearly desirable both from an ecological and economic point of view, for any habitat can only support so many animals. Any surplus causes starvation or increases stress on the other species.

Suggestions have been made from time to time that appropriate browsers be introduced to the rangelands of the American Southwest. Up to some 10,000 years ago there were a number of native browsers feeding on various desert shrubs as well as grass. The Shasta ground sloth, for example, ate yucca, creosote bush, and agave. But the sloth and most of its fellows seem to have been eliminated by superpredation (see Chapter 15) 10,000 years ago and their niches have remained unoccupied. The only major replacement, the cow, can utilize only a small fraction of the total biomass available. The desert shrub community that covers 1 million square miles in the Southwest regularly produces 1200 pounds of dry matter per acre per year, twice that produced by most grasslands. If the range were stocked with an exotic browser that could utilize this production, the range could support as many as 15 to 30 million new browsers. If the new species were carefully selected, there need be no competition with the cattle that now stock the range, or with the native bison, antelope, elk, sheep, and deer. A sustained-yield harvest based on a herd of 30 million eland (one possibility) would provide an enormous quantity of protein from forage that is presently unused. In view of the problems of mammalian introduction in other countries, red deer in New Zealand particularly, great care would have to be taken before such an experiment were allowed. But if properly done it could be considered a form of environmental restoration, for it has only been 10,000 years since many of the species now considered exotics were artificially eliminated from the Southwest.

Another time-buying factor is the continuing surplus of food produced by the United States, Canada, Australia, and New Zealand. Until recently, up to 60 percent of the United States wheat crop was distributed abroad under Public Law 480, the Food for Peace Program. The USDA estimates that by 1984 demands by hungry nations for this surplus will exceed the supply. More pessimistic estimates of this meeting of supply and demand suggest 1975 as the critical year. In any case, if in the remaining ten years the recipient countries cannot increase their food production to at least the self-sufficiency level or if they cannot control their population growth, there seems no alternative to mass famine as Malthus predicted in 1798.

Future Technology

With great application and perhaps a bit of luck, the world can avoid the famine predicted for it in the next two or three decades. But what of the year 2000 and beyond? What hope does technology provide for the accommodation of two, three, or four times the present population of the world or even a more slowly growing population over a long period of time?

Only about 40 percent of the earth's total land area is capable of productive use, at least in the direct sense of cropland, grazing, or timber production. The other 60 percent is tied up in ice and snow, mountains, and desert. Of the potentially cultivatable land only 25 percent could reasonably be brought under cultivation and only one-third of this land is in Asia, where the demand is most critical: six out of every ten humans live in Asia.

Increase of cultivated land

For a better idea of why it is so difficult to bring new land under cultivation despite its apparent suitability, let us look at a commonly cited example. The terra firma soils which occupy most of Amazonia bear a luxuriant rain forest (Figure 22.5). But as we saw in the discussion of soil (Chapter 3), the forest, once established, accumulates and then recycles most of its mineral nutrients, using the rather infertile soil mostly as a means of absorbing moisture. When the forest is removed, the action of sunlight, temperature extremes, and alternation of wet and dry conditions causes rapid erosion, laterization, or both. If Amazonia were, as naïve dreamers like to claim, the future bread basket of the world, Brazilians would have discovered the fact

Fig. 22.5 • As lush as this rain forest appears, the soil beneath it is relatively unfertile, for most of the nutrients are already incorporated into the vegetation and are quickly recycled. (*Eli Morowitz*)

already and would have cleared and farmed most of it. Today this enormous area produces less than one-half the food eaten by its 2 million inhabitants. Far from being the economic mainstay of the Brazilian economy, it contributes only 1 percent of the gross national product while requiring 3 percent for its maintenance. It is extremely unlikely that the Amazon basin will ever be successfully used for conventional crop farming.

One other example of attempting to bring new land under cultivation was the great east African groundnut scheme. Much of east Africa is a savannah, scattered shrubs and low trees mixed with grass. In 1947 the British, then administrators of the territory, decided to open over 3 million acres of the countryside to the cultivation of groundnuts, or peanuts, as they are known in the United States. Research was done to assure that peanuts would grow satisfactorily in the climate and soil of east Africa. Then 30,000 acres were bulldozed and plowed, with great difficulty as it turned out, for the roots of trees and shrubs in areas of low rainfall are long and ropey and exceedingly difficult to remove. This accomplished, it was found that the stones in the soil, which were extremely hard and sharp-edged, damaged farm implements, compacted under the tread of tractors, and made the soil almost impervious. In addition, alternation of heavy rain and drought scuttled crop after crop. In 1951, after investing over $100 million, the British admitted failure.

As a general rule, any area of the world not already grazing cattle, growing trees, or producing crops is not capable of doing so using current technology. With much research and development, some of these marginal areas may yet be made productive, perhaps in ways beyond our present imagination. But we cannot tacitly assume that presently unproductive land is simply waiting for the touch of man to be brought into production.

There are large tracts of desert around the world that are neither too sandy, too mountainous, nor too cold to be farmed, if water could be supplied. The problem is not water supply per se, for in the next few decades desalinization will probably reduce the cost of fresh "manufactured" water from its present $1.00 per 1000 gallons to the point where economics will allow its use in irrigation (25 cents or less per 1000 gallons). However, once desalinated, water must be moved against gravity, over mountain ranges, and up high plateaus, for few deserts are located at sea level, even if they are adjacent to the seacoast. All of the long-distance water supply systems of large cities— New York's from the Delaware River and the Catskills, Los Angeles' from the Sierra and Owens Valley—operate on a gravity flow system. The billion-dollar California water scheme, which is bringing water from northern California to southern California, can compete with local desalinization only because the transport of this water is partially free due to gravity. If some way is developed to overcome the cost of moving desalinated water up from sea level, it may well be possible to bring water to the desert, but that is only a part of the problem.

Irrigation

However sweet, water carries some dissolved salts in solution that are left behind as the water evaporates in the hot desert sun. Unless the soil is periodically flushed and drained to remove these salts, they accumulate, making agriculture progressively more difficult (Figure 22.6). Furthermore, many deserts, otherwise suitable for

Fig. 22.6 • Sugar beets grow only sparsely along the salt-encrusted ridges between irrigation furrows in this field. Irrigation waters containing these salts rose to the ridge surfaces through capillary action and there evaporated, leaving their salts behind. (*State of California, Dept. of Water Resources*)

irrigation, lie in basins having no drainage to the outside. How would the water used to flush accumulating salts out of the fields be evacuated? Could we afford the cost of pumping it against gravity, back to the sea? Perhaps some deserts can be made to bloom, but not in the foreseeable future.

Assuming marginal lands and deserts could ultimately be cultivated, what crops would be planted? We have seen how few of the edible plants have been exploited; most plants, particularly those which have been intensely bred, are native to and grow best in the temperate zones of the world. But any new land being brought under cultivation is likely to be in the tropics. What is needed is an extensive research program that will screen the thousands of potentially edible tropical plants, isolating those that seem to offer the best possibilities: suitability to soils characteristic of the new lands, high protein content, resistance to drought and pests, and acceptability to the people in need of these foods. These initial efforts must be followed by an intensive selection and breeding program to combine the most useful features into new crop plants. Since the development time is long in plant breeding and selection programs, present efforts must be greatly accelerated if suitable crops are to be ready to take care of any sudden technological advances in marginal land use.

What kind of harvest *can* we expect of the world ocean? Altogether, the ocean "fixes" around 18 billion tons of carbon each year

The ocean in the process of photosynthesis. If this seems high, remember that 90 percent of the world's vegetation is in the sea. The open sea, about 90 percent of the total, produces fifty grams of carbon per square meter per year. Shallow coastal waters and estuaries produce about twice this, and local upwellings (areas where nutrient-rich deep water comes to the surface, causing a sharp increase in productivity) may produce up to six times that of the open sea.

With our present technology we are harvesting in most instances only the fish at the other end of this production (Figure 22.7). In the open ocean food chains may have five links between the algae producers and the fish that are harvested, coastal waters usually have three links, and the areas of upwelling only two. Because energy is lost between each link, the shorter the food chain, the greater the amount of energy that can be harvested.

When the yields of the three types of oceanic water are compared, the open ocean appears to be a biological desert, producing only a small fraction of the fish harvested and with little likelihood of any increase. In contrast, the upwellings that form only 1 percent of the ocean surface produce *one half* of the world's harvest of fish.

Annual production of fish is about 240 million tons, but not all of this can be harvested. Some fish must remain to breed and, since man is not the only fish predator, others are taken by sea birds and mammals. Perhaps 100 million tons are available to man on a sustained yield basis. In 1967 the fish catch was 60 million tons. At the

Fig. 22.7 • This etching by Brueghel embroiders a bit on the "big fish eat little fish" aspect of the aquatic food chain. (*Academy Collection, Philadelphia Museum of Art. Photograph by A. J. Wyatt, Staff Photographer. Courtesy of the Pennsylvania Academy of the Fine Arts*)

current rate of increase, 8 percent per annum, the ocean will be yielding as many fish as it can *within ten years*. Indeed, some parts of the ocean are already overexploited: the area between the Hudson Canyon and the Nova Scotia shelf produced 1 million tons of fish in the early 1960s but the catch has been falling off ever since.

Quite irrationally, at least one half of the fish caught commercially are thrown away because of finicky public attitudes about which fish are edible and which are not. Since fish contain all the amino acids necessary for man to manufacture protein, a huge amount of valuable protein is being wasted. An efficient system of trawlers working from a mother-ship base could easily deliver all the fish caught to the factory ship, which could clean and freeze the most desirable fish, then process the trash fish into fish protein concentrate. The Russians are exploring this possibility; indeed it seems too efficient and profitable a possibility to ignore.

Another promising possibility is the raising of some types of seafood under controlled or semi-controlled conditions, opening the way for more convenient and efficient harvesting and processing than is possible today with wild populations of fish and shellfish.

Of all the thousands of edible species of plants and animals in the sea, surprisingly few are eaten and even fewer (50 out of 20,000) are systematically cultivated. A few species of seaweed are grown in Japanese estuaries and surely far more could be grown in other countries. But seaweed is not to everyone's taste, although we all eat considerable quantities as alginate, a stabilizer that is finding its way into more and more processed foods (see Chapter 12). Oysters have been raised for centuries and have proved to be a valuable and nourishing food capable of yielding 6000 pounds per acre (Figure 22.8). But increasing pollution of estuaries is destroying more beds than are currently being developed. Mussels, a favorite shellfish in the Mediterranean countries, are almost totally ignored as a food supply in North America. Considering their superb flavor, far superior to most clams, and their extreme abundance along rocky coasts, mussels are one of many unexploited food sources in North America. Some forms of shrimp, milkfish, and mullet are raised in ponds with considerable success, particularly in Asia. Conch, abalone, lobster, blue crab, and other species of shellfish could be raised in tank farms or salt water ponds. Some recent work done in Massachusetts has reduced the time required for lobsters to reach sexual maturity from eight years to eighteen months, suggesting that lobster farming may soon become a practical operation. Research is also progressing on the utilization of thermal effluent for rearing food species (see Chapter 6).

Fig. 22.8 • Tarred sticks are used as a substrate for growing the Sydney rock oyster near Sydney, Australia. (*Australian Dept. of Information—photo by James Fitzpatrick*)

While the annual production of fish and shellfish may seem enormous, most of the true production of the ocean lies, like the greatest mass of an iceberg, out of sight. When we catch fish, we are harvesting only a tiny portion of the food chain energy represented by that fish. Behind every ounce and a half of tuna that we eat, there is one pound of mackerel, ten pounds of herring, 100 pounds of zooplankton, and 500 pounds of phytoplankton. Phytoplankton consists of the unicellular algae that begin the whole food chain by fixing carbon into food by photosynthesis. Rather than accumulating all the loss of energy involved in long food chains, it would make far more sense to harvest algae directly. On an average the algae in one acre of sea surface could yield about three tons dry weight, compared with the average yield of twenty pounds of fish per acre of ocean.

The problem is that harvesting algae is not easily done: getting those three tons of algae out of the water is one major problem, converting them into palatable protein is another, and persuading people to eat this protein in preference to fish is yet another. It would be more convenient to grow algae in tanks under optimum conditions than to try straining them out of huge quantities of water, but pilot operations of this type have never been as productive as expected. Apparently much more research is necessary before algae can either be harvested or raised directly as a major source of protein.

The potential yield resulting from shortened food chains applies to the land as well as the sea. Beef as a harvester of energy is incredibly inefficient; the yield of protein is forty-three pounds per acre per year. If the protein is in the form of milk, the yield is increased slightly to seventy-seven pounds per acre per year. Soybeans can yield 450 pounds of protein per acre per year and processed alfalfa, 600 pounds. When the demand for protein becomes great enough, eating beef as a major source will become an increasing luxury, and then disappear altogether. Two-thirds of the earth's population has already, by necessity, made the shift.

New technique
in crop production

Even within the spectrum of plants already utilized, we can do much better than the 4.5 million calories per acre per year provided by wheat; potatoes give 8 and sugar beets 13 million calories per acre per year. The difference is due to the small fraction of the wheat plant we actually eat, less than 30 percent. In contrast, 50 to 65 percent of the sugar beet or potato plant is directly edible. When the pinch becomes sharp enough we may have to shift to crops that are more completely edible, or we may breed new varieties of grain that have edible stems as well as heads.

The short-circuiting of food chains is but one way to eliminate as much waste as possible by going to the primary producer. The next logical step would be direct synthesis from an abundant organic source. Three approaches have already been worked out on a pilot basis. Roughage waste from cereal crops, corn cobs and straw, when treated with a hot acid, forms a molasses-like substance that can be fermented to produce yeast protein. Although the expense would be ten times as great as using molasses from sugar cane or sugar beets, it could be done. Conversion of most of the woody residues from today's crops could lead to a doubling of our present food supply.

Direct synthesis

Another approach treats soybeans with a mild alkali that aids in the extraction of the soybean protein. This extract can be processed like synthetic fabric fibers; when extruded, the extract coagulates in a suitable bath into fibers ranging from one- to thirty-thousandths of an inch in diameter. These can be oriented in parallel or at right angles, forming a mat (Figure 22.9a). The resulting material is tasteless but

Fig. 22.9 • (a) Soybean protein can be processed like synthetic fabric fibers. (Ralston Purina Co.) (b) When oriented at right angles to form a mat and artificially flavored and colored, luncheon meats can be synthesized: left, "smoked beef"; center, "corned beef"; and right, "smoked turkey." (*Worthington Foods*)

can be appropriately flavored and colored to resemble in taste and texture the familiar forms of meat: chicken, beef, sausage, and luncheon meats (Figure 22.9b). Although at present more expensive than the real thing, something true of every pilot project, it probably tastes as good as many meats currently available and could serve the valuable function of weaning the meat-eating public to an acceptable substitute when the real thing becomes too expensive and scarce a source of protein.

The most revolutionary approach to direct synthesis of food involves crude oil. Anyone who has spent time in the tropics realizes that there are microorganisms that will attack anything, no matter how indigestible or poisonous it may seem to us. A few years ago a French research group found some microorganisms that could use petroleum as a food source and thus be harvested as a substantial yield of protein. One enthusiastic estimate suggests that the equivalent of all animal protein needed by the world today could be produced from but 3 percent of the annual crude oil production of the world; however, tooling up from a test tube to the scale necessary to supersede beef as a protein source presents extensive technical problems. In the meantime, the world's oil supply is being rapidly depleted.

Direct chemical synthesis of food, though feasible, is still some years and perhaps decades away.

Where does the world stand then when present and future technology is turned to the task of producing food? If the rate of world population growth could be reduced to 2 percent at the same time the rate of food production everywhere could be raised by the same amount, the population of the earth could double by the year 2000 without mass starvation.

Balancing serendipitous technological developments against natural disasters, we are still left with a big if. While many countries have made great strides toward balancing their population growth and food production rates by decreasing the birth rate, by increasing food production, or both, many more have made no progress in either.

As late as it is, we still have before us the choice of voluntarily limiting our population and our consumption of raw materials. Before long, however, limited availability of food and natural resources will remove that choice forever.

FURTHER READING

Bardach, J. E., J. H. Ryther, and W. O. McLarney, 1972. *Aquaculture: the farming and husbandry of freshwater and marine organisms.* Wiley, New York. Complete review of the state of the art.

Chapman, V. J., 1970. *Seaweeds and their uses.* 2nd ed. Methuen, London, England. Comprehensive.

Dasmann, R. F., 1964. *African game ranching.* Pergamon-Macmillan, New York. Experimental evidence that native game can out-produce domestic stock and still conserve the landscape.

Ehrlich, P. R. and A. H. Ehrlich, 1970. *Population, resources, environment.* W. H. Freeman, San Francisco. An excellent resource book.

Gutcho, M., 1973. "Textured foods and allied products." *Food Technology Review* No. 1. Noyes Data Corp., Park Ridge, N.J. Fascinating discussion of the techniques currently used to manufacture food products.

Holden, C., 1971. "Fish flour: protein supplement has yet to fulfill expectations." *Science* **173,** pp. 410–412. Developing a technique is obviously only the first step to making it widely available.

Idyll, C. P., 1970. *The sea against hunger.* Crowell, New York. Examines each of the major sources of food from the sea.

Martin, P. S., 1969. "Wanted: a suitable herbivore to convert 600 million acres of western scrubland to protein." *Nat. Hist.* **78**(2), pp. 34–39. Martin presents an interesting case for introducing exotics to fill empty niches.

493

THE PEOPLE PROBLEM

Paddock, W. C., 1970. "How green is the green revolution?" *BioSci.* **20**(16), pp. 897–902. Paddock views the green revolution with a rather jaundiced eye.

Rhyther, J. H., 1969. "Photosynthesis and fish production in the sea." *Science* **166,** pp. 72–76. The most up-to-date survey of the fish resources in the world ocean.

23

POPULATION CONTROL

OF ALL THE PROBLEMS discussed in this book, overpopulation is the one problem the entire world must share. But we have become so concerned about the quantitative aspects of the billions of people we fatalistically anticipate in the next few decades that we forget the critical importance of the *quality* of life on earth. For hundreds of millions living in underdeveloped countries, life is so impoverished that quality, in our affluent terms, is unknown.

There are signs posted all over downtown Edinburgh, Scotland, that read "The amenity of our streets is commended to your care." The Scots sign is thought-provoking far beyond its literal intent, for the amenity of future life on earth *is* in our care. Amenity is being able to show your children skunk cabbages pushing up through the corn snow of March, going swimming and being able to see your toes, visiting a National Park rather than a national parking lot. It is very easy to pick some point in time past and say "those were the days," regretting the subsequent loss of amenity, for the passage of time never seems to add to the quality of life, only the quantity. Like the realtor who advises buying land now for it will never be cheaper, we must make a stand now, for the quality of life will only lessen in the future.

But an even more important aspect of amenity is personal freedom. The Wild West of the frontier days was wild because of the freedom, not simply opportunity, to lead your life as you wished. As populations grew, the freedoms began to wither away. For a while one could always follow a frontier north, south, east, or west; but there are no more frontiers in our country. Backpack into the most remote wilderness and you will surely run into another family doing the same. As the number of people increases, the number of problems arising from interactions increases, more laws are passed, more freedoms are limited. If you feel hemmed in now, sharing North America with 300 million other North Americans, think of the restrictions necessary to allow 600 million or a billion North Americans to live peaceably together.

The only way to preserve what amenity and personal freedom remain is by population control. Despite centuries of chamber of commerce propaganda, growth is not always coupled to prosperity; neither is it necessary or inevitable. Its control is possible with the means at hand, but first we must overcome our superstitions, religious objections, political barriers, and just plain ignorance. Not until *all* the people of the world fully understand the consequences of unrestricted growth can present solutions be applied, and better ones developed.

MOTIVATION AND POPULATION CONTROL

However desirable as a national goal, neither population growth nor population decrease can be imposed. People respond not to external coercion but to internal motivation; the most remote goals can be achieved if people are properly motivated. Conversely, government policies can be effectively undermined if motivation is lacking, as Prohibition conclusively demonstrated in the United States two generations ago. Thus, family planning has no hope of success unless people are motivated by moral, economic, legal, or social means.

Family planning is not synonymous with birth control, for birth control is specifically designed to limit the number of births. Since the purpose of family planning is to plan for the size of a family, help is also given to couples who have problems conceiving children. But far more people are desirous of controlling births than eliciting them, and family planning is commonly used as a euphemism for birth control. For when questioned, most women express the desire to control the size of their families and not simply to give birth to child after child throughout their childbearing years. To help such women, many underdeveloped countries—India, Pakistan, Turkey, Colombia, Morocco, Kenya, and Jamaica, to name just a few—have set up birth control programs and clinics.

Since the family planning movement dispenses birth control information as well as information to help overcome sterility, there has been much opposition to family planning activities in Roman Catholic countries. While the Catholic Church is still unrelenting, in 1961 the National Council of Churches sanctioned the practice of birth control, moving some of the religious deterrents to a realistic program of population control. Attitudes on birth control are sometimes further complicated by nationalist or political groups who feel that birth control is being applied toward a political or even racist end; some black groups in the United States and Hindus in India have made this charge. But as long as birth control information and devices are made available without coercion to everyone who wants them, such a charge is unwarranted.

Plainly, the social and moral attitudes within our society and throughout the world will have to undergo considerable modification before birth control becomes a way of life for the entire population. At present, young single women are under considerable social pressure to marry and raise a family, despite education or challenging career possibilities. While greater economic and social opportunities have become available to women quite recently, through the activities of such groups as the Women's Liberation movement, much more will have to be done in this area if population growth is to be effectively checked.

Population Policies

Governments have been directly influencing population growth for years. Governments could, however, discourage population growth by the same means. This might be accomplished by rewarding childless individuals with at least equal, if not lower, tax burdens than parents, hence removing the attractive tax deduction reward for having children. Other proposed schemes involve setting up an inverted tax scale, increasing the tax with every child. Delayed marriage, already beginning to take place as a result of increased educational opportunities and requirements, might be further encouraged by government subsidy. The government of India is considering a law that would raise the minimum marriageable age, by which it hopes to shorten the childbearing years of its citizens.

However these are *material* rewards for remaining single, delaying marriage, or having few children. What about the psychological rewards? Social pressure, which usually tends to push people into marriage and then into raising families, may at times have a negative effect on population growth. During the Depression the birth rate fell dramatically in most countries and the one- or two-child family became the norm.

Finally, those parents who can financially and emotionally support more than two children should be encouraged to increase the size of their families not by further natural births, but by adoption. Circumstance and accident will always result in parentless children who need love and protection as much as one's own. At the moment the family is the only institution capable of fulfilling the needs of *all* of society's children.

In Europe birth control has varying degrees of official recognition. In the Scandinavian countries and Great Britain it is legal; in France, Belgium, and the Netherlands it is considered semi-legal, that is, a couple may practice birth control for medical reasons. However, in Italy, Portugal, Spain, and Ireland, still predominantly Catholic countries, birth control remains illegal. Neither the United States nor Canada has a national policy on family planning, nor are such plans likely to be developed in the near future. Sweden and Great Britain officially encourage family planning, but most other Western

European countries have no defined population policy. France is an exception; it has an active government-sponsored program providing subsidies for larger families, with increasing benefits as the number of children increases. The dissemination of appropriate propaganda also helps the French government to express its approval of larger families. Ironically, France has been plagued by periods of population decline over much of the twentieth century.

Japan presents another type of government policy on family planning. Japan's 102 million people are jammed into a land area about the size of Montana but much more mountainous. In 1968 the population density was 1333 people per square kilometer, compared with 565 for the Netherlands. As a result there has been societal as well as governmental pressure for a reduced birth rate. This has by and large been accomplished over the last couple of decades. Legalized abortion and widespread use of contraceptives have reduced the birth rate from 34.3 in 1947 to 18.8 in 1970. For years now, small families have been the socially accepted practice in Japan. However, with increasing industrial output the country is experiencing a labor shortage—rather, a shortage of cheap labor, for more and more young people are going on to school and postponing their entering into the labor market. When they do, it isn't as cheap factory labor. As a result, a subtle change in government policy is taking place. Instead of removing the heavy subsidy on rice farming and retraining the farmers released for industrial jobs, a return to larger families has been suggested as the easiest solution to the problem.

This last example is interesting because it illustrates the major problem with population control today. Countries have no coordination of their political and economic policies. When one country experiences a decline in population it does not draw on the population resources of another country; instead, it chooses to view people as a natural resource and to seek to increase the size of this resource. But people are not a natural resource; rather, people consume natural resources. If we have overpopulation in any one country, that country must tap the resources of the countries which are still underpopulated to meet its consumer needs. If we are to have realistic population control we must change the view that population is related to specific political economies.

BIRTH CONTROL METHODS

Once the economic and social motivations for smaller families are established, the means of achieving the goal of fewer births can be implemented. Ideally, people should have children only when they want them. This is the essence of family planning. Too many families have children who are unwanted, and often resented as well. The unwanted child often matures into an unwanted adult, maladjusted and a burden to society.

The means of avoiding childbirth are many, running the gamut from celibacy to abortion, but the most reliable method is sterilization. The term sterilization has an ominous ring to it because of the implied permanence of the operation. Often confused with castration by the uneducated, sterilization of the male simply prevents the release of sperm. This is accomplished in a minor operation, vasectomy, that severs the sperm ducts and can be performed in a few minutes with a local anesthetic. Females may also be sterilized, but the operation to tie off the oviducts or fallopian tubes which conduct the egg from the ovary to the uterus requires internal surgery. Sterilization has *no* physiological effect on the sex drive, although if the significance of the operation is not clearly understood, there may indeed be psychological effects on the libido. Studies of Japanese couples where the male was sterilized showed that in only 3.5 percent of the couples was the frequency of sex reduced; in contrast, 28.5 percent reported an increase, probably due to the release from anxiety over possible conception.

Sterilization

The major drawback of sterilization is its implied permanence. Actually, due to improved surgical techniques, over 50 percent of sterilized males and females can be restored to fertility. Still, many doctors in the United States hesitate to sterilize because of the ever-present threat of lawsuits from patients who have changed their minds.

A way around this problem that is being followed by many is the deposition of a supply of sperm in a sperm bank by a husband before sterilization. Then, should conception be desired at some time in the future, the sperm can be withdrawn from the deep freeze and used to inseminate the wife artificially.

In India, sterilization is encouraged by cash incentives, and over 4 million people, mostly males, have been sterilized in the past ten years. With the possibility of new techniques for temporary sterilization, such as blocking rather than severing the fallopian tubes or sperm ducts, this contraceptive technique may become as accepted in developed countries as it is in some underdeveloped ones.

Of all the modern contraceptive techniques available, the rhythm method, although widely practiced, is the least effective. This is most unfortunate since it is one of the few contraceptive techniques endorsed by religious bodies otherwise opposed to contraception, notably the Catholic Church.

The rhythm method

In theory, if a woman knows when she has ovulated and does not have intercourse until the egg has left her body, conception can be avoided. Unfortunately most women are not aware of when they ovulate, although by monitoring of vaginal temperature and glucose levels the time can sometimes, but not always, be pinpointed. To make things worse, at least one-sixth of all women in their reproductive years have irregular periods that make calendar watching useless.

Then too, the rhythm of a period can be upset by tension, anxiety, and illness. Finally, expecting an uneducated and unsophisticated person to avoid conception by using the rhythm method is not so much ingenuous as it is cynical.

There are many mechanical contraceptive devices available, for example, the condom, the diaphragm, and the IUD (intrauterine device). IUDs are made of stainless steel or nonreactive plastic and may be just about any shape (Figure 23.1).

Fig. 23.1 • The Lippes loop (top) and Daikon shield are two of a great variety of intrauterine devices currently in use. (*Planned Parenthood–World Population*)

When an IUD is inserted into the uterus it serves as a mild irritant which interferes with the implantation of a fertilized egg into the uterine wall that is necessary to begin embryonic development. While this irritation is tolerated by most women, some 30 percent are made uncomfortable or bleed, necessitating removal of the IUD. Another 10 percent eject them spontaneously. A second generation IUD, T-shaped and copper, has reduced side effects to less than 1 percent. IUDs made of copper seem to be the most effective. Apparently some of the copper dissolves, is absorbed, and serves to attract leucocytes or white blood cells, which in some way interfere with survival or development of the freshly fertilized egg. The great value of this device for those women who are receptive to it is that once inserted by a physician it can protect against conception indefinitely. Generally after two years, 55 to 60 percent of the women fitted with IUDs are still successfully wearing the device. The remainder must use alternative methods. The IUD has become quite popular in both developed and underdeveloped countries as well; over 2 million IUDs are in use in India alone.

The first contraceptive pill, marketed in 1959, contained two hormones, a progesterone and an estrogen, which prevented ovulation. However, the estrogen mimicked some of the effects of pregnancy: nausea, bleeding, weight gain, and nervousness, and more seriously, the tendency of the blood of pregnant women to form blood clots. The incidence of deaths from this side effect is low—3 per 100,000 users—much lower than the death rate from normal pregnancies.

Unfortunately the possibility of side effects was not thoroughly explored at the time the pill was introduced. So when Senate hearings in late 1969 publicized some of the problems of contraceptive pills, many women gave up the pill in panic, without being made aware of the odds, which as far as side effects are concerned, favor the pill over normal pregnancy. Still, millions of American women are taking the pill with close to 100 percent effectiveness when the sequence is closely followed. The real problem with the pill is its cost, which, unless subsidized, makes it available only to the higher income groups. Another problem interfering with its use in underdeveloped countries is the irregular pattern of intake that is required.

A second generation of chemical contraceptives is being researched and will probably be available in a few years. By enclosing a prescribed dose of progesterone in a permeable plastic ring which can be inserted in the vagina much like a diaphragm, a small but steady dose of hormone can be absorbed. Progesterone alone seems to have an effect on the sperm, making it unable to fertilize the egg. Further, progesterone, unlike estrogen, does not prevent ovulation. The ring lasts for three weeks and is replaced after menstration if contraception is to be continued. Another approach is to impregnate with progesterone a small piece of plastic, perhaps the size of a pencil lead, then insert it just under the skin. Theoretically such an insert could regularly meter out the prescribed dose of progesterone over a year or perhaps even the reproductive life span of a woman. If pregnancy were desired, the implant could be easily removed and conception could take place on the next cycle.

Another group of compounds, the prostaglandins, have some promise of providing agents that can produce labor, induce abortion, and make possible contraception after fertilization, the long sought "morning after" pill. Prostaglandins, normally found in very small quantities in cells throughout the body, seem to act as enhancers or suppressors of hormones. However, they are not themselves hormones.

A somewhat more distant prospect of drug control over contraception is provided by a hormone awkwardly called luteinizing hormone-releasing hormone/follicle-stimulating hormone-releasing hormone (LH-RH/FSH-RH). Its function is fortunately more simple than its name: it controls the synthesis and release of luteinizing and follicle-stimulating hormones produced by the pituitary gland. These two hormones then cause the ovaries and testes to secrete their respective sex hormones. LH-RH/FSH-RH could be used to disrupt the normal menstrual cycle and prevent ovulation; the pituitary gland might be prevented from releasing LH and FSH, thus preventing ovulation; if LH-RH/FSH-RH can be neutralized then it might be possible to immunize women for several months; ovulation might be induced in the middle of the menstrual cycle, allowing for intercourse without danger of pregnancy during the rest of the cycle.

Lest women feel that they are forever to bear the brunt of contraceptive application and potential side effects, the possibility of male contraceptives is becoming more feasible. Apparently antigens exist in sperm and testes that can cause antibodies to form in the female, interfering with reproductive processes. An enzyme, lactate dehydrogenase (LDH-X), has been isolated that might be injected into a male whose body would produce antibodies to the enzyme, thereby interfering with sperm production. Still another enzyme, sorbitol dehydrogenase, can, if inhibited, bring sperm production to a halt. While male contraceptives are clearly possible, it may well be at least a decade before any one of these possibilities is worked out.

While new oral or implanted chemical contraceptives look promising, there is a testing period of eight years required by the FDA to anticipate possible side effects. This is done to protect the public, as is the FDA's charge, but every drug, including aspirin, has side effects. Considering that during an eight- to ten-year development-testing program some underdeveloped countries are well on the way to doubling their population, this is a long time to wait. Many women are far better off taking the pill than running the risk of complications in pregnancy, or worse yet, being unable to sustain the life they have created.

The effectiveness of four of the major contraceptive techniques is listed in Table 23.1, which will allow you to take your own measure of the risk of pregnancy.

Table 23.1
Comparative Effectiveness of Various Contraceptives∗

METHOD	PREGNANCIES PER MILLION USERS
pill	5,000
IUD	30,000
diaphragm	120,000
rhythm	240,000

∗ Djerassi, C., 1969. "Prognosis for the Development of New Chemical Birth Control Methods." *Science.* 166:468–473.

THE ABORTION QUESTION

For centuries man has practiced abortion, despite censure on religious, moral, and legal grounds. The widespread practice of abortion to terminate unwanted pregnancies has recently been brought more forcefully to the attention of governments, since abortion rates are particularly high in those countries where birth control is illegal. In Italy, for example, there is probably one abortion for every live birth and in Santiago, Chile, 40 percent of all hospital admissions involve abortion attempts. Where contraceptives fail, abortion should be considered as a method of birth control, but its legitimacy or acceptance is a far more complicated question.

Biologically, egg and sperm cells are neither more nor less "human" than any other cell in our bodies. Hence it makes little biological difference whether an egg and sperm are killed by their failure to be united or shortly after their union by removal from the body naturally or artificially. The question arises then, at what point does this fertilized egg become human? Medically and legally, humanity is recognized after 5 months. Should an embryo die before this point it is regarded merely as tissue and disposed of like an excised gall bladder or tonsil. After 5 months the embryo becomes a fetus, and under common law it has a legal identity. Should a natural abortion or stillbirth take place after this point the body is buried and the woman is regarded as having given birth.

Since most induced abortions are performed within the first 3 months of pregnancy, a good two months before the law formally recognizes civil identity, there should be no legal interest in such abortions. This has not been the case, however, and the reason is historical. In the last century, when most anti-abortion statutes were written, abortions performed at any time by anyone were dangerous. This was a time when surgery was quite risky because of the almost total lack of sanitation in the operating room. Abortions were regarded as fifteen times more dangerous than natural childbirth. To protect women from this risk, abortions were declared illegal with almost no exceptions.

Today an abortion performed by a physician during the first trimester (pregnancy is usually divided into three periods of three months) is quite safe, involving only one-eighth of the danger of complications associated with live birth, with no risk of subsequent sterility.

In recognition of this change in the medical status of abortion, anti-abortion laws have been under attack in the courts on the basis that the right and choice of abortion should reside with the woman who, after all, must bear the responsibility for her actions. Medically and legally, there is no reason why a woman should not have an abortion in her first trimester of pregnancy if she so desires.

In response to this pressure the United States Supreme Court in 1973 handed down a decision that makes unconstitutional state laws forbidding abortion during the first trimester. As a result of this decision, abortion laws are presently in a state of flux in the United States. Most of the countries of western and southern Europe and South America still have restrictive abortion laws. India is considering the legalization of abortion as Japan did just after World War II. Apparently shocked by the huge increase in abortions that followed liberalization, the Japanese government subsequently embarked on a massive program to encourage the use of contraceptives and is even considering reinstating controls on abortion. Similar public reactions are seen in other areas after abortion laws have been relaxed. But much of the increase in abortion simply reflects the shift from furtive, unreported abortions often resulting in injury or death, to hospital- or doctor-supervised abortions that are reported as a routine statistic along with chicken pox or measles. In the United States, for example, there used to be at least 1 million illegal abortions per year. Now that abortion has been legalized nationally, these formerly illegal abortions will doubtless be viewed by outraged moralists as a "rising tide of immorality." Forgotten will be the millions of abortions which have always taken place, legally or illegally.

The chief opponents of legalized abortion through the years have been religious bodies, particularly the Roman Catholic Church. The Church has maintained, not as dogma but as current teaching, that the soul is present from the moment of conception and that any unnatural interference with its progression through to birth is a sin. This concept is called immediate animation. But the Church's view on the matter has not always been so; no less an authority than St. Thomas Aquinas held that the soul developed gradually as the embryo slowly became human, a view congruent with contemporary legal opinion. But after the death of St. Thomas this mediate animation concept fell out of favor, although it has never directly been eliminated as an option. It is quite possible that in future years the Church may again alter its position in view of the seriousness of overpopulation and its threat to *all* of man's institutions.

What then of the role of abortion as a form of birth control? Despite its present ease and increasing legality, no one would seriously recommend abortion in place of one of the contraceptives previously discussed. It is far more desirable to prevent a conception than to terminate one. Yet an unwanted pregnancy is often a traumatic experience even if resolved by abortion. Abortions can perform a valuable service to society by allowing those women who are least fit physically, emotionally, and financially to avoid the birth of an unwanted child. The ultimate social and moral cost to a society that tolerates the production of unwanted children is surely far greater than any circumstances surrounding an abortion.

FUTURE POPULATION AND ECONOMIC GROWTH

Birth control measures can be effective in reducing population growth: Japan cut its birth rate drastically by passing a broadly permissive Eugenic Protection Act in 1948, which made abortion available to everyone, while South Korea and Taiwan have also reduced the birth rate by use of the pill and IUDs. Ceylon, Pakistan, and India are, after several years of somewhat disappointing and grossly underfunded efforts, also beginning to make progress.

India plans in the next ten years to reduce its birth rate from forty-one to twenty-five per thousand, which will require that 40 percent of married couples of childbearing age use contraceptives, an incredible task. But if and when it is accomplished, the Indian rate of growth will still be 1.5 percent, compared with 0.78 percent for the United States, 0.29 percent for Sweden, and 0.18 percent for Luxembourg.

Without any effort to control population there will be 7 billion people in the year 2000. With a decreased rate of 2 percent yearly growth for all nations there would still be 5.6 billion people by 2000. Two billion more people can probably be accommodated on earth, but adding 4 billion more staggers the imagination, for the earth is barely able to support its present population; indeed most men live at a subsistence level far removed from the embarrassing affluence of western Europe and North America. When you consider that Americans, who account for 6 percent of the world's population, are absorbing a disproportionate amount of the world's raw materials to maintain their affluent standard of living, it is not surprising that the underdeveloped nations have thought it hypocritical for us to urge them to control their populations while developed nations continue to grow, albeit more slowly, and use still more of the world's resources. Our insistence upon pollution control is also misinterpreted by nations just beginning to industrialize. Underdeveloped nations should not be denied the benefits of technology simply because of the problems it has generated. Just as developed nations have shared their medical and food production expertise, they can also share the birth and pollution control technology they have gained through the hard experience of overpopulation and badly polluted environments. Underdeveloped nations, then, could be spared the ill effects of development by careful planning while enjoying the benefits of the rising standard of living that development allows.

If the population of the world were frozen at its present level there might be some hope that technology could in the next several decades begin to feed and house the world's population adequately and perhaps even begin to close the yawning gap between the few who have and the many who do not. But any increase in population

505

beyond present numbers, 1 or even 2 percent, so widens the affluence gap as to make it virtually impossible for the average Asian or African *ever* to have the good life which we as Americans now consider our inalienable right. A reduction in the growth rate is not enough. Population growth must cease altogether, not just for India, China, Egypt, or Jamaica, but for the United States, Russia, New Zealand, and Denmark. For it is not the Indian, Chinese, or Egyptian peasant who is making unprecedented demands on world resources, generating sonic booms, or releasing radioactive wastes; it is the middle class of the world. Caught up in a cycle of more-demanding-more, *they* must bear the responsibility for much of the world's pollution and share the responsibility for the population control that must come.

Since the only rationale for continued growth is its traditional association with progress, the first step in making the transition from an expanding population to a zero growth population is the realization that growth and progress are *not* inseparable. Indeed, if we take the stance that progress has a qualitative as well as a quantitative aspect, then growth is clearly *inimical* to progress. For the same population growth which traditionally provides new markets, more profits, and greater consumer affluence at the same time has eroded our personal freedoms and steadily debased the viability of our environment. Once we realize that our definition of progress must be broadened to include the achieving of quality in our future, we can begin to plan our economy to accommodate this change. This will certainly mean a shift from a growth to a steady-state economy.

Since much of the gross profit of industry is reinvested in expansion to meet the demands made by population growth, a steady-state economy should free some of this capital for by-product utilization, recycling of wastes, pollution control facilities, modernization of physical plant and the research and development necessary to supplement diversification. While the details of this shift from a growth to a steady-state economy are best left to economists, the positive impact of such a shift on the environmental problems that we have discussed in this book would be impressive.

With a stable population and a steady-state economy, the world's industry should be fully occupied for decades in satisfying the minimum demands of the world's people for decent housing with electricity, pure water, and sanitary facilities that most Americans now take for granted. This is an enormous market which has barely begun to develop, but when it is satisfied, diversification to meet the increasing leisure needs of the earth's now stable population should assure continuing profits for industry and a better life in both quantitative and qualitative terms for all men.

FURTHER READING

Blake, J., 1971. "Reproductive motivation and population policy." *BioSci.* **21**(5), pp. 215–220. Explores some of the problems of achieving zero population growth.

Callahan, D., 1972. "Ethics and population limitation." *Science* **175**, pp. 487–494. A wide-ranging discussion of the ethics involved in population control.

Demarest, R. J. and J. J. Sciarra, 1969. *Conception, birth and contraception: a visual presentation.* McGraw-Hill, New York. An excellent graphic review.

Galon, L., 1971. "The new mystery—maybe miracle—drug." *The New York Times Magazine,* 5 December. A summary of prostaglandin research.

Hardin, G. J., 1969. *Population, evolution, and birth control.* 2nd ed. W. H. Freeman, San Francisco. Superb collection of basic sources. The section on abortion is especially well done.

Heller, A., 1972. *The California tomorrow plan.* William Kaufmann, Inc., Los Altos, California. An enlightened development plan for California is contrasted with what will probably happen.

Jackson, H., 1973. "Chemical methods of male contraception." *American Scientist* **61**, pp. 188–193. Examines the current possibilities.

Kapp, K. W., 1963. *Social costs of business enterprise.* Asia Publishing House, Bombay, India. Considers the costs which have been externalized by industry and hence borne by the public at large.

Marx, J. L., 1973. "Birth control: current technology, future prospects." *Science* **179**, pp. 1222–1224.

Mishan, E. J., 1970. *Technology and growth; the price we pay.* Praeger, New York. An iconoclastic view of the economic cost of growth.

Murray, B. G., Jr., 1972. "What the ecologists can teach the economists." *The New York Times Magazine,* 10 December. Looks at the ecological/economic concepts of growth, movement of materials, and competition.

Shepard, P. and D. McKinley (eds.), 1968. *The subversive science.* Houghton Mifflin, Boston. A superb collection of writings about the ecology of man.

Wagar, J. A., 1970. "Growth versus the quality of life." *Science* **168**, pp. 1179–1184. A critical look at the growth myth.

EPILOGUE

ALL ANIMALS AND PLANTS, whatever their habitat or form, follow one inexorable rule: their populations grow, slowly or rapidly, to some point of equilibrium with their environment, then level off in a change from steady growth to a steady state. This point of equilibrium, known as the carrying capacity of the environment, may be determined by food supply, moisture, nesting sites, competition, or various combinations of these, but it is inevitably reached. If the equilibrium is disturbed, the excess population is lopped off by the environment; directly, as in a patch of seedling ponderosa pine thinned by fire and competition for moisture and nutrients; indirectly, by social stress that in overcrowded animal populations may break down normal reproductive behavior.

For thousands of years, man was little different. While he had tools that certainly differentiated him from other animals and unquestionably gave him a competitive edge, the margin was not marked and his numbers reflected fairly closely the carrying capacity of his environment. But now this has changed, as we saw in the final chapters of this book. Man has manipulated his environment until it has reached a dangerous point of disequilibrium.

Most of the environmental problems we face today result directly from increasing population. Man demands more products that pollute more water and more air, makes more wastes than can be disposed of, takes up more and more suburban space, and puts impossible demands on overexpanded cities. A stabilized population would not solve all these problems overnight, of course, but it would give time to work out more permanent solutions.

Achieving a steady-state population is not of itself enough. A stable population, it is true, will allow a greater allocation of resources to attacking environmental problems, but far too much emphasis has been placed on *technology*. The truism runs, "the technology that has caused the problems can solve them," but as fast as old problems are solved, new ones appear. Clearly technology is not responsible for our current environmental woes any more than a hammer would be for a bruised thumb. The basic cause is man's lack of ecological sensitivity. Until man can become aware of his fellow organisms and the correspondence between their well-being and his, until the environment is regarded as a responsibility rather than an economic opportunity, population control can only partially solve environmental problems.

The first step toward developing an ecological sensitivity is to inform yourself. Just reading this book will give you an overview of environmental problems. Use the bibliographies provided to get more details about areas that interest you, read a good daily or Sunday paper, or a weekly news magazine. Appendix 1 provides a list of periodicals that treat man-environment interactions. Subscribe to one of these or read them occasionally in the library. Use your information in town meetings and public hearings in your community. *No* community, however idyllic, is without environmental problems. Interest yourself in these problems, then interest others; organize and make your feelings and opinions known to public officials and businessmen. Using just this approach the League of Women Voters with only 150,000 members has had an effect quite out of proportion to its rather modest membership.

If working with others, joining organizations, and trying to influence public opinion is not your way, you can still exert a telling effect by a number of purely individual actions. Some environmental problems seem to defy the individual; suggestions on how to deal with these have been made throughout the book. But there are many opportunities for each of us to make a small effort that will have a cumulative effect. Throwing a beer can out of a car window is a trivial action; 100 million such acts become a national problem. Similarly, the short list of unilateral opportunities for action that follows may seem trivial, but there is no better way to develop an ecological sensitivity than to think about the environmental implications of ordinary daily activities and then exercise the free choice you have in these areas to preserve environmental values.

We all function in many roles that have their cumulative environmental effects.

As a home owner or apartment occupant:
1. Repair and keep in repair leaky faucets. The steady drip of a faucet can easily waste four gallons of water a day.
2. Keep a bottle of water in the refrigerator for a cold water supply. Running the tap until water is cold may waste gallons for every glass.
3. Don't burn household trash in the backyard; use waste collection services. Solid waste disposal is easier to control than the acrid smoke from backyard incinerators.
4. Put lawn clippings and fallen leaves into a compost pile or into the garden rather than burning them. This will recycle some of the mineral nutrients and reduce air pollution at the same time.

5. Avoid persistent and broad spectrum biocides around the house and garden; try for some natural control of garden pests by encouraging their natural enemies. Replace some of your lawn with shrubs that will attract birds and give them food, shelter, and nesting sites.

As an automobile owner:

1. Keep your car properly tuned to reduce unburned hydrocarbon emission. Maintain pollution control devices.
2. Replace faulty, noisy mufflers and use your car horn only when absolutely necessary.
3. Leave your car whenever possible; walk wherever you can. Take some form of mass transport to work; if it is dirty or inefficient, bring pressure to bear on public officials to improve service. Vote for bond issues designed to upgrade or expand mass transit facilities.
4. Keep a litter bag in your car at all times and use it.

As a consumer:

1. Be reasonable in your demands on power, particularly during peak periods. Turn off air conditioners and outdoor lights when they are not necessary.
2. Use nonleaded gasoline in your car as it becomes available.
3. Read the labels on the food you buy. While it is impossible and unecessary to avoid all additives, the fewer in the diet the less risk to your health.
4. Refuse to buy clothing, accessories, or novelty items made from endangered animals; if in doubt, check it out.
5. Be aware of environmental values when you buy or build a house; make your requirements known to builders and realtors.
6. When you buy a car consider safety features and pollution control devices as well as luxury features and color schemes.
7. Use returnable bottles when you have a choice; avoid "overpackaged" food and goods when possible.

As a spouse and parent:

1. If you are single, seriously consider whether marriage is for you. Everyone is not suited for the responsibilities, as the divorce rate suggests, and there are professions which offer rewards for single people commensurate with marriage.
2. If you marry, limit the size of your family to two children. If you want more, adopt as many as you have financial and emotional resources to support.

Time is growing short. Unless population growth is reduced to zero by the year 2000, problems which are just beginning to concern us now—air pollution, water and food supply, waste recycling, the inequitable distribution of affluence, and the lack of ecological sensitivity—will become impossible to solve.

The usual moral drawn from the fable of the tortoise and the hare is that steady efforts to solve problems will ultimately win out. But if we let the hare symbolize man's yearning for a "good life" and the tortoise represent steady population growth, another moral can be drawn: despite the slow pace of the tortoise there is a point beyond which the late-starting hare cannot possibly catch up, no matter how hard and fast he runs. We are probably within a few decades of a point in time where the sheer mass of people on the earth will make continuing degradation of the environment irreversible for the affluent *and* the deprived. While the cost of "catching up" is already staggering, it is not yet beyond man's reach. But if we should falter now, not only will the universal goal of peace and happiness for mankind slip from our grasp, but man's very humanity will become extinguished in a struggle for mere survival.

Appendices
Index

APPENDIX 1

Keeping informed of the current developments in man-environment interactions requires more than reading an occasional newspaper. There are many periodicals which, wholly or in part, are concerned with man and his environment. The following list is by no means complete; rather, it includes several points of view. Many of these publications can be found in local libraries, most in college or university libraries. The asterisk indicates organizations whose main thrust is toward active participation in preserving the landscape or improving the quality of man's environment.

Professional Journals, but Often Containing Articles of Interest to Everyone

AMERICAN FORESTS
The American Forestry Association
919 Seventeenth Street, N.W.
Washington, D.C. 20006
monthly

BIOLOGICAL CONSERVATION
Elsevier Publishing Co., Ltd.
Ripple Road, Barking
Essex, England
quarterly

BIOSCIENCE
American Institute of Biological Sciences
3900 Wisconsin Ave., N.W.
Washington, D.C. 20016
to members of AIBS only, bimonthly

ENVIRONMENTAL POLLUTION
Elsevier Publishing Co., Ltd.
Ripple Road, Barking
Essex, England
quarterly

ENVIRONMENTAL SCIENCE & TECHNOLOGY
American Chemical Society Publications
1155 Sixteenth Street, N.W.
Washington, D.C. 20036
monthly

JOURNAL OF RANGE MANAGEMENT
The American Society of Range Management
P.O. Box 13302
Portland, Oregon 97213
bimonthly

JOURNAL OF SOIL AND WATER CONSERVATION
The Soil Conservation Society of America
7515 Northeast Ankeny Road
Ankeny, Iowa 50021
bimonthly

JOURNAL OF WILDLIFE MANAGEMENT
The Wildlife Society
Executive Director, Suite S-176
3900 Wisconsin Avenue, N.W.
Washington, D.C. 20016
quarterly

NATURE
Subscription Department
Macmillan (Journals) Ltd.
Brunel Road, Basingstoke
Hampshire, England
weekly

POLLUTION ABSTRACTS
Dept. Y–5
P.O. Box 2369
La Jolla, California 92037
bimonthly

SCIENCE
American Association for the Advancement
of Science
1515 Massachusetts Avenue, N.W.
Washington, D.C. 20005
weekly

515

Periodicals Intended Primarily for Nonspecialists

AUDUBON
* National Audubon Society
1130 Fifth Avenue
New York, New York 10028
bimonthly
beautifully illustrated

THE CANADIAN FIELD-NATURALIST
The Ottawa Field-Naturalists' Club
Box 3264, Postal Station "C",
Ottawa 3, Canada
quarterly

ENVIRONMENT
Scientists' Institute for Public Information
P.O. Box 755
Bridgeton, Missouri 63044
10 issues
probably the broadest coverage of all these magazines

THE FUTURIST
World Future Society
P.O. Box 19285, Twentieth Street Station
Washington, D.C. 20036
bimonthly
exceedingly interesting speculations about the future

THE LIVING WILDERNESS
* The Wilderness Society
729 Fifteenth Street, N.W.
Washington, D.C. 20005
quarterly

NATIONAL PARKS MAGAZINE
* National Parks Association
1701 Eighteenth Street, N.W.
Washington, D.C. 20009
monthly

NATIONAL WILDLIFE
* The National Wildlife Federation
1412 16th Street, N.W.
Washington, D.C. 20036
bimonthly
also publishes an annual Conservation Directory listing organizations, officers, phone numbers, and publications; a very useful compendium

NATURAL HISTORY
The American Museum of Natural History
Central Park West at 79th Street
New York, New York 10024
10 issues

NEW SCIENTIST
New Science Publications
128 Long Acre
London, WC2E 9QH
England
weekly

OCEANS
Oceans Publishers, Inc.
7075A Mission Gorge Road
San Diego, California 92120
monthly

SCIENCE AND PUBLIC AFFAIRS
Bulletin of the Atomic Scientists
935 E. 60th Street
Chicago, Illinois 60637
10 issues
primarily concerned with atomic energy and disarmament but increasingly oriented to broader environmental issues

SCIENCE NEWS
Science Service, Inc.
1719 N. Street, N.W.
Washington, D.C. 20036
weekly

SIERRA CLUB BULLETIN
* Sierra Club
1050 Mills Tower
San Francisco, California 94104
monthly

APPENDIX 2

THE FOLLOWING ORGANIZATIONS and those marked with an asterisk in Appendix 1 are representative of the many environmentally concerned groups which seek to improve the deteriorating relationship between man and his environment. For details of their programs, write to the organization directly.

Air Pollution Control Association
4400 Fifth Avenue
Pittsburgh, Pennsylvania 15213
educational group involved with the art and science of air pollution control

American Institute of Architects
1785 Massachusetts Avenue, N.W.
Washington, D.C. 20036
professional organization that promotes excellence in urban design, community planning, and preservation of historic architecture

American Shore and Beach Preservation
    Association
P.O. Box 1246
Rockville, Maryland 20850

Appalachian Trail Conference
Box 236
Harpers Ferry, West Virginia 25425
manages the vounteer maintenance of the Appalachian Trail, and can provide guidance to groups that wish to establish and maintain trail systems

Conservation Foundation
1717 Massachusetts Avenue, N.W.
Washington, D.C. 20036

Defenders of Wildlife
731 Dupont Circle Bldg.
Washington, D.C. 20036
mostly concerned with humane treatment of wildlife and elimination of unnecessarily cruel methods of trapping and capturing

Ducks Unlimited
P.O. Box 66300
Chicago, Illinois 60666

Environmental Defense Fund
P.O. Box 740
Stony Brook, New York 11790
an organization of lawyers and scientists which undertakes legal action to bring about amelioration of environmental abuses

The Environmental Law Institute
Suite 614-E
1346 Connecticut Avenue, N.W.
Washington, D.C. 20036
conducts a program of research in environmental law and provides a document service to lawyers and scholars

Friends of the Earth
620 C Street, S.E.
Washington, D.C. 20003
lobbies for ecologically sound land use, restoration, and preservation

Izaak Walton League of America
1326 Waukegan Road
Glenview, Illinois 60025

League of Conservation Voters
324 C Street, S.E.
Washington, D.C. 20003
nonpartisan group promotes election of ecologically aware politicians whose actions are monitored by analysis of roll-call votes on environmental issues

National Trust for Historic Preservation
740 Jackson Place, N.W.
Washington, D.C. 20006
will provide advice and technical assistance to groups wishing to restore and preserve sites or buildings of historic significance

517

Nature Conservancy
1522 K Street, N.W.
Washington, D.C. 20006
main interest is preserving examples of various environments, grasslands, forests and deserts, for educational purposes

Scientists' Institute for Public Information
30 East 68th Street
New York, New York 10021
coordinates the utilization of scientists in public information programs

Trout Unlimited
5850 E. Jewell Avenue
Denver, Colorado 80222

Water Pollution Control Federation
3900 Wisconsin Avenue, N.W.
Washington, D.C. 20016
concerned with the application of the best engineering principles known to the design, construction, and operation of waste water collection and treatment systems

World Wildlife Fund
910 17th Street, N.W.
Suite 728
Washington, D.C. 20006
seeks to preserve endangered species throughout the world

Zero Population Growth
330 Second Street
Los Altos, California 94022

Many of the regional offices of the Environmental Protection Agency have compiled lists of the environmental organizations within their areas. To obtain such a list contact the Public Affairs Director in your district.

REGIONAL OFFICE	AREAS SERVED
Boston, Massachusetts 02203 (617) 233–7223	Connecticut, Maine, Massachusetts, New Hampshire, Rhode Island, Vermont
New York, New York 10007 (212) 264–2515	New Jersey, New York, Puerto Rico, Virgin Islands
Philadelphia, Pennsylvania 19106 (215) 597–9370	Delaware, District of Columbia, Maryland, Pennsylvania, Virginia, West Virginia
Atlanta, Georgia 30309 (404) 526–3004	Alabama, Florida, Georgia, Kentucky, Mississippi, North Carolina, South Carolina, Tennessee
Chicago, Illinois 60606 (312) 353–5800	Illinois, Indiana, Michigan, Minnesota, Ohio, Wisconsin
Dallas, Texas 75201 (214) 749–1151	Arkansas, Louisiana, New Mexico, Oklahoma, Texas
Kansas City, Missouri 69108 (816) 374–5495	Iowa, Kansas, Missouri, Nebraska
Denver, Colorado 80203 (303) 837–4905	Colorado, Montana, North Dakota, South Dakota, Utah, Wyoming
San Francisco, California 94111 (415) 556–6695	Arizona, California, Hawaii, Nevada, American Samoa, Guam, Trust Territories of the Pacific, Wake Island
Seattle, Washington 98101 (206) 442–1203	Alaska, Idaho, Oregon, Washington

INDEX

Illustrations are identified by **boldface** numbers, explanations of terms by *italicized* numbers. When illustrations or explanations occur within a discussion of several pages, the entire discussion is identified.